WHAT I TELL YOU THREE TIMES IS TRUE

Published by

Librario Publishing Ltd

ISBN: 1-904440-38-X

Copies can be ordered via the Internet
www.librario.com

or from:

Brough House, Milton Brodie, Kinloss
Moray IV36 2UA
Tel /Fax No 01343 850 617

Copyright © 2004 Ian Parker
All rights reserved.
No part of this book may be reproduced, stored in a retrieval system, or transmitted by any means, electronic, mechanical, photocopying, recording, or otherwise, without written permission from the publishers.

WHAT I TELL YOU THREE TIMES IS TRUE

CONSERVATION, IVORY, HISTORY & POLITICS

IAN PARKER

Librario

For Christine, with love and thanks for putting up with a hell of a lot.

> "Just the place for a Snark! I have said it twice:
> That alone should encourage the crew.
> Just the place for a Snark! I have said it thrice:
> **What I tell you three times is true."**
>
> Lewis Carroll 1876 – *The Hunting of the Snark*

ACKNOWLEDGEMENTS

Since the first draft of this book appeared in 1994, many people have commented upon it and the numerous subsequent texts. Their advice was valuable and, *inter alia*, resulted in a 300,000 word text falling below 200,000 words. Had most had their way, it would have been shorter still. Yet whether or not I took all of their advice, I am deeply grateful that whatever its stripe, it was given. Listed alphabetically I pay special tribute to Tony Archer, David (DB) Brown, Stan Bleazard, John Caithness, Jonathan Cobb, Bill and Wendy Cooper, Philip Coulson, Tony Dyer, Fred Guilhaus, Michael Gwynne, Anthony Hall-Martin, Chris Huxley, the late Peter Jenkins, Jonathan Kingdon, Dick Laws, Chrissy Bradley Martin, Rowan Martin, Martin Meredith, Dieter & Norbert Röttcher, the late Peter Saw, George Schaller, David & Jane Stanley and Stuart Whitehead. I am also indebted to the late Peter Davey and Dave Richards for help with photographs. There is also a deep appreciation for those from whom I acquired such insights as I may have into the subjects about which I have written, foremost among them the Wata – the Elephant People – and the traders who not only gave information, but also a large measure of trust. Finally, and above all, there is Christine, my wife who had to live with the consequences of a spouse who tried to unravel what many did not want unravelled. To all, my thanks are no mere genuflection: they are real.

CONTENTS

Introduction		11
Prologue		15
Chapter 1	Vasco da Gama's Legacy	23
Chapter 2	Foundations	31
Chapter 3	Mau Mau and Gamekeeping	37
Chapter 4	Conserving	41
Chapter 5	The Quintessential Game Warden	51
Chapter 6	And There Were Others	61
Chapter 7	A Band of Gentlemen	73
Chapter 8	Natural Gentlefolk	83
Chapter 9	The Partners	93
Chapter 10	A New Approach	99
Chapter 11	Rotten Apples in the Barrel	109
Chapter 12	On Tusks	117
Chapter 13	The Battle of Wounded Foot	129
Chapter 14	Somalis	149
Chapter 15	Soldiering Again	163
Chapter 16	The Dol Dol Anti-Poaching Safari	175
Chapter 17	Double-O Six and Seven-Eighths	183
Chapter 18	The First Ivory Consultancy	199
Chapter 19	Sudan	207
Chapter 20	Hoisting Hunters with Their Own Petard	221
Chapter 21	Spades Are Spades	229
Chapter 22	Ebur	237
Chapter 23	A Trip to Washington	251
Chapter 24	The South African Connection	259
Chapter 25	A Brokering Service	269
Chapter 26	Mogadishu and Ethiopia	273
Chapter 27	Burundi	285
Chapter 28	The Emerging Overview	305
Chapter 29	CITES - The Unworkable Treaty	313
Chapter 30	A Prince and a President	327
Chapter 31	Burundi Again	333
Chapter 32	Sometimes You Get Lucky	343
Chapter 33	The Dream Team	357
Chapter 34	There is Something Wrong With Conservation	365
Chapter 35	What I Tell You Three Times is True	375
Chapter 36	The Tyranny of Freedom	385
Chapter 37	Farewell the Pleistocene	393
Epilogue		401

INTRODUCTION

"Who the hell do you think you are?" The question raises a point. Who does one think one is – other than oneself? In my case as good an answer as any came from a particularly boorish professional hunter. Having told his wife he was going to reconnoitre for elephant, he had flown to Europe to meet a lady. When his trackers arrived and told his wife they had located good elephants and could the Bwana now come and hunt them, the cat was out of the bag. When he did return, the lady held her peace and let him hang himself describing the weary miles vainly following spoor, before she drew his attention to their travel agent's latest account. The off-the-cuff explanation he then gave was ingenious. As Kenya was approaching independence he thought the family might have to emigrate. He went, so he said, all over Europe looking for a prospective new home. Not wanting to unsettle her betimes, he had kept his trip secret (that being the only truth).

It was then I had arrived for dinner. The better to convince his wife, he started to give me a lecture on Europe's finer points, and he wagged a finger at me,

"You; you ignorant Kenya hick, you don't know what art is." This was thin ice indeed. "Why, when I was in the Louvre in Rome..."

He fell through the ice and there is no point going on. Yet the scene remains vivid and I have treasured "you ignorant Kenya hick" ever since: so succinct and certainly as accurate as any other description bestowed on me.

Possibly hicks should not write books because, despite my best efforts, this is a long, rambling, disjointed dissertation. Indeed, without reading it, an Australian publisher in marsupial mode, described it as non-fiction, something between enviro- and *realpolitik*, written by someone unknown from Kenya. Consequently he did not need to read it. Others in London were less descriptive but equally pessimistic.

I wrote it for a variety of reasons. First, writing helps my thinking. Second, to record for posterity issues about ivory, conservation, politics and history. I wrote autobiographically because many events involved me personally and my experience strings them together. Also, all historians have unique personal baggage and biases, which influence the tales they tell. It is only fair that readers should know what these are, the better to make sense of what is written.

I started my career thinking I was a conservationist, but soon found that I was simply not an '-ist'. History, people and their behaviour became more interesting than wild animals. The natural world took up much of my time and gave much enjoyment, but more and more its fascination lay in what it revealed about humans. My life spans the demise of the British Empire. Conservation and particularly matters to do with ivory provided a surprisingly good synopsis of that process in Africa where good governance disintegrated into chaos in country after country. It also gave insight into Britain and the developed world's contribution to what has happened. Trends in conservation have their parallels in other aspects of human affairs but, because motives to conserve are deemed in some universal way ethical and 'good', they are not quite so closely curtained from scrutiny.

As in African history, so in this book, ivory is a thread running through the fabric, whether in the distant past or in the cataclysmic present. David Livingstone's inherent dishonesty, Lord Puttnam's interests and the Mad Mullah of Somalia's actions may, at first glance, appear to have little connection to conservation in Africa, yet in the compass of my life ivory brought them together. No wildlife product in history has so influenced a continent's political fortunes. The first civilisations wanted it; Indians, Arabs and Portuguese schemed and bickered among themselves to get it; tusks underwrote the 18th/19th century calamity of slaving in central Africa; ivory led the United States of America to establish the industrial world's first formal diplomatic links with Zanzibar in 1839[1]. It gave financial incentive to partitioning Africa into colonies. More than any one product, ivory influenced Africa's present political shape.

For seven short colonial decades (1890-1960) its influence receded. Africa's economies were absorbed into Imperial economies and metropolitan currencies became her currencies. Memories of ivory's great days flickered on in places like eastern Kenya where the Wata and the Kamba unlawfully continued their hunting and trading traditions, but its value was never forgotten. With independence one after another of Africa's nations was cast off from the industrialized Imperial economies to survive as best they could on their own. In less than a decade ivory came back into its own, regaining all its earlier influence. With political freedom came freedom to use and abuse resources. Where governance and economies remained cohesive, as in Kenya, those with lawful authority exercised that freedom over the elephant resource, mediated principally through the Game Department and its successor, the Wildlife Conservation & Management Department. Elsewhere, as in Tanzania and Zambia, ivory gave some independence to rural people generally, and was a means of avoiding the economic tyranny their new leaders inflicted in the name of freedom. It was available for the taking by anyone who could hunt.

Ivory financed the Somali guerrillas' way into Kenya, enabling them to pursue their historical agenda for expansion. Referred to as 'poachers' by conservationists, themselves and government alike, it was never that simple. The term hid the underlying political goal. Those who used it were concealing something altogether more awful.

In the Southern Sudan, likewise, the Northerners came into the south on ivory collecting expeditions with trains of baggage camels to carry the booty. As they went deep into tsetse country they took along veterinarians to administer prophylactics and cure tsetse-borne trypanosomiasis that would otherwise have destroyed their transport. On the one hand it was buccaneering for profit, but the longer term, evolutionary end was getting at places to which Northerners felt they had right. And having prospered, even temporarily, conviction that an area was *theirs* rose by orders of magnitude and the distant goal of having it all became that much more worth pursuing. The recent oil finds in the south immeasurably hardened a resolve set by ivory two centuries earlier.

Though negative for elephants, Kenyatta's use of ivory – awarding it to reward selected allies – was politically practical. After all, whatever reservations one might have had about such tactics, the product – a politically and economically stable Kenya – was an African success story.

---

[1] P. E. Northway 1954. *Salem and the Zanzibar-East African trade, 1825-1845*. Historical Collections Vol. XC April. Essex Institute, Salem, Mass.

Burundi's position in Africa's unrecorded 'black' economy exposed ivory's use both as a route round government mismanagement and in interstate trading. Looked at from economists' rather than conservationists' standpoints, this use was both understandable and more a 'good' than a 'bad' thing. Ivory opened a door into South Africa's politics, exposing a wider political and economic net across the continent. It was impossible to follow the ivory thread without being drawn into the vortex of African politics. In Kenya it illuminated the intertribal power game, which is their cornerstone. In the Sudan I was nearly burned by the north-south conflict. In South Africa it exposed, *inter alia*, how sanctions were being broken and led into the realms of espionage. In Marxist Ethiopia taxi drivers begged me to curb irreverence for Marx, Engels and Lenin frowning down from billboards. In Mogadishu it gave a foretaste of what was just around the corner for the crumbling wreck of Somalia – the non-state – and to Rwanda and Burundi, the states that should never have been.

Ivory's use reflected people's circumstances. This held good across the breadth of my experience from the manifest unfairness of the Wata predicament, through Obote's men taking large volumes of ivory out of Zaire, Idi Amin and his men pillaging Uganda's national parks as they raped every resource, things going wrong in Kenya, and the traffic in tusks and rhino horns out of Angola into South Africa and beyond: all made that point.

In Tanzania, Zambia and the Sudan it was sensible to use ivory when the national currencies were worth so little. While it did not benefit elephants, it kept international trade flowing in times of economic chaos and unquestionably benefited many people. Because of this, history will judge the Asian entrepreneurs more kindly than contemporary conservationist critics. If they broke rules, were those rules sensible?

South Africa's officially organised ivory and rhino horn trafficking arose out of the virtual war situation. Just as those who beat their breasts and wailed about Somali poaching in Kenya ignored the condition of which it was a symptom, so it was with South Africa. In judging those involved my reaction has been, 'There, but for the grace of God, go I'. The solution lay not in conservationist lament, but in stopping the war. And in a crystal clear glimpse of reality, the matter of ivory showed that even in Britain's and the USA's policies, expediency and not ethics ruled international affairs. As their reaction to my work showed, containing Kenya in the western camp by overlooking corruption was a greater need than stopping incontinent use of elephants. If I learned one thing from following ivory, it was that ivory *was* politics.

Yet it will not remain so. To have fulfilled the rôle it did these past three millennia, it had to be a *great* resource: it had to be very generally available and above a certain threshold of abundance. With elephants available in their millions and present in all but the four smallest sub-Saharan states it was indeed a great resource. Once it falls below that threshold – which may now be the case – it can no longer serve this historical rôle. Just as silver was the 'working bullion' before the nineteenth century's great gold strikes, something else will have to take its place. The paradox, where elephants were concerned, was that their great abundance greatly complicated conserving them. As they become fewer, so it will be easier.

And then there is perversion of commonsense conservation into the cause of conservationism: an irrational faith as bedevilled with fundamentalism as any other religion in modern times. In its madness it destroys what it wishes to preserve. Be warned, that my views of conservationists are often at odds with their self-perceptions.

This is the rich and immensely varied tapestry upon which I endeavour to throw light. The conflict between brevity and coherence has troubled me throughout. Be warned, therefore, that some aspects of my tale may seem irrelevant when they first appear. Hopefully, as it unfolds, relevance should emerge, and casting so wide a net eventually make sense. Perhaps the goal was overly ambitious though, like Thoreau, I see low aim rather than failure as the crime. Maybe one should not expect otherwise from an ignorant Kenya hick.

PROLOGUE

The high sun muzzled the dry Nyika woodland and, but for the soughing wind, all was quiet. Mammals and birds rested in the thin shade of leafless branches. Only the tireless *tumnale*, tiny stingless 'eye' or 'sweat' bees, toiled on oblivious of the heat. Their coming and going from the small waxen tube at the entrance to their hive in a rotten *Commiphora* branch caught the man's attention. For the past two hours he had been sitting and waiting. Now, taking a knife from his belt, he reached up and jabbed it into the soft wood beside the bees' entrance and prised the blade sideways. The branch split and a portion broke off exposing a chamber full of small spherical wax cells. Some were filled with a dough-like yellow pollen cake. Others held a clear sweet honey. The treasure was pillaged and gone in two mouthfuls. The knife, cleaned by rubbing in the earth, was back in its sheath and the small pale brown man lapsed back into immobility, waiting and watching.

Fifty yards away in the evergreen branches of a *hadama* tree (*Euphorbia robecchii*) lay a *wakala* (a savannah monitor lizard). Spreadeagled, withered and motionless, seemingly dead, it had been there for two months hardly moving. Even the shower of earth thrown up from an elephant below as he dusted himself failed to make it stir. Yet beneath its sagging, wrinkled skin, it breathed ever so slowly, awaiting the rains.

Under the *hadama*'s umbrella five bull elephants made a single solid grey mass as they dozed. From time to time one shifted weight from one leg to another, or moved slightly to keep in the creeping shade. Occasionally the silence was broken by a long sonorous sigh, or a sound like a slack sail filling suddenly as a great triangular ear slapped against neck and shoulder. At wide intervals there was a rush of damp, soft thuds as an elephant emptied its bowels and the warm, horse manure-like, smell wafted downwind to the man.

His high cheekbones and slightly slanted eyes would not have seemed out of place in the Kalahari but curly hair, not in tight Bantu clock-springs or Bushmen's peppercorns, betrayed more northern blood. He wore nothing but a short cotton loincloth or *kikoi*, held in place by the raw leather belt that threaded through his knife sheath. A pair of rawhide sandals completed his dress. Beside him, propped against the tree was a quiver, which held seven arrows, each just over a metre long. They were unusually robust. Each fletched with four flights made from vulture feathers identified the man as a Watic or member of the Wata. Their fore-ends were bound in thin strips (*chora*) of soft dikdik skin to protect the broad metal arrowhead and the seven-inch shaft into which the head was tightly bonded. This shaft was covered with a layer of black, tar-like poison containing the virulent cardiac glycoside – ouabain. In turn it fitted into the main arrow shaft.

Beside the quiver was a massive bow, as long as the man was tall. In cross section it was not round but D-shaped, the shank of the letter forming the bow's outer face or 'back' and the arc the inner side or 'belly'. The depth of the D near the bow's mid-point was nearly two inches. In cross section the bow was so crafted that the softer sapwood of the stave from which it was carved ran along the back of the bow while the belly was the heartwood. Instead of forming a single arc, two arcs met at the centre and in archer's terms formed a double-convex bow. Although the man did not know it, the same technology was used in the great longbows of medieval Wales and England.

Early that morning he had scanned the riverbank where elephants had come to drink in the night. The most recent tracks were left by a group of five bulls an hour before dawn. If he was lucky they would not be far away. Quiver across his back and bow horizontal in his right hand, he hitched up his belt and set off on the tracks at a fast walk. Scarcely had he gone two hundred paces when a rufous-necked bustard gave its loud, mournful whistle off to his right and slightly behind him. This was propitious, but almost immediately there was another equally good omen as a Nubian woodpecker called close by – again to his right. Both calls gave the same message: today he would kill a bull elephant. Had the calls been to his left, his victim would be a female. The closer and louder, the more certain the message.

An hour had passed when a bustard called again on his right and simultaneously he saw fresh elephant droppings ahead. Coming up to them he shed a sandal and thrust his toes deep into the moist mass. The heat told him it had been dropped within the past fifteen minutes. Not a hundred paces later he saw the first elephant and within a minute of this he had seen two more. They were spread out, roughly in line abreast, feeding and moving slowly across the early morning wind drift that was still too light to be called a breeze. It would gather strength soon enough, and by mid-morning would be a steady strong wind that blew hard until well after dark. It was always thus at this time of the year and had been since the southeast monsoon first blew. The ever-unpredictable rains might fail or bring floods, but the desiccating south-easter that sapped moisture and life out of the coastal hinterlands was, like nothing else, regular and reliable. The hunter had angled off to one side as he closed with the elephants to be down wind of them when the wind picked up.

The country was fairly open and he was in no hurry. Soon he was level with the great grey animals and moving with them at their slow, erratic progress some fifty paces or so from the nearest and perhaps twice this from the furthest. The closest was a *tofa* with tusks whose lengths were around four *mikono* – the biblical cubit of distance between elbow point and fingertips outstretched. They would weigh around eighty *ratili* each, the measure's name stemming from the Arabic *rotl*, a weight fractionally less than the English pound, and which had, with the cubit, been a standard measure for the three thousand or more years that aliens had traded down Africa's eastern seaboard. A *tofa* was a worthy prize in anybody's eyes. The two more distant elephants were smaller, both *boro* with tusks of three *mikono*. He could easily close with the *tofa* and take it.

He was at the point of withdrawing two arrows from his quiver when two woodpeckers flew into the broken *Commiphora* not five paces to his right. First one called then, before it had ended, the other repeated the succession of loud, high-pitched, accelerating notes. Scarcely had the second finished when not far away another called and as it ended, yet another further off joined in. All had been on the right and slightly to the front of where he was facing. The man pondered. In years of hunting the birds had never called so close as the first two and never had they been answered by two more on the right. Any one, on its own, would have been a benison on his decision to take the tofa so close by. Yet two birds so close and four calling in all was extraordinary. If he had learned anything down the years it was to treat the unusual with caution. All too easily one missed the detail which turned the seemingly benign malign. A frisson like the lightest of spiders danced fleetingly down his back.

Instead of closing in on the *tofa*, he merely continued to keep abreast of the elephants' slow movement. He was certain five elephants had been in the group that left the river. With three in sight, where were the other two? Had they broken away further back, or were they still in front? He paid more attention to the way ahead. The last thing he wanted was for any elephants in front to angle across his path and be downwind of him, get his scent and flee. And though they might be far out of sight, they would communicate in the mysterious way that elephants have and the three still upwind would run too.

The sun had climbed a full ten degrees higher and he was beginning to think that there were no other elephants when he heard a branch break ahead. From the sound it was a big branch or even a tree trunk, which meant there was a good chance that the elephant responsible would stand and feed off it for a while and allow the following three and their shadow to close the distance.

Soon, like tall anthills, two wrinkled elephant sterns could be seen through the brush before him, the occasional tail movement confirming what they were. As always, the hunter's eyes strained for a glimpse of their ivory. What ivory did they carry? The first was another *boro* but with one tusk broken. He didn't get a chance to see the second's tusks until all four of the other elephants were interposed between him and it and initially he could not believe his eyes. He had never seen their like. A bull bearing ivory of five *mikono* or more – an *usho* – was every hunter's hope, but this one was well over five *mikono*; certainly six but more like seven. Not only were they long, but thick: where they emerged from the lip they were as thick as the hunter's thigh and they held this thickness far down their lengths. It was a bull such as few ever saw and then only once in a lifetime. No wonder the birds had been so insistent.

The great *usho* stayed out of the hunter's reach on the far side of the herd as the elephants fed. For a certain kill he wanted to be within fifteen paces of it and preferably less than ten before loosing his arrow. He must bide his time. The sun rose higher and as if upon a command the bulls stopped feeding and moved into the *hadama*'s shade. Knowing that they would rest here through the hot hours, the hunter sought shade too and took up his vigil, close enough to move forward quickly in case the *usho* came to the near side of the group and gave an opportunity to arrow it. This had not happened.

The sun was well past its zenith and its heat perceptibly less when the elephants stirred. One after another they released cascades of urine and quite suddenly moved out from under their resting tree. Indeed the move was sufficiently sudden for the hunter to think momentarily that, somehow, they had scented him. As they moved he rose, slung his quiver across his back and picked up the great longbow. Watching the herd he was relieved to see the initial determined march slow and the bunched animals spread apart and start feeding. Again they angled across the wind so that he could safely keep abreast of them. Seemingly deliberate, the great elephant stayed on the far side.

Thus it was for the next hour. Then the pattern changed. The *usho* stayed overlong to eat his fill of a bush and the rest moved forward so that although still the furthest upwind, there were no other elephants between him and the hunter. Reaching over his shoulder with his left hand, the man grasped the cap on his quiver and pulled it off, letting it hang loosely on an attaching strap to the main tube. His fingers gripped the flights of two arrows and drew them out. Eyes on the elephant and working by feel, he untied the end of the leather binding that shielded the head and first ten inches of one arrow and then the other. In turn

Then, as he drew on the string he swung it down until the bow was at its fullest extent as the arrow pointed directly at its target – a hand's breadth forward of the rear leg, low in the left abdomen.
(Sculpture by Rob Glen: proportions
from life exact).

he pulled the strips and the arrows had spun in his hand as the leather came away, exposing their broad, razor sharp heads and eight inches of shiny black *Acokanthera* poison or *hada*. Holding both arrows along with the bow in his right hand, and still watching the elephant, he tucked the thin parchment binding strips into his waist belt, then felt backward with his left hand groping for the quiver cap, pushed it back firmly onto the quiver body.

The *usho* moved to catch up with his companions and the hunter sensed its unease if they moved too far away. To locate them it angled downwind to pick up their scent trails, coming closer to the hunter. But it didn't come close enough. When it caught up with its companions the *usho* was no longer the furthest upwind from him. Instead there was only one other elephant between them and the hunter and, experienced as he was, the man's pulse quickened.

For the next hour, as the elephants and their observer progressed, the two nearest to the man moved almost in unison. It was as though the younger animal deliberately shielded the *usho* from any downwind approach. Not for the first time the man felt misgiving. Perhaps the omen-birds had meant the *tofa*? It had, after all, been a prime target and easy to approach: there had been no calls since he had turned down the opportunity. But surely the birds had referred to the *usho*? They had been so loud and close and at least four of them. He would wait. If the opportunity he sought didn't materialise, there was tomorrow, other days and other elephants. And so he waited, seeing but unseen.

Without warning the nearest guardian elephant suddenly stopped, moving its trunk delicately over the ground. Finding the right spot, a huge leg swung forward and scuffed the ground with the nailed front edge of its foot and the elephant started to dig, not as other animals do, by pushing the earth backwards, but elephant style by pushing it forwards. For all the inappropriate shape of an elephant's foot as a digging tool, three ponderous forward scuffs made a trench a yard long and nearly a foot deep, exposing a small, wizened, potato-like root. Delicately the elephant picked it up, manipulated it briefly between the two fingers of its trunk and placed it in its mouth. It chewed slowly and the wet, sucking, lip-smacking sounds carried clearly to the watching hunter. While the elephant returned to sniffing the ground and digging for more of the morsels, a gap gradually widened between it and its companions. Soon it was so far behind them that the great tusker was no longer shielded from a downwind approach.

Recognising his moment, the bowman nocked one arrow onto the bowstring and clasping the other against the bow-stave in his left hand, he closed with the elephant. Crouching, gliding, almost flowing he came closer and closer. His goal was a six-foot shrub behind which he could straighten up. He calculated the elephant would pass within ten paces of it. A sound from his right made him look back to see the elephant that had stayed digging was now moving to catch up. It came as if guilty of abandoning its guarding and there would be a race between the *usho* passing the hunter's ambush point and the laggard coming across his tracks and giving the alarm. Moving in on his quarry across the path of the oncoming rearmost elephant, he was committed.

The bowman arrived at his chosen shrub and was still in a crouch when the *usho* decided not to go past it, but feed off it. The hunter straightened up as the elephant turned towards him. Neither had seen the small grey antelope crouched under the bush watching both. As the man stood its nerve broke and the dikdik tore off giving its shrill, penetrating alarm whistle "wheez, wheeeez, wheezi weez weez!", right across the elephant's bows. The huge animal shied to the right exposing the whole of its left side. With head high, the great sweep of its tusks forwards and upwards made a magnificent frieze against the late afternoon sky. And the elephant's eye fell on the man pulling in his mighty bow not seven paces away. With a wrench of his back and shoulder muscles and the solid thews of his arms, the double arcs bent backwards to their fullest extent forming a single curve. The draw started with the bow held over the man's head and the arrow pointing skywards. Then, as he drew on the string he swung it down until the bow was at its fullest extent as the arrow pointed directly at its target – a hand's breadth forward of the rear leg, low in the left abdomen. Held on target for a fraction of a second, the arrow was released. So close was the elephant that the sound of the bow and the arrow's stripe were as one and the missile went so deep that only its feathered flights protruded from the elephant's wrinkled side.

The elephant roared at the violent blow and fiery pain as it wheeled and fled. The other elephants listened for a split second and they, too, were in headlong flight, each in the direction they had been facing and not, as did females and young, bunching first. The hunter sat down abruptly. A thudding heart and weak knees surprised him, recalling memories of long ago when aged thirteen he had arrowed his first elephant. He wondered how birds knew what would happen for he had no doubt that the elephant was now his. The arrow strike had been perfect. Had he been with others the aftermath would have resounded with tension-dispersing laughter and talk. Yet he was alone and silent. It took five minutes to collect himself and take up his victim's tracks.

In the first two hundred paces of the elephant's flight the fire in its abdomen mounted fiercely. After this the pain diminished but the world started to spin. It staggered, fell to its knees, rose and continued with bowels and bladder emptying almost continuously out of control. It could not draw enough breath and its heart slowed to heavy thudding beats that were becoming irregular. Soon it could not run and then no longer walk. It took all its failing senses to stand and it was swaying and trembling when the hunter came up with it. He stood and watched not bothering to place his second arrow. Indeed after observing it for a few moments he withdrew one of the leather *chora* held in his belt, rebound the arrowhead and poison shaft of the arrow in his hand and returned it to his quiver. The elephant's trembling became progressively more violent before it lost all control and crashed over on its side. For five minutes or so it moved spasmodically before its heart gave a last contraction and stopped. It was dead six hundred and twenty-three paces from where the arrow had entered it. Had the Watic a watch he might have noted that it took twenty-four minutes from arrow strike to death.

The small brown bowman walked slowly round his victim before unsheathing his knife and cutting off the end of the elephant's tail with its coarse wiry hairs and sticking it in his waistband. Not until the tail end had been cut off was the elephant technically taken. Then he went to the other end of the carcass and, in awe, touched the uppermost of the two great tusks. First his hand just rested on it, getting the feel, Then he placed both hands about it where it emerged from the elephant's head. It was thicker, far thicker, than he had originally judged. Then in time honoured fashion he measured it, placing his elbow on that point in the skull where the tusk would begin and laying forearm and outstretched hand along the tusk, then moving elbow to that point where fingertips had reached and repeating the process down the entire length of ivory. It was seven *mikono* long and the partner tusk on the underside no shorter. They were the biggest tusks he had ever seen and probably the biggest seen by any of his people in generations.

Then he sang. In a high African tenor he sang the song that he had first sung with his father years ago beside his first elephant. With that first elephant, excitement and boyhood had minimised the deeper ritual meaning of this song. Now it was a hymn of the deepest significance, the spiritual high point of his life, a salutation and bonding with the elephant, the traditions that bound their peoples – the elephants and his – in a common destiny. As the last cadence floated into the distance, he had stood, head bowed, overcome with emotions beyond describing.

Night was close and he would have to make his way back to his camp through the hostile dark. On the morrow he would come with his people to take the tusks and dry the meat. First, however, there would be ceremonies to perform, ceremonies of thanks and expiation. Yet nothing would ever again catch the moment as he sang alone by his elephant. Later the dances that formed part of these ceremonies would give way to singing, merriment and laughter. And then around the fires that would burn about the elephant's remains for the next few days while his people ate and the meat dried, he would re-enact the hunt in its entirety. Every move, every nuance of the past twelve hours would be acted, recounted and analysed. His watchers would absorb this so accurately that they, in turn, would be able to retell the hunt as though they, themselves, had been present. It would become history and part of their lore to be passed down the coming generations for as long as they were Wata – the Elephant People – into the mists of the future. And as always they would wonder at the ancient wisdom in the omens that had been proven once again.

With head high, the great sweep of its tusks forwards and upwards made a magnificent frieze against the late afternoon sky. (Sculpture by Rob Glen)

Vasco da Gama

CHAPTER ONE

VASCO DA GAMA'S LEGACY

When the Imperial British East Africa Company (IBEAC) was founded and granted a royal charter to administer what became Uganda and Kenya, its founders believed it would be profitable from trading one commodity above all others: ivory. That alone, committed the Company to conserve elephants and from the colonial era's outset, elephants and ivory have been at the core of conservation policies, not only in Kenya and Uganda, but across Africa. The IBEAC was not original in its hopes over ivory. The commodity, along with gold, has been sought by all of Africa's invaders down the millennia. Given their historical, political and commercial importance, elephants, ivory and its production in past centuries should have been well studied. Strangely, this was not the case. The depth of our ignorance has been embarrassingly ever more apparent. It wasn't as if information was lacking, for it was not. Had it been used, it would have influenced politics overall and African conservation in our times would have been approached very differently.

* * * *

Arabs, Indians Persians and the Chinese have traded with Africa's east coast for thousands of years. It is also likely that Helladic Greeks and then Romans were in contact with it. The Dark Ages expunged such knowledge Europeans may have had and by medieval times the region was unknown to Europe.

In May 1487 King João of Portugal sent two spies into and beyond the Muslim realms of the Middle East to find out who produced the silks and spices that were in such high demand throughout Europe. Today, with refrigeration and modern transport, fresh food is available everywhere all the year round and it is difficult to comprehend the monotony of medieval winter diets. Spices gave variety to otherwise boring food and consequently were in demand to a degree now beyond recall. Produced in the east, they were traded overland becoming more expensive at every stage, through Asia Minor to Europe. If a sea route could bypass these steps, spices would be cheaper and those who carried them would be rich.

One of the spies – Pera da Covilha – got as far as India and found out where and how spices were grown. On his way home he took the opportunity to sail by dhow from Oman down the eastern coast of Africa as far south as Sofala in present Mozambique. He learned that further south still the coast was both navigable and swung from a predominantly southerly orientation to a more westerly alignment. He also learned that the Arabian and Indian mariners came down the coastline to get ivory – which was available all along the seaboard – and for gold from the African interior in what is present-day Zimbabwe.

When this information eventually reached the Portuguese Court it tied in with that brought back nearly a decade earlier by Bartholomew Diaz. Sailing down the west coast of Africa, he had rounded Cape Agulhas and reported that from there on the coast ran east and north of east. It needed little imagination to assume that Diaz and Covilha had described the same coast from opposite ends and that once ships rounded the Cape, the way was open to India.

When Vasco da Gama set sail in 1497 for the east coast of Africa and India, it was not quite the step into the unknown that school history books had me believe. He must have had a pretty fair idea of where he was going and what he would find. The real hero was surely Pera da Covilha who, alone and disguised, travelled through strange lands collecting information? Through da Covilha's work, da Gama expected to come to Sofala, knew of all the ports he would find to its north and that crossings to India were made annually by fleets of dhows, and that all the eastern coast was under Islam.

Within a decade of da Gama's voyage the Portuguese ruled from Sofala to the Straits of Hormuz. While their motive was trade, pure and simple, they justified their actions in the name of Christianity. It made them ethically comfortable – doing God's work – but also ensured the Pope would favour their claims to sovereignty over any that might arise when other Christian powers woke up to the new route's possibilities. And if one wonders by what right the Pope ruled on such issues, suffice it that in those days when Europe was still Catholic, his word *was* law. Some, like the Argentinians, still believe it – when it suits them – and rest their claim to the Falkland Islands on it. Spreading Catholicism was good business five hundred years ago!

* * * *

The Portuguese were as aggressive against the vessels of other Christian nations as they were towards those of Islam, forbidding them to trade between the eastern African coast and India. At sea they were initially strong enough to enforce such rules, but even at the height of their power, the Portuguese were not numerous enough to secure the African interior militarily. Other than where the Zambezi basin was concerned, they never tried to colonise the east and central African hinterlands.

The Zambezi basin was different because it contained the Mwene Mutapa (Monomatapa) empire, successor to the Zimbabwe civilisation, which was a source of gold. To forestall other European powers getting a trading foothold with it, Portugal proclaimed Mwene Mutapa a vassal state. Several military expeditions went into the area to support this claim over the next two centuries, but never subjugated Mwene Mutapa's people. Until the twentieth century, Portuguese rule away from coastal Mozambique was at best loose. Catholic Christianity was taken into the interior and trading posts were established far up the Zambezi. Yet the furthest inland that posts and missions were held for any time was the eastern highland edge of modern Zimbabwe.

In the interior the Portuguese presence was made not so much by officials directly under the Crown in Lisbon, as by private individuals and freebooters – *sertanejos* – who carved niches for themselves well away from the coast and outside government control. Such men married local women and established dynasties of mulatto descendants who were 'Portuguese' only when it suited.

In effect there were two Portuguese presences: colonial civil servants with the missionaries whose interests were the national and clerical exchequers, and the *sertanejos* who traded on their own accounts, dealing in gold, ivory and slaves. Yet for all that the Portuguese were harsh rulers, both civil servants and *sertanejos* went to some lengths to preserve political stability in the Zambezi basin and African interior generally. Royal edict demanded that the Portuguese be nice to the natives, not so much through innate kindness or concern for their welfare as from understanding that trade with the interior depended on political stability. Thus an order to Governor Frois in the late 1720s reiterated what had worked successfully up till then; he must –

> "... not allow any European nation whatever to hold trade or commerce with the Negroes of the coast, ..."[2]

nor permit –

> "any of the said nations to establish themselves in the land . . for which it is very necessary that no offence should be given to the Kaffirs inhabiting the said shores."

Ivory, one of the two commodities that the Portuguese wanted, was heavy and bulky and could only be carried to the coast by people. For them to do this freely, there had to be peace so that caravans could move unhindered. The historian Edward Alpers described events[3]. The trade had three components: Portuguese, African and Indian. The Portuguese produced little that either India or Africa wanted. The Indians produced cotton cloth and goods which Africans desired and which they would exchange for gold and ivory. The Portuguese had to insert themselves into this commerce. Their vessels outbound from Europe collected gold and ivory in Africa and took them on to India where they traded them for spices, silks and cotton cloth. On the return run they dropped off the cotton cloth in Africa to buy more ivory and gold and took the spices and silk on to Europe. The cycle would then be repeated. The system worked for as long as they could monopolise shipping and swap cotton goods for the ivory and gold.

Naturally it was in the Indians' commercial interests to cut out the Portuguese middlemen and they sought permission to trade directly in Portuguese-controlled East Africa. Portugal's African colonists and the Catholic Church worked equally hard to keep Indians out of territory that they controlled. In the first century or so of Portugal's rule in the Indian Ocean, they succeeded. Yet as time wore on and the cutting edge of Portuguese imperialism grew blunt, Indians worked their way round the barriers.

Using tactics that are as characteristic of them today as they were three centuries ago, the Indian merchants addressed the people who mattered: those at the top. The most senior Portuguese to whom they could appeal was the Viceroy who, conveniently for them, was based in Goa on the Indian sub-continent and not in Africa. One of the attractions of being appointed Viceroy of Portugal's Eastern African and Indian possessions was that it gave the incumbent opportunity to acquire wealth. Understanding this, Indian merchants loaned capital to successive Viceroys. Once loans were made the recipients could be approached for favours – among them permission to trade in Africa. Thus while the colonists in Africa wanted none of it, the Viceroys tended to favour the Indians. Consequently the African

---

[2] In Alpers E.A., 1975. *Ivory & Slaves in East Central Africa*. Heinemann, London.
[3] Alpers E.A., 1975. *Ibid.*

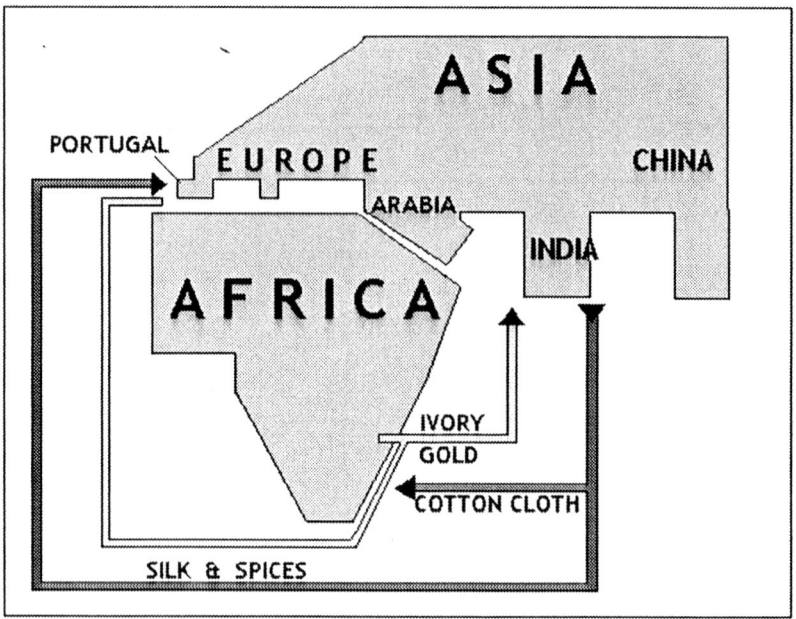

A diagram to illustrate the pattern of Portuguese trade maintained by naval might: ivory and gold from Africa to India, cotton cloth from India to Africa and silks and spices from India to Portugal.

colonists found it progressively more difficult to compete against Indians bartering their own cotton cloth directly with the ivory caravans and gold carriers coming out of the hinterland.

Through the seventeenth and eighteenth centuries the Arabs and Islam gradually recovered ground from the Portuguese, first in the north and then progressively moving south. By the latter half of the eighteenth century only Mozambique remained Portuguese. The Arabs allowed Indians into their recaptured ports so when Kilwa was lost by the Portuguese, the natives of the Zambezi basin could deal with Arabs and Indians there as readily as with the Portuguese in Mozambique. Naturally they sold to whoever offered the best deal. No longer controlling the sea-borne traffic, Portuguese fortunes waned.

If the natives no longer brought ivory and gold to the Portuguese, but sold them instead to the merchants in Kilwa, there was now no reason to be nice to them. Coincidentally the French were opening Mauritius and Reunion for large-scale sugar production, and needed abundant labour which, in those times, meant slaves. The nearest source of slaves was Africa. Thus as the Portuguese grip on the ivory and gold trades reached low ebb, Mozambique found itself the focus of a strong new demand for slaves. And if there was no need to be nice to the natives, why not sell them as slaves?

Slavery was new neither to Africa nor the Portuguese. Yet until the late eighteenth century it had not figured as a major disruptive force in the interior of central and south-eastern Africa. Once the Portuguese took to it on a large scale, however, slaving quickly changed the *status quo*. Lacking manpower to carry out slaving raids on any scale, the *sertanejos* fanned quarrels between African communities. Lending arms to the stronger sides, they took payment in captives. Soon the hinterland was aflame from Delagoa Bay in the south to the Rovuma River and beyond in the north. In the chaos which followed, the people in Zimbabwe stopped mining gold. The political stability needed for long distance ivory portage vanished. Caravans could no longer move safely.

This set off a second commercial crisis. When caravans ceased, ivory and gold stopped arriving at the coast. Even the traders in Moslem controlled Kilwa were affected. Their solution was logical. If the commodities were not coming to them, they must go to where they were produced. Thus heavily armed Arab and Swahili caravans, financed with Indian capital, started probing inland. They found what they wanted easily enough, but their problem was how to get tusks back to the coast. Ivory is heavy. There were few navigable rivers. Tsetse flies prevented using draft animals. Tusks had to be carried by people.

Because of the political instability caused by Portuguese slaving, willing porters were unavailable. Men would not take the risks of long journeys through other tribes' lands. So the Arabs and Swahili took the labour they needed by force. By seizing people, chaining them together, making them carry the tusks acquired, the ivory questers got their goods to the coast. Arriving there they increased their profits by selling their forced carriers and greatly improved the outcome of trading ivory.

Thus the ivory seekers multiplied the disruptive influences that commercial Portuguese slaving had started. Once established, their system became a self-fuelling juggernaut that rolled northward from hinterlands of the mid-Zambezi and around Lake Malawi, through what was to become southern and central Tanganyika, around Lake Tanganyika and deep into the Zaire basin. And as it spread it became progressively more efficient and co-ordinated – as new commercial systems tend to.

By the third decade of the 19th century this huge commercial enterprise was largely controlled by the Sultan of Oman, who in 1823 had moved his capital to Zanzibar, no doubt the better to look after his interests. The coastal ports from Kilwa in the south to Mombasa in the north had become staging posts for ivory and slaves on their way to Zanzibar. And this vast enterprise was, in turn, responding to a fast-rising demand for ivory created by the European and North American Industrial Revolution.

As happens when looking at history from one angle, the foregoing simplifies reality. Yet, in a nutshell, prior to the arrival of the Portuguese, Central East Africa had had centuries, if not more than a millennium, of trade contact with Asia. It imported trade goods and exported gold and ivory along with other natural or grown products. The Portuguese captured and dominated this commerce in the name of Christianity. They were initially successful, but could not handle Indian competition, which set a chain of chaotic events in motion.

* * * *

The wave of slaving for ivory that the Portuguese started, but which the Arabs and Swahilis expanded across Tanganyika (the mainland component of the Republic of Tanzania), and what is now Zambia and the Democratic Republic of the Congo, did not influence northernmost Tanganyika or Kenya. Apparently there were two reasons for this. First there were the Masai whose vast lands lay across the routes to the interior and whose martial prowess was such that forcing a way through Masailand called for military capacities the slavers did not have. Second, as the flow of ivory out of northernmost Tanganyika and

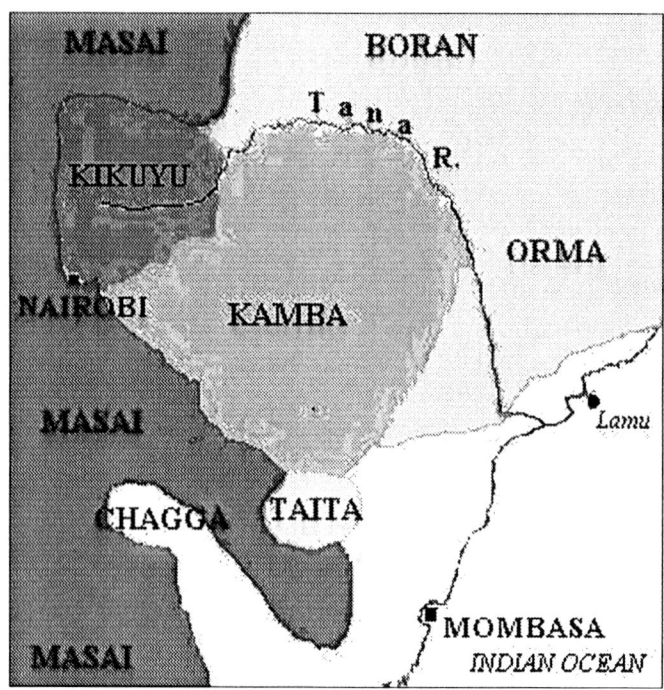

From the 16th to the 19th century, the Kamba traded ivory out of central Kenya and traders did not have to go inland for it. When the Kamba system faltered in the 19th century, traders had to go inland to get the ivory and had to pay tribute to the Masai if they wished to avoid Kamba land. (Photo Peter Davey)

Kenya to the coast was sustained while it collapsed elsewhere, there was no need to go into the interior to get it. All the hinterland people traded ivory either directly or indirectly, but none more so than the Kamba of east central Kenya (Ukambani). They specialised in hunting elephants and selling tusks, often in co-operation with the coastal Giriama tribe.

Responding to growing nineteenth century demand for ivory, the Kamba spread their collection net ever wider across neighbouring tribes. Naturally it was not long before some of these neighbours tried to cut out the Kamba middlemen. Just as naturally, the Kamba closed the route through Ukambani to outsiders. Ivory supplies from Kamba neighbours started to dry up and around 1840 the hitherto reliable Kamba suppliers began to fail their markets.

Emboldened by their successes further south in Africa, the coastal caravaneers resolved this by going inland and getting the tusks that were not coming to them. As the Ukambani route was closed, they could either by-pass the Kamba to the north, through dry, inhospitable country peopled by warlike Orma, or to the south through the more equitable Masai plains. The caravans chose the latter and instead of using force, they paid the Masai *hongo* (tribute or blackmail) and were allowed to pass through. From the coast they went to Kilimanjaro then keeping west of Ukambani, to Kikuyu country where they could get provisions before working deeper into the interior.

The caravans were as unable as leopards to change their spots, however, and kept an eye open for intertribal quarrels that could be turned to advantage as had happened so devastatingly further south. By the late 1870s and early 1880s their behaviour had been such that they were not allowed into Kikuyu country and could only barter at the borders. Von Höhnel documented this state of affairs[4] when he and his leader Teleki were on their epic walk to Lake Rudolf (now Turkana).

The tense relations between the caravans and the Kikuyu presaged chaos. In the 1880s the Masai lost most of their cattle from plagues of bovine pleuropneumonia and then repeated waves of rinderpest. With this loss went their military strength and an inevitable decline of respect from the coastal traders. Had it not been for the coincidental imposition of British and German rule over Eastern Africa at this point, the ivory/slaving chaos might well have infected Kenya as it had elsewhere.

When the first whites ventured inland, they came with the same Swahili porters who accompanied the ivory caravans. Relying on their porters for information about the people they met, the British learned that the Kikuyu were hostile and treacherous. This was reinforced by Kikuyu reluctance to let caravans into their country. At the same time the Kikuyu, observing that white men came with the distrusted coastal caravaneers, assumed that the whites would be equally avaricious and unreliable. The attitudes of both groups towards one another were thus strongly influenced by the company in which the whites arrived and were strained and rather hostile from the start ... but that is another story.

* * * *

Before the white man's coming the people of inland Central and East Africa had had long commercial intercourse across the Indian Ocean. Pre-eminent in this trading was ivory: as no other animal product it had influenced the region's human history. Had the first IBEAC administrators known the complex history of the ivory trade, or that through it the people of the continental interior had sustained long contact with the outside world, they may not have acted as they did. They would have realised that many Africans had travelled to the coast and met men from other lands and it is possible that some had also been overseas.

Benighted and isolated in the continental interior the tribesmen may have been – but only in relative terms. A persistent myth exists across the interior of modern Kenya. Ascribed to the prominent Masai seer – Batiaan (aka Mbatia wa Gaterimu[5]) – he foretold of red (white) men coming from the east and conquering the land. Among the invaders' innovations would be an iron snake or railway. Today other tribes lay claim to the same prophesy. Now attributed to mystic abilities, these forecasts probably had a far more prosaic basis. Most of Eastern Africa's ivory went via Zanzibar to Mumbai (Bombay) in India. The Arab and Swahili caravaneers would have known this. As seafarers it is likely that some had voyaged to India and, if they had not, they certainly knew the Indian ivory merchants who bought their produce in Mombasa and Zanzibar. Among these traders many would have made the trip to India. And having done that, they would have seen the railways that the British Raj

---

4 Höhnel L. von, 1894. *Discovery of Lakes Rudolf & Stefanie*. Longmans, London
5 *verbatim* from the late Dr Louis Leakey who, among other things, was an authority on the Kikuyu and who claimed that the Masai Batiaan was in fact a Kikuyu from the Chania valley.

introduced on a large scale in the last three decades of the nineteenth century. The information about white people, their power and their mechanical abilities would have been carried into the African interior along with men seeking ivory. It is inconceivable that their African suppliers would not have asked where all the ivory was going and for what purpose and equally inconceivable that they would not have been told. The forecast that the white people and their contraptions would penetrate the African interior was thus based on common sense. There would have been nothing mystical about it until white romancers gave it a twist in keeping with their perceptions of benighted Africa.

Perhaps if the authorities had known a little more about the history of the ivory trade, they would have appreciated the degree to which much of eastern and central Africa had been in contact with the outside world. They may never have tried to monopolise it in the name of the Crown. Yet speculation aside, in the name of ivory Vasco da Gama left quite a legacy.

Prior to the 1880s, the Masai had sufficient military strength to deter any passage through their lands. (Photo Peter Davey)

CHAPTER TWO

FOUNDATIONS

A maternal forebear, Dirk Pretorius, landed at the Cape in 1671. His daughter Anna Elizabeth married Dutch immigrant Christian van der Schelde from Zeeland on 1st April 1731 in the Cape Town Dutch Reformed Church. Their daughter Maria married Christiaan Paulsen from Husum (not far from the modern German/Danish border) who had come to the Cape in 1743 as a 21-year-old soldier in the service of the Dutch East India Company. They had ten children. One daughter, Anna Elizabeth after her grandmother, married one Arie Steyn (also written as Styne), a Dutchman from Aalsmeer in Holland. He captained a Dutch frigate – the *Meermen* (Mermaid) – stationed at the Cape to protect ships of the Dutch East India Company from the British Royal Navy and other marauders. As the single church in the Cape, the records of the Dutch Reformed Church vouchsafe this solidly Afrikaner root.

The purity of the line was tainted when their daughter – Maria – married a British Huguenot called James Fichat who, as a Lieutenant in the Royal Marines arrived in the Cape in 1807. Unlike many fellow Huguenots he did not become an Afrikaner, but stayed staunchly British. The records show that at the time of his death in 1825 James Fichat had a farm – The Grove near Veldhuyzen – of ninety-nine morgen, had a son and three daughters, and owned one prize negress, seven slaves, three horses, three oxen, five donkeys, six pigs, two wagons and a cart. A morgen was equivalent to about two acres, the term deriving from the Dutch word for morning and originating as a measure of the land a diligent ploughman could plough in a morning. How a prize negress differed from a slave I know not; nor do I know what prize she was awarded.

A lawyer-descendant – one Sydney Fichat – who was my grandfather, arrived with his family in Mombasa toward the end of 1903. Fed up with Rhodes and sickened by the second Boer War, like many who then came to the British East Africa Protectorate (which only became Kenya in the 1920s), he was making a new start. These South African settlers were responding to a call by the Protectorate's Commissioner – Sir Charles Eliot – for whites to take up land. I doubt if they knew the invitation was motivated by need to make the Uganda Railway pay[6]. I doubt, too, whether my grandfather realised that Britain's Imperial India looked upon East Africa as *lebensraum* for Indians. Both factors influenced subsequent events in the region.

Trained in South Africa's Roman Dutch law, Sydney Fichat was not allowed to practise as a lawyer in the British East Africa Protectorate, which operated under Anglo-Indian rules. In quick succession he worked for the *East African Standard* while it was still owned by the Jeevanjees (suggesting he either had no racial prejudice or was so broke that he set it aside). The venerable historian Edward Rodwell told me that Sydney Fichat repeatedly stole electric light bulbs from the Jeevanjees. Why, or even if it was true, I have never established. Fact or merely Rodwell's sly humour, it drifts as an historical mote of dust in the chasms of

---

[6] Hill M. F., 1950. *Permanent Way: The Story of the Kenya & Uganda Railway.* East African Railways & Harbours, Nairobi.

my mind: pointless but nonetheless there. Fichat grew potatoes near Kikuyu and then became a successful estate agent, publisher and general businessman. He was politically active and gained notoriety when, with 'Cape to Cairo' Grogan and another settler they flogged a native for not showing respect to a white woman, causing a furore that reached back to Whitehall. For a while he was secretary to the Colonists' Association led by Lord Delamere. When he died of cancer in the 1920s his family was rooted in Kenya.

His partner – Stella Marie Perossi – was a strong character for whom life held no grey. Her relations with others reflected this and they either liked or disliked her and *vice versa*: there was no middle ground. A person of conviction and contradiction, she was a suffragette, champion of African rights, guardian of the underdog but believed that all should know and accept their stations in life (underdogs should be well looked after, but were underdogs nevertheless). A stickler for the *right* way of doing things, she was not above correcting a diner in a restaurant on the etiquette of eating peas, notwithstanding that she had never met him before or that he sat two tables away from her. She gained public attention by supporting a move in the Legislative Council to deport white women who cohabited with black or brown men. She added a rider, however: providing that the same applied to all white men who cohabited with black or brown women. The move failed!

She and her two elder daughters were nursing volunteers in the First World War and among the very few white women who served in the African Carrier Corps hospitals. After the war all three went to Britain and gained formal nursing qualifications, she as a midwife and her daughters as nurses. Later she opened Nairobi's first curio shop, ran two hotels (both then respectable but much later became brothels) and farmed coffee. Indomitable, prickly, didactic, British to the point of paranoia, she was the very stuff of Empire. All through her life she was devoted to 'her' Africans – who knew her as 'Njeri'. She died in 1971. Seemingly so transparent, Stella Perossi nonetheless took secrets to her grave that would have surprised her contemporaries. She may have been British through her birth in England, but was not English as her father was Italian and her mother Belgian. To her grandchildren's consternation and in a practice well ahead of their times, she and Sydney Fichat never wed.

My mother was born in Nairobi in 1908, the last but one of six children. The sixth died of dysentery in 1913 and was buried in the cemetery next to Nairobi's Uhuru Highway, under a small gravestone that gave only her name: Una. As a boy I saw it, but when I went there recently it was gone and most other graves had been vandalised. Graveyards, windows to so much European history, play a lesser rôle in Africa. Perhaps because she grew up under so strong an influence as my grandmother and was the youngest of five children, my mother was gentle and unassertive. With a strong sense of humour, she was a gardener with the greenest of fingers and a great compassion for the underdog. As the young wife of an Assistant District Commissioner (ADC) in Nyasaland (now Malawi) in the 1930s, her outspoken criticism of chaining prisoners in gangs raised many senior eyebrows.

My father was an administrator in the Colonial Service who served in Nyasaland. His forebears were Scots with a family seat at Fairlie on the Firth of Clyde. Their earliest overseas venture was in Jamestown, Virginia, where one James Parker dealt, *inter alia*, in slaves. He was a Royalist and for siding with the Crown in the American Revolution and then serving on Cornwallis's intelligence staff, lost his American fortune and had to start anew. With partners he planted sugar in Guyana and the West Indies and their Liverpool based Company – Sandbach, Tinne & Co – had its own shipping line, and a near monopoly on Demerara sugar until bought out in the middle of the twentieth century.

I came to Kenya on the outbreak of the Second World War when my father joined the RAF and my mother returned to her family home for the duration. My parents' marriage became a war casualty and I have lived in Kenya ever since. In retrospect, I suppose that through antecedents, birth and upbringing I am an Imperialist and by definition a colonialist. When I was young, neither term was pejorative. Yet times change. Then abortion was a sin and capital punishment just redress for murder. Now abortion is a right (of the mother's that is – but not of the unborn child's), hanging is out and imperialists and colonialists are wicked.

Mark you, I neither feel wicked nor do I think that the British Empire was an unmitigated evil. To the contrary, it gave good governance in many lands that both before and after it, experienced lawlessness and corruption. The Commonwealth itself is explicit acknowledgement by former British possessions that the Empire had its good points. By inclination I want to believe that, had it persisted, all would still be well: but this is perhaps no more reasonable than wishing for perpetual youth.

* * * *

My earliest conscious recollections are overwhelmingly visual and best likened to short film clips. They must have formed when I was between two and three years old in Nyasaland. Perhaps most vivid is one that involves a man-eater. In keeping with the time and place, I had a personal 'minder', a youth to look after and entertain me out of doors. My companion of the same age, Brian Wilcox the District Doctor's son, also had a minder. We both had pedal cars, which we could of course pedal independently if that was our wish. Yet both had broom shafts tied to the rear bumpers with which our minders could, upon command, push us. The times, recall, were colonial!

Our minders took us for daily drives in and around the station. On this morning we turned right out of a driveway onto a dirt road. From the sunlight in my recollection it must have been early and our turn was to the south. The road ran level for a way then dipped into a narrow cutting, crossed a wide, flat valley bottom or *dambo* of open grassland, then rose slightly to disappear into a wall of woodland. At the bottom of the cutting to the right of the road was a pool surrounded by a jostling herd of cattle.

We were well down the cutting and near where the road came out onto the grass plain when bedlam broke out. Cattle came away from the pool in a headlong stampede towards and past us, then away up the road along which we had come, urged on by yelling herdsmen. Dust swirled and the air stank of cow dung. When it blew away all that remained of what, seconds before, had been a crowded scene were two tots in pedal cars and their minders.

Across the valley floor a small yellow animal walked along the edge of the bush. It was small only because it was far away – perhaps some six hundred yards or so. Yet from its gait, general appearance and my subsequent familiarity with lions, I have no doubt the yellow animal had indeed been a lion. Apparently a man-eater was at large and whether or not this was it, our minders turned us around and pushing us ahead of them fled pell-mell in the wake of the fleeing cattle.

* * * *

The butterfly we knew as a 'Daddy Christmas' – scientifically Papilio demidocus.
(Photo Peter Davey)

Like many, perhaps the majority, of Kenya's 'Settler' boys of my generation I was enthralled by natural history. This is hardly surprising given our rich environment. Encouraged both at home and at school we dived into this great cornucopia, collecting butterflies, moths, insects, birds' eggs, feathers and skins. Some of us also collected snakes – though it was not so widely approved by parents. The outcome was a general natural history knowledge that set us apart from subsequent generations. Between the ages of eight and twelve I think we knew more about local natural history than any section of society – other than African hunters – with which I have been involved since. Individuals unable to name common birds and butterflies were exceptions. Some names were strictly our own. The common citrus swallowtail butterfly was a 'Daddy Christmas' and not until collecting butterflies years later with my son did I appreciate this must have derived onomatopoeically from its specific scientific name *demidocus*.

The war may have made toys scarce and enhanced small boys' predilections to collect. Yet chasing insects and other life was also hunting: an activity for which boys have a natural bent. Taking butterflies and birds' eggs prepared us for bigger things: hunting mammals. As is still the case, the boarding school year fell into terms and holidays. While at school I was with lads of my own race and collecting natural history specimens was stimulated by the competitive spirit characteristic of such institutions. In the holidays, however, on the family coffee farm the only other white boy was my brother, James. Before I was twelve the three year gap between us separated our interests and abilities so markedly that I associated more with the farm labourers' sons of my own age. Like my white peers they, too, enjoyed hunting. If our interests differed, they were more interested in whether or not the quarry was edible. Some tribes, notably from western Kenya, considered even small birds as *mboga* (a vegetable) to be eaten with their staple diet of *ugali* – maize meal cooked like Italian polenta. These we killed with 'catties' (catapults) made from forked sticks and rubber strips from car tyre inner tubes, or by making snares from a cow's tail hairs. A single cow's or a horse's tail hair is strong enough to snare a sparrow-sized bird.

We were eager for bigger game but until I was old enough to have a gun we ran down our quarry with dogs. Bushbuck and grey duiker were flushed often enough, but we never caught one. Nevertheless the chases were exciting. We were more successful with scrub hares (*Lepus saxatilis*). Nine out of ten times they got away. If pressed these hares go to ground and occasionally one would do so where we could get at it. Tearing around the countryside together we learned about one another as well as nature. Many of my school contemporaries had similar experiences.

By nine years old many or even most white boys (and many white girls) had access to an air rifle. By twelve we had graduated to ·22 rifles and smaller bore shotguns. Those not allowed such weaponry hung on the words of others blessed with more reasonable parents. The less fortunate envied lads who had bagged an antelope. And from an early age we developed a frontiersman proficiency with arms. To this day, let a man or woman lift a firearm and we will know on the instant whether or not they know guns. It is in the way they touch it, lift it and instinctively locate its point of balance.

As we approached puberty and secondary education, horizons broadened and other opportunities beckoned. A few, of which I was one, retained our interest in birds and animals and if there was a common reason for this, it was John George Williams, Ornithologist at the Coryndon Memorial Museum (now Kenya's National Museum). A short, ebullient Welshman, he imparted enthusiasm for zoology as no-one else I have met. Mark you, the time he gave to youngsters rewarded both him and the museum richly. By harnessing our interests he controlled and honed our collecting habits. Collecting for the museum served a useful purpose and we were introduced to formal biology. Through John Williams a whole cohort of my generation were directed into wildlife-based careers.

Grey duiker which we chased with enormous enthusiasm and no success.

* * * *

Natural history was but one part of a general schooling between 1941 and 1949. The overall ambience was very British. The cane – Victorian antidote to all small boys' problems – was used a great deal. Counselling (wimpish word) was unheard-of. The war heightened senses of honour, duty and jingoism and I recall a padre telling his congregation of small boys that God had appointed the British to lead the world to better things. Such certitude made us proud. In as far as we knew anything, the British Empire was definitely a good thing.

The idea that all things British were wonderful slipped a bit when I went to England's Cheltenham College in January 1950. I was not the only Kenya colonial there. The College's Imperial military connections and the fact that Kenya had many soldier settlers created strong ties between school and colony. In the colonies we had a sense that the most British of the British – and therefore the most superior of a superior race – would be in Britain. Somehow the people in the 'home country' would be on a higher plane than us colonials. I went there expecting to be something of a bumpkin and was slightly shocked to find that this was not so.

Britain was emerging from the trauma of a World War that, in relative terms, had hardly affected Kenya. It was demoralised, near broke and floundering in costly social experiment. Pride in being British, senses of honour, duty and responsibility seemed on a lower plane to that of colonials. Dispossession and nationalisation were bearing down on those whose sons were our contemporaries. Masters as well as fellow pupils brought up racialism: the general drift being that we colonials should not be living on what must surely be African land? This philosophical hostility, for all that it was low key, confused boys brought up believing in the Empire's unity.

A profound personal impression came from a visit to the College 'Mission': a charity funded by the College to give recreational facilities to boys in what today would be called a disadvantaged part of town. Cheltenham College lads were encouraged to go there to both broaden their own outlooks and benefit those among whom we mixed. I went once out of curiosity, not charity, and was shocked. I had not met semi-literate and ignorant white folk before. For all that it sounds insufferably arrogant, so absolute was my belief in white (and British) superiority, that until the visit I did not appreciate whites could seem more ignorant than African labourers back in Kenya.

Amongst Kenya's white colonists traditional British class distinctions counted for little. The College Mission exposed the nether pole of England's social stratification which, given my attitude at the time, reinforced the opinion that far from the Kenya mob being poor country cousins, it was 'Poms' who fell short of expectations.

This attitude was not unique to a small clique of boys at one English school, but is widespread in the former colonies and Dominions: Rhodesia, South Africa, Australia and New Zealand. And if the difference between the colonial (Imperial) and the 'home country' attitudes needs a highlight, one set of statistics makes it: the proportion of able-bodied men who rallied to the colours in the second Boer War and two World Wars. Proportionate to overall numbers, far more colonials took up arms than did those in Britain itself. Perhaps this should not surprise. After all, those who went out to make the Empire as rulers, merchants, settlers and colonists were self-selected for their Imperialism. The less adventurous stayed at home, gave up the Empire, then cut their long-standing trade links with the remaining Dominions for a place in the communal European bed. Getting into bed to profit is an old profession, but many a colonial who fought for King and country has since crossed the bourn wondering why he rallied to the flag. The Australian phrase 'whingeing Pom' and that country's drift to cut ties are the outcome of these feelings.

I run off at a tangent. Parents and masters said that I should try for Oxford to read zoology, and had it not been for Mau Mau I might have done so. Certain that my country needed me and not wanting to miss out on any action, I threw in the academic towel, returned to Nairobi, reported for duty in December 1953 and was mortified to be told that, at seventeen, I was too young. To fill in time I worked on Glanjoro Farm owned by Cenydd and Alison Hill just outside the town of Nakuru. Cen was a hard taskmaster who believed that before one could lead or direct subordinates, one should be able to do the job at hand oneself. Thus, despite being white, I started my working career with a hammer and stone chisel knocking holes into concrete under the African farm mason's eye. Cen's attitude was not atypical of that prevailing generally and I recall this start to make a point. In his eyes, I had no innate superiority over other workers. If I was to lead I must first know how to work.

The Hills were mixed farmers who ran a large dairy herd. Working with the cattle entailed rising before dawn and not being through until well after dark. Cen ensured that young hired hands had no free time. Though I have never worked as hard since, neither have I enjoyed a job more than those farming days employed for a munificent five pounds a month. When I left to join the Kenya Regiment, however, I knew that pulling cows' tits was no way to earn a living. No other job is so unremittingly routine.

CHAPTER THREE

MAU MAU AND GAMEKEEPING

The eastern wall of the Rift Valley presented the Uganda Railway's builders with a formidable problem. To get from the rim to the valley floor two thousand feet below, the line would have had to follow a gentle gradient cut into the valley wall with viaducts spanning the many ravines that incised it. Today the rails follow such an alignment, but in 1899 the engineers were under pressure to get railhead to Lake Victoria as quickly and cheaply as possible. They came up with an unusual, albeit temporary solution. Trains would descend to the Rift floor by winch down a gradient of 45°. Arriving at the valley rim, trains were lowered coach by coach and wagon by wagon on a winding lift. The weight of one wagon going down was counterbalanced by another coming up on a parallel track. Today the alignment of this ingenious system is followed by the high-tension electricity line that brings Kenya power from Uganda's Owen Falls dam.

I knew nothing of this in January 1955, on my first anti-Mau Mau patrol as a private in the Kenya Regiment's 'O' Company. Having combed a patch of forest our platoon took a brief rest. Deployed in a circle more than 50 yards across – everyone facing outward – the forest was so thick that only the person on either side of one was visible.

I happened to be watching my neighbour, a forester in civilian life called Basil Twitchin. Startled to see him raise his Patchett sub-machine gun I looked in the direction of his aim and briefly glimpsed an African in a shabby mackintosh plunging away through the greenery. Basil fired two short bursts and I snapped off a rifle shot. All movement in front of us stopped and the silence was profound. Our platoon commander and his sergeant were quickly with us. While the rest of the platoon stayed still, the four of us moved forward and some fifty paces to our front came across a very ordinary looking Kikuyu sitting down. There was nothing sinister about him.

He did not seem frightened – just resigned. With hindsight he must have been in deep shock. There was no blood that we could see, but when asked if he had been hit, he indicated his shoulder. A bullet had passed through a rolled blanket, then the point of his left shoulder, through the top of his chest and out under his right arm-pit, seemingly having missed all the vital blood vessels and organs in that vulnerable area. Such penetration was more likely to have been a rifle bullet than a 9mm Patchett slug. Further examination revealed two of Basil's bullets had hit low on the spine and were still lodged within. Though it was these two that must have brought him down, the man seemed unaware of them. He spoke rationally for a few minutes and then died. His fingerprints were taken for identification and his corpse was left in the forest gloom.

The event had been anti-climactic. A small, inoffensive-looking person had had the misfortune to bump into us and died. To sustain a sense of righteousness one had to remind oneself that it was (presumably) such as he who had hacked women and children to pieces, had taken vile oaths to kill all who did not conform to their views and deserved no mercy. Anyway, he had brought it on himself as he must have known that anyone found in the forests was liable to be shot on sight. I doubt that such thinking was different to that in which most new soldiers rationalise killing others. In books and films first sight of a dead body at close quarters is traumatic. None of us rookies seemed much affected. One or two looked a little green, but most soon became pretty callous. The only time I recall being sickened – as were several companions – was when a white policeman's wife kicked a dead body as it lay in the police station precincts. Somehow that was deeply shocking.

My first patrol was one of several which ended in the same way. Few prisoners were taken and bodies were left where they fell. The only section of the security forces that took prisoners regularly were members of the Police Special Branch who needed them for intelligence. Some of these prisoners were 'turned' and went back amongst their erstwhile comrades in 'pseudo-gangs' to either capture or kill them. British troops were supposed to play by the rules and treat the wounded according to the Geneva Convention, but this seldom happened. Prisoners were rarely taken and the wounded died of their wounds. The Mau Mau had even less interest in taking prisoners.

After several weeks in the forests of the east Rift Valley wall north-west of Nairobi, our Company went to Narok and to what is now the Masai Mara Game Reserve (much of what is now open plains was then thick scrub and woodland) and through the beautiful country above the Nguruman escarpment on the Rift Valley's western rim. We went to hunt for one of the last aggressive Mau Mau gangs still operating. Led by a half Kikuyu half Masai we knew as Mugurori, it had recently killed some game scouts and wiped out a British army survey team, in both cases capturing its victims' weapons, and was now well armed. We failed to find the gang, which was later destroyed by a police pseudo-gang.

From Narok I was seconded to the Administration as a temporary District Officer: my full title was District Officer (Kikuyu Guard) and we were known as DOKGs. Initially as second in command to Mike Whittall, my duties were military and we led a platoon-sized Tribal Police Combat Unit based in a 'fort' high up in the Aberdare forests above Fort Hall (now Muranga) District. There were four such units, each led by two young whites – either as DOKGs or Kenya Regiment soldiers on detachment. Later I took over a unit and my second-in-command was one John Henry (Jock) Anderson. Our year in the forests was idyllic. We fished for trout, shot game for the pot and I collected birds for the Museum in Nairobi. We emerged frequently from our retreat to attend parties and play occasional hockey. As long as we patrolled the forest, we were doing what was demanded of us and no-one interfered. By 1955 the few remaining Mau Mau rebels were far better in their bushcraft than we were and incredibly adept at concealment. We had very few contacts and our rôle at that stage of the rebellion was not so much a high kill rate as to keep constant pressure on those still in the forests. If they kept coming across our footprints, that alone sustained the psychological stress of knowing that they were being hunted.

Like all idylls it ended. The Mau Mau in the forest were so few that it was no longer worth keeping units such as ours in position. Jock had served his two years of military service and returned to civilian life. I elected to stay on for longer than my compulsory two years and was assigned to administrative duties in the locations outside the forest.

Momentous changes were afoot. Government took advantage of the State of Emergency to force the Kikuyu into accepting changes in land tenure that they had rejected earlier. Their traditional communal land ownership was a barrier to progress. In Kiambu District, which was typical of Kikuyuland generally for example, there were 37,134 smallholdings of which 17,316 were less than four acres in extent; 10,359 were between four and seven

As DOKGs we led a platoon of Kikuyu Tribal Policemen after Mau Mau gangs in the forests: I.P (left) and Mike Whittall. (Photo M. Whittall)

acres; 7,975 between seven and 29 acres and only 1,484 more than 30 acres. The average holding was 5·3 acres but divided into eight fragments – often widely separated. The traditional sense of tenure had evolved around shifting agriculture and that while occupying and tilling land gave a person rights to it, actual ownership was vested in the community or clan. Rising human numbers and limited land was preventing shifting agriculture. Plots that were small and widely scattered were difficult to farm efficiently. Government wanted to bring fragments together, issue the owner with title deeds for the consolidated block and abolish communal ownership. Thereafter further subdivision into units of less than four acres would be discouraged. On an owner's death his farm was supposed to be left to one son who should pay each of his brothers the value of an equal share within five years. Inherent in this thinking was acceptance that those forced off the land would have to seek employment in a field outside agriculture.

Elsewhere in Kenya the Kalenjin people accepted this land consolidation, but Kenyatta and the Kikuyu had rejected it for short-term political expediency. Once in power Kenyatta reversed his opposition and endorsed the idea. By giving peasants private title to their land they acquired the collateral to raise capital. Today it underpins a prosperity and concern for land that elevates Kenya above all other African countries.

Outside the forests many of my duties were preparing the way for land consolidation. At the same time there was a parallel programme to repatriate the thousands of Mau Mau interned in concentration camps for the duration of the Emergency. It was interesting, but after a few months at it I was thinking about a more permanent career.

By 1956 I may not have had a university degree, but had been educated in an altogether broader field of human affairs. Those of us seconded from regular army duties to outstations, lived alone with Africans and were among the last to have that Imperial experience which made the British Empire what it had been. By virtue of one's whiteness and nationality one was expected to lead. As eighteen- and nineteen- year-olds we were responsible for the lives of others on a scale that is difficult to convey to people of those ages today. We were not intrinsically more capable, but merely conditioned. One did what was expected because it was expected, without sensing anything remarkable about it. Indeed it was all a bit of a lark. This only stands out now because the expectations are gone. What counted was mental attitude. The experience gave us a self-confidence beyond academe to impart and I never regretted coming back to Kenya to fight Mau Mau instead of going to university.

* * * *

In mid-1956 two posts in the Kenya Game Department were advertised, inviting applications. Though I knew nothing of the game laws and had only the vaguest idea of what game wardens did, it must have to do with wild animals and I applied for one of the posts. There were, I believe, over eighty applicants and I was very surprised to be selected. The other winning applicant – David Brown – was also from the Administration (which may be a clue), only he was a pukka District Officer and not a temporary DOKG as was I. The Chief Game Warden, Willie Hale, was an ex-Administration man, having previously risen to the rank of District Commissioner. That, too, may have counted heavily where David and I were concerned. I had had two of Kenya's better known trout streams (the North and South Mathioya) in my area of Fort Hall and Willie Hale loved trout fishing. Because of the Mau Mau Emergency Regulations he had to have special clearance to fish these streams and I had to look after him while he did so. This, possibly, influenced my fortunes.

CHAPTER FOUR

CONSERVING

An Irishman might say this chapter was written with the foresight of hindsight. It breaks the chronological narrative to establish perspective for conservation in Kenya between 1900 and 2000 and to give an idea of the field I entered in 1956.

At the end of the 19th century there were no 'ecologists', 'conservationists' or 'environmentalists', yet I doubt that the world was a worse place. The desire to preserve wildlife was strong, but it supported few specialists except gamekeepers. The gentry of Britain and Europe went to some lengths to looked after nature. Their interests were not as holistic as is the general case now and they had a category of which one hears little today – vermin – for animals which they felt should not be preserved. The plant counterpart to vermin is weed. Annual sales of rat poison and weed killers are billions of pounds more than is spent on conservation – proving both are still abundant and important.

In Britain gamekeepers were the only professional cadre devoted to preserving wild animals. In Germany foresters looked after wolves, wild boar, red deer, roe deer and bears as part of their duties. Nevertheless across Britain and Europe it was part of the landowners' general ethic to understand the basics of game preservation. In keeping with the times, issues were seen in simple and certain terms. The rules of gamekeeping revolved around how to produce a maximum of desired species (e.g. pheasants, partridges, red grouse, boars and deer) and keeping predators as few as possible.

Europe's acquisition of African estates was paralleled by fascination with their flora and fauna. The gentlemanly interest in natural history fostered at home was a feature of the Europeans in Africa. Hoping for immortality by having an animal or plant named after them, many travellers collected as and when they could. It all helped foster a deep and growing sense of responsibility for the planet's wildlife. As soon as the Dutch established themselves at the Cape in 1652 they took steps to preserve wildlife. The author and biologist, Clive Spinage, listed some of them.[7]

In 1657 Governor Jan van Riebeeck forbade shooting birds. Further decrees on hunting were issued in the same year and then again in 1667, 1668, 1680, 1687, 1751, 1771 and 1792. In 1680 Governor Simon van der Stel brought in laws that prohibited unlawful hunting, introduced a licensing system and a ten month closed season with no hunting. Before 1800 the rules decreed fines, flogging and forfeiture of firearms for poaching. The sale of poached game products was banned, dogs had to be controlled and it was illegal even to disturb wild animals. Long before the British bought the Cape from the Dutch for £6 million in 1814, the Dutch residents – the Afrikaners – led the world in this colonial endeavour.

It has become conservation dogma that because the world's first national park was proclaimed in the United States of America (Yellowstone in 1872), modern conservation

---

[7] Bothma J. & Rabie A., 1983. *Wild Animals* in *Environmental Concerns in South Africa*. Juta & Co. Ltd., Cape Town. (For a summary see Spinage C., 1991. *History & Evolution of the Fauna Conservation laws of Botswana*. The Botswana Society, Occasional Paper #3. Gaborone, Botswana.)

must have started there. In truth, over the past four millennia and possibly longer, wildlife and forest sanctuaries have been proclaimed spontaneously in many cultures around the world including Africa, and the idea is not a uniquely American contribution to land use. The USA may have used the name 'national park' first, but that is all. The concept of conserving, which is what counts, is far older. Where white people are concerned, South Africa was ahead of the USA by three centuries.

The Afrikaners' steps to preserve game are interesting as for the past two centuries the British maintained that the South African Dutch were game exterminators. Afrikaner game laws show that the British were wrong. For the time being, however, game laws were an early feature of Africa's control by whites. Whether or not the Portuguese preceded the Dutch in making conservation rules I don't know because I do not read or speak Portuguese.

Yet while white interest in preserving African fauna can be traced at least as far back as the mid-1600s, it reached a higher pitch after the continent's 1884/85 partition into European estates. An excellent exposition of what happened in East Africa was assembled by historian Nora Kelly[8]. Rules drawn up by the Imperial British East Africa Company (IBEAC) in 1887/88 provided for both conserving and using wildlife. They derived from more than just a conservation stimulus. Believing it would make the Company profitable from the outset, the IBEAC declared a monopoly over trading elephant tusks. It also realised that the equable, healthy climate of highland British East Africa, with its open country and visibly abundant wild animals would attract wealthy sportsmen and laid down licence fees that such hunters would have to pay.

Herman von Wissman, Governor of German East Africa (Tanganyika), was keenly interested and it was his brainchild to establish uniform conservation measures across Europe's African territories. Whitehall picked up the idea and the first international conference on preserving Africa's fauna was held in London in 1900[9]. While it achieved little on the ground, it confirmed substantial official interest in conserving at the end of the 19th century. The Convention emerging from the conference opens *In the name of Almighty God* – the ultimate patron. It implies a greater certainty about the order of things prevailed in those days. Yet it also underwrites a spiritual dimension to conservation that recurs widely across the centuries.

Surprising many modern conservationists, the London Convention listed animals which might not be killed under any circumstances, decreed the protection of immature and female animals, ordered the setting aside of reserves, called for the strict regulation of trade in game products and many other restrictions besides. In a nutshell, in principle little new emerged over the subsequent hundred years.

The colonial government that took over when the IBEAC went broke, strengthened and expanded the Company's conservation rules. In the British East Africa Protectorate (later Kenya) an administrative officer, Mr Richard Crawshay, was initially responsible for enforcing the game laws. Shortly thereafter Arthur Blayney Percival was appointed the country's first official game ranger. However, the true founder of the Game Department was

---

[8] Kelly N., 1978. *In Wildest Africa: the Preservation of Game in Kenya 1895-1933*. Ph.D. Thesis, Simon Fraser University, Canada.

[9] Anon, 1900. *Convention for the Preservation of Wild Animals, Birds, and Fish in Africa*, signed at London, May 19, 1900. Her Majesty's Stationery Office, London.

Blayney Percival, Kenya's first game ranger.

Sir Frederick Jackson who, as a senior administrator and an exceptional naturalist, laid all the plans and secured all the agreements for a special conservation department which was created in 1907[10]. Blayney Percival became this department's first official. For six years he had worked alone as a member of the Administration. A field man who disliked office routines, he did not want to be the new departmental head. Thus in 1907 when the Game Department was formed, Lt Colonel J. H. Patterson of *Man-eaters of Tsavo*[11] fame was made Game Warden of the Protectorate. Percival became senior assistant game ranger and two assistant game rangers were appointed: Major C. J. Ross DSO, and a Mr G. H. Goldfinch.

Patterson was neither a good naturalist (one has only to read his *Man-eaters of Tsavo* to appreciate that), nor a particularly good hunter. An abrasive and high-handed martinet, he had made a strong impression when as the Uganda Railway Administrator of Nairobi in 1899 he burned the fledgling town's shanty quarter to the ground to stop a plague outbreak. It did that, but landed the Railway with numerous claims for lost property.

---

[10] Parker I.S.C. & Bleazard S., 2002. *An Impossible Dream.* Librario, Scotland.
[11] Patterson J. H., 1907. *The Man-eaters of Tsavo.* MacMillan & Co. London.

Patterson lasted only a year as Game Warden before resigning under a cloud. Going on safari to inspect the boundaries of the Northern Game Reserve he invited a husband and wife to accompany him. On the trip the husband was shot dead, allegedly by his own hand. Inevitably rumour had it that Patterson, lusting after the wife, had done the deed. Later, after leaving Kenya, he married her giving the rumour added substance. But Kenya, then as now, has been an exceptionally fertile bed for rumour, so no historian should treat the tale as anything else.

Percival acted as Warden until a man named R. B. Woosnam was appointed to replace Patterson in 1910. Thoroughly professional, *Appendix II* of his *Annual Report for 1910-1912*[12] (the years were combined) is a surprisingly modern dissertation on *The Question of the Relation of Game and Disease*. Despite his veterinary professionalism, however, there are sections in Woosnam's reports that either contradict one another or do not come to grips with reality. In *Appendix I* to the combined reports for 1910/11 and 1911/12 he wrote –

> *"Now it is a remarkable and conspicuous fact that the game and the natives in British East Africa have never interfered with each other, there has never been any collision between them, and they have always been able to co-exist in perfect harmony."*

Given Woosnam's knowledgeable comments on disease, this implausible statement is surprising. By the time that he wrote it, the need to protect crops and property from wild animals was already apparent and in neighbouring Uganda had become a critical issue[13]. His sentiment was rooted in what he wanted to believe supported by some simplistic logic. He argued that as both game and people were present when white man arrived, and as both had presumably existed together for a very long time, *quod erat demonstrandum*, they must live harmoniously. If this was not so then surely the natives would have got rid of the wild animals long before white man arrived?

Suffice it that Woosnam did not see the natives as a threat to conservation. His *bêtes noires* were the white settlers as this quote from the same source makes clear –

> *"It must however always be remembered that . . the white colonist has been an unqualified drawback and evil to the game . . . "*

Another quote shows even more starkly his failure to grasp the issues –

> *". . in the East Africa Protectorate there is a very large native population who are bound to be provided for somewhere, and this means that there must be set aside large areas of land for Native Reserves. In these areas the game and the natives can and will live and both be preserved together."*

---

[12] Woosnam R. B., 1913. *Annual Report of The Game Warden, British East Africa Protectorate.* Government Printer, Nairobi.

[13] Swynnerton C.F.M., 1923. *Report on the control of elephants in Uganda.* Government Printer, Entebbe.

Woosnam obviously had conservative views on how fast the natives were likely to progress. Yet where land is put to arable use, game is eliminated regardless of the colour of the people involved. Even where pastoralists were concerned, Woosnam was wrong, though he was not alone in this perception.

Writer after writer still holds that Kenya's Masai live in harmony with the wildlife around them. The assumption arises from observing that *nomadic* pastoralists occur on the same ranges as herds of wild animals. If they share the same ranges, they must surely get on?

Zoologist Mike Norton-Griffiths[14] showed up the fallacy of this argument. Analysing aerial surveys over more than 250,000 square kilometres (100,000 square miles) of East Africa, he observed that the contrary was true. Over 93% of the counted wild animals were found in areas where, at the moment of counting, there were no people or their stock. They might be quite close to one another, but they did not mix as, for example, do zebra, Thomson's and Grant's gazelle and wildebeest. Clearly they were avoiding one another. However, because the humans were nomadic, the wild animals could use areas as and when people moved off them. The two groups shared ranges and resources, true, but not the same space at the same time. This sharing collapses when people and stock become sedentary. When they stay in one place, wild animal use of it drops dramatically.

Conflict between pastoral herds and wild fauna was not just over direct access to resources. The transmission of disease is strong ground to keep them apart. Predation has always been a source of outright conflict. Large tracts of Masailand are today strangely deficient of predators. This came about through widespread poisoning by the Masai with an arcaricide that cattlemen used to keep their stock free of disease-bearing ticks between the 1950s and 1970s. To lions, leopards and hyaenas the ingredient Toxaphene (a trade name) was both attractive and lethal. They willingly drank it from the plunge dips through which cattle are forced to swim, and eat meat into which it has been injected. This quiet poisoning campaign by Masai ran for many years, giving the lie to peaceful Masai co-existence with wildlife. They looked on predators as stock thieves in the same way as most stockmen.

All this is perfectly obvious, but when Woosnam wrote his reports the level of general conflict between wildlife and people was at its lowest for a century before or since. The great plagues of bovine pleuropneumonia and rinderpest in the 1880s had reduced cattle numbers, which, in turn, must have lessened competition between stock that survived and wildlife. All the people had been afflicted by smallpox epidemics and the massive famine of 1897-9. Their numbers, too, were reduced.

His claim that the game and the natives would be preserved together has a certain grandeur. It says much about conservationist attitudes towards people both then and now, but also stresses the huge cultural gap perceived between the primitive African state and that of the civilised Europeans.

Woosnam said that his policy would have three pillars. The first was to prevent trade in game products. The second was to introduce game laws to the natives that, combined with other civilising measures, would encourage them into lawful commerce, which excluded wildlife. The third was the proper supervision of white sportsmen coming to Africa to

---

[14] Norton-Griffiths M. Pers comm.

shoot. This was as good a description of Departmental policy over the next fifty years as was ever issued: childish in both its simplicity and deficiencies.

Counting Woosnam himself, there were only four officers in the Game Department, who were helped by a staff of less than fifty African Game Scouts. At that time the British East Africa Protectorate was over 700,000 square kilometres (273,000 square miles) in extent. Divided into four equal ranges this would have given each officer 146,240 square kilometres to look after. To give readers some perspective, this is bigger than England or New York State or North Island, New Zealand. To add to this perspective, an estate in England of say 5,000 acres (c.20 square kilometres) would, at the turn of the century, have had at least one gamekeeper, but more probably a head keeper aided by one or two assistant keepers to look after the game on it. Thus while experience in developed, law-abiding England indicated manpower of an order of at least one person to 20 square kilometres was needed, in wildest Africa, the staffing level was one man to 5,700 square kilometres. Woosnam accepted that he was understaffed, but his comment in 1913 that "one more additional ranger is badly needed," merely showed how far he was from reality. In recent times the subject has received some rational consideration. For example, Rowan Martin of Zimbabwe, engineer, biologist and, *inter alia*, a piper of merit, produced Martin's law[15] which presents conservation's manpower needs as a formula. It is one among several to choose from; but all make the point that with the wisdom of hindsight, Woosnam was sadly distant from reality.

The game laws did, however, make provision for creating honorary game rangers or wardens. It was a device that allowed the head of the Game Department to confer all the considerable legal powers of a game ranger upon any one he felt warranted such responsibility. Theoretically this cadre of amateurs increased the strength of the Game Department enormously. For extended periods of the colonial years there may have been substantially more than one hundred honorary wardens on the books. Yet, as a body, they never fulfilled their theoretical promise.

From the outset there was an element of romantic delusion in perceptions of game rangers. Their duties called for men who were competent with arms, self-reliant, knew the ways of animals and how to hunt them. Although only one of Woosnam's 1910 staff had been a professional soldier, the military was the obvious field in which to find men with such qualities. It is thus not surprising that a high proportion of rangers and wardens in colonial game departments and national park services had been soldiers. The association was helped by the tradition that hunting was manly and army officers were encouraged (if not expected) to participate in it. Military service in India gave many the opportunity to indulge in *shikar* and get acquainted with big game. It was further helped by Britain's class structure in which an officer was assumed to be a gentleman, and part of a gentleman's background was knowing something about game and gamekeeping.

The connection between familiarity with wilderness and hunting skills with martial matters was not peculiar to British Africa, past or present. Davey Crockett, Daniel Boone and the American West's frontier scouts were closely linked to the army and played an

---

[15] Fact Sheet 6, African Resources Trust, Harare. (N_s = A where A is expressed as km^2)

influential rôle in breaking Red Indian opposition to white arrivals. The concept, probably correct, is that if one can hunt animals one can hunt humans. The converse of the idea that soldiers make good game rangers is that game rangers make good soldiers, which is supported historically. In times of crisis the military has always turned to hunters, rangers and wardens, whether in the American west, the last century's two world wars or in the independence wars up and down Africa over the past four decades. In tacit acknowledgement of their military value, such men were usually made officers. Given this situation it was not surprising that on the outbreak of World War I every ranger in Woosnam's Game Department including himself enrolled in the army. Neither is it surprising that they were employed in scouting and intelligence gathering.

Most recently in the war for Zimbabwe's independence that ran from 1974 to 1980, the military-gamekeeping-hunter connections were restated. National parks staff were organised into volunteer tracker units (VTUs) where wardens and their game scouts tracked guerrillas and led the Rhodesian army into 'contacts' with them.

* * * *

Woosnam's policy statement had two crucial deficiencies: (i) knowledge of wildlife's rôle in African lives and that in certain situations they depended on it, and (ii) appreciating the conflict between wildlife and people. The early white gamekeepers thought that so 'civilised' an idea as preserving the wild had yet to dawn in the African mind. And it followed that if the idea was lacking, there could be no indigenous conservation systems. This was incorrect. Africa has conserved for reasons that varied from reserving hunting rights for a king to ensuring the welfare of the spirits of the departed. Shaka Zulu, the King of the Barotse and the Mwami of Burundi had game sanctuaries where they, alone, might hunt. King William I of England setting aside the New Forest as his own hunting preserve was a direct analogue.

The Chewa and Manganja people of Malawi had a complex religion in which the spirits of the departed played a strong rôle in the lives of the living and for this reason had to be kept happy. Associating undisturbed nature with tranquillity, grave sites were protected from fire and disturbance and consequently became small nature sanctuaries scattered across the land. For the Mijikenda people of coastal Kenya certain hilltop sites are sacred. These *kayas* are not allowed to be disturbed and have become unique samples of the region's flora. The Ogiek forest hunters of Kenya's Mau range have a complex system for dividing wild resources among themselves. Not only does this allocate families' and clans' hunting rights to specific areas, but sets seasons when certain animals may be hunted. Whatever the rationales, all were conservation systems.

The early whites knew Africans hunted, but made no effort to establish why they did so or to measure its importance. That some Africans had to hunt to survive was given little thought. Section 36 of the 1909 Game Ordinance made some allowance for those who traditionally lived by hunting to go on doing so, but it was dropped from subsequent game laws.

No concessions for Africans to use wildlife were made, nor was the possibility entertained that they might want to hunt for the same recreational reasons that white men hunted. This was not basically a racial attitude, but more an extension of Britain's social class system in which the landed gentry were seldom in favour of their own peasantry being allowed to hunt. Nevertheless, racial attitudes did exacerbate this inequitable outlook. In Kenya, white security influenced a classic Catch 22 situation. Blacks should not have guns because they might be used against whites. The game laws stipulated that game could only be taken with firearms. If the natives had no guns, then they could not have game licences. Game licences cost so much that Africans could not afford them. If they could not afford the licences, then no ground existed for relaxing the prohibition against them having guns. Whichever way one approached the problem, Africans were effectively barred from hunting.

Africans had produced ivory for millennia, which proved a hunting competence. That they had traded it with men from overseas – including the British – was evidence it had a rôle in their economies. These facts were abundantly documented but simply overlooked when forming game policy. Small wonder that Africans in Kenya so generally disregarded conservation laws. Consequently villagers shielded poachers and game scouts have had to be armed throughout the twentieth century. Woosnam's policy made it inevitable.

The first white colonists found that the Game Department treated them as it did the natives and denied them rights to control or get rid of game on their land. Referring back to this pre-1909 situation, at a meeting of the Convention of Associations in August 1913 Delamere, the Settlers' leader, said[16] –

> "We have seen game laws introduced here by the Colonial Office which can only remind one of the game laws of the old feudal days in Europe."

The situation Delamere was harking back to had in fact been changed in 1909, when the new game laws allowed landowners to destroy animals causing damage or being a nuisance to them. Woosnam conceded in the following year –

> "it was no longer possible to continue the compulsory preservation of the large herds of gregarious species of game on the areas which were occupied by the colonists."

He then went further –

> "If success is to be attained, it is essential that the regulations dealing with the preservation of game should meet with the approval and support of the body of colonists who have to live under such regulations ... They are beginning to realise that under the present Game Ordinance the game on their land is their own property and, in many cases, of considerable value to them, and the game on private land will be far more efficiently protected in this way than by compulsory Government protection. For each landowner who wishes to preserve the game on his land at once virtually becomes a game ranger. This is the explanation why there is any game existing in Europe today."

---

[16] in Hill. *Ibid.*

What Woosnam had not foreseen was that when Government took control of wildlife matters, it had to assume responsibility for their depredations upon agriculture. This gave rise to 'control work', the euphemism for killing animals to protect crops or exact revenge on them for damage done. As was also the case with its sister departments in Uganda and Tanganyika, in their control work the Game Departments, guardians of the wild, paradoxically killed more wild animals than any other body in the land in the twentieth century. Woosnam had not foreseen that control shooting would constitute the fourth and most robust pillar of policy.

Woosnam was unfortunately killed in the Great War. Yet the policy he outlined was to last out the colonial era and beyond. The Game Department was amalgamated with Kenya's National Parks organisation in 1976 to form the Wildlife Conservation & Management Department (WCMD). In 1989 the WCMD became a quasi-government body, the Kenya Wildlife Service (KWS), under a board of trustees. The longest serving and most impressive of Woosnam's successors was Captain Archie Ritchie (1924-1948) who proposed and was instrumental in the formation of national parks. With this striking exception, none who followed Woosnam redressed the basic deficiencies of his policy. Consequently, throughout its history, Kenya's conservation authorities have presided over a more or less constant decline in the country's wildlife.

Creating national parks was the one step that rectified defects in Woosnam's prescriptions. The Royal National Parks of Kenya Ordinance was enacted on 26th June 1945. Inherent in the national park concept was knowledge that Pleistocene wildlife cannot be preserved amongst Holocene people or on their agricultural land. For this reason and once they were formed, the national parks became Kenya's primary conservation devices and the Game Department became of lesser importance.

Yet while national parks were an altogether more sensible way of preserving wildlife than the game laws which preceded them, they embodied many of the weaknesses in colonial thinking. No consideration was given to what Africans, the majority of the country's people, thought of the idea. The boundaries of every park dispossessed some local people of some asset. In the most extreme cases they had to vacate homes which lay within the new parks' borders. In some places communities lost grazing rights they had held for centuries. Hunting grounds were lost, but as traditional hunting was illegal, the hardship was difficult for the policy makers to grasp. At the other end of the spectrum of losses, locals were denied access to resources such as honey, fuel, bush foods and traditional medicines.

* * * *

Thus from the outset of colonial times the British East Africa Protectorate and subsequently Kenya had, on paper, conservation policies and laws to back them up. Yet, with hindsight, both policies and supporting laws were practically and philosophically exceedingly defective.

George Adamson, quintessential game warden and romantic.

CHAPTER FIVE

THE QUINTESSENTIAL GAME WARDEN

Broadening the perspectives of conservation in Kenya, this and the following two chapters will recall some of the personalities who were prominent in this field. George Adamson was arguably the best known game warden of twentieth-century Africa. Mark you, his fame came not from his work as a warden, but for his rôle in the lives of two formidable females: his wife Joy and the lioness Elsa. Subsequently he took up with other lions and ended in a blaze of glory (no pun intended) when shot by Somali bandits. I doubt George ever claimed to be the quintessential game warden, though he certainly liked the image and it is how he was widely seen.

George hated office work, but enjoyed writing and wrote well, though his spelling was a trifle phonetic. As he described it, being a game ranger was as romantic in real life as it was portrayed in the media. This even came through in his monthly and annual reports which made light reading. Those for the years 1938-49 survive and can be examined for objective evidence of how good a warden he really was before he took up with lions and the media.

Born in 1906 in India, educated at Dean Close School in Cheltenham, George drifted through a variety of interesting jobs. He joined the Game Department in July 1938 to look after the Northern Game Reserve. From the outset he referred to himself as 'Assistant Game Warden – Northern Reserve' at a time when most in the Game Department called themselves Game Rangers.

In 1897 Kenya's first game reserve was proclaimed[17] and encompassed most of 'Kenia District' – an ill defined but nonetheless huge area that extended from the Aberdares and Mount Kenya, taking in the Laikipia Plateau northward into the unexplored north of the territory. With the arrival of white settlers, their need for land and a better idea of the land and its people, the Kenia Game Reserve was found too unwieldy and impractical. It was abolished and in 1900 two vast new game reserves – the Northern and the Southern – were proclaimed to replace it. The Northern encompassed a chunk of what is now northern Kenya. The Southern was equally impressive taking in all of modern Kajiado District. The Northern Game Reserve was home to the Samburu tribe, at least three hunter/gatherer 'Dorobo' tribes as well as some Turkana and Rendille. The Southern was occupied from end to end by Masai.

No shooting was allowed in either Northern or Southern Game Reserves and before the creation of national parks they were 'absolute sanctuaries'[18] the nearest to that concept that the Game Department had come up with. However, other than banning hunting, the Department and its officers had no control over the people who lived in the game reserves. To those who did not know this, the two reserves looked impressive on paper. They contained thousands of people and yet bigger numbers of livestock. Government had a greater

---

[17] Foreign Office communication to the Commissioner for the British East Africa Protectorate, Sir Arthur Hardinge, dated December 30th 1897 (FOCP 7032/181).

[18] Woosnam. *Ibid.*

responsibility to look after the humans and their development than it did to preserve animals and both Northern and Southern Game Reserves lacked conservation teeth.

Reduced to basics, George's job as assistant game warden was to stop anyone hunting in his reserve and to protect the people who lived in it from the depredations of wild animals. Later, in 1942 after a brief period of military service, he was made responsible for game affairs in all northern Kenya east of Lake Turkana and in addition all land north of the Tana River down to the sea and the Somali border. With never more than 40 Game Scouts and often fewer, he was supposed to both protect people from wild animals and enforce the game laws over 217,600 square kilometres (85,000 square miles). True, the Administration and Police with their greater resources were supposed to help him generally, but the task was too big for the available manpower.

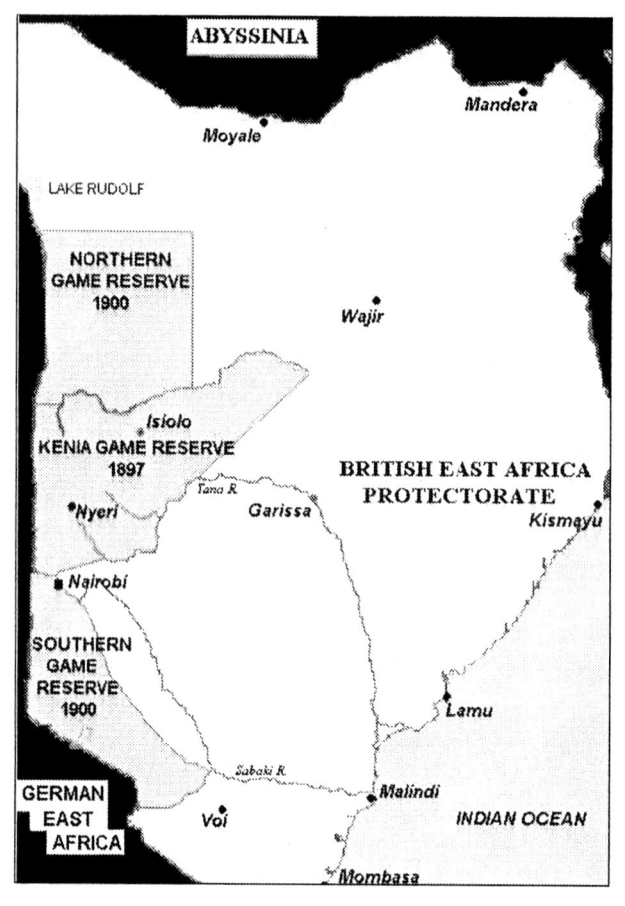

The British East Africa Protectorate's game Reserves in 1900.

We can analyse George Adamson's reports to see what he did. To protect humans and stock in the eleven years between 1938 and 1949 he killed –

- 50 lions (averaging between four and five a year) for man-eating or stock-raiding,
- 52 elephants (averaging between four and five a year) for crop-raiding or killing people,
- 3 rhino (one every three to four years),
- 5 buffalo (one every two years),
- 4 leopards (one every three years),
- 2 wild dogs (because he did not like them),
- 1 crocodile and countless hyaena.

What is striking about these figures is actually how few animals were accounted for. They are less than the rates for some of the country's bigger ranches. The figures above relate to animals that he knew were killed. It is badly deficient in one aspect. Far more than hunting them down and shooting predators, George Adamson used poison, principally strychnine. Many poisoned animals will have died uncounted. As a poisoner George Adamson had no rival in the Game Department. He used it routinely on hyaenas for which he seemingly had a pathological dislike. No-one with experience in Kenya's Northern Frontier District (or NFD as it was usually known) will deny that hyaenas do a lot of damage, but George's attitude was extreme. In March 1939[19] he wrote –

> *"There are certainly far too many in many places, wherever I find them troublesome I always put down poison."*

His technique was to put lethal doses of strychnine in lumps of meat that were then scattered about the place where the animals were causing trouble. After a 'good' night between ten and twenty dead animals would be found. As not all died in the vicinity, the actual deaths will have been higher than recorded and included an unspecified number of dogs, jackals, leopards and lions, all of which scavenge and take baits. In addition to his own poisoning George also made strychnine available to the District's administrative officers to hand out to local herders. In May 1943[20] he reported –

> *"I have asked the DO Maralal to restrict the issue of poison to natives. I consider that the poisoning of lion has been overdone in the past year."*

George thus never knew how many animals died of his poisoning or from the poison that he distributed to the local people through the Administrators. Like most game wardens he shot many antelope for the pot and took out elephant licences for the ivory which he

Hyaena, for which Adamson had an irrational loathing and poisoned widely throughout his career.
(Photo Peter Davey)

---

[19] Adamson G., 1939. *Monthly Report to the Game Warden for February 1939.*
[20] Adamson G., 1943. *Monthly Report to the Game Warden for April 1943.*

then sold to augment his salary. So, too, did his wife of whom George wrote[21] to the Game Warden on the 30th May 1946 –

> *"Joy ... got her first elephant a few weeks ago, an eighty pounder and very proud of it she is too and will not hear of selling the tusks."*

Between 1938 and 1949, George Adamson and his men arrested 447 Africans (averaging 41 a year) for poaching, three Afrikaners, two British Army officers and twelve Italian ex-prisoners of war. The majority of African cases were petty in that they had either killed animals to eat or possessed some part of a game animal – an ostrich feather for example. The Afrikaners had killed an elephant without a licence, the officers had shot a couple of buck for the pot while the Italians had been collecting ivory to work into artefacts. Somalis comprised less than 3% of Adamson's arrests. His poaching cases averaged around four a month if his time on military service is excluded, and reveal much about his inclinations.

In 1942 the area for which George was responsible was increased by 600% and took all districts north of the Tana River. There was a small increase in the men under his command, which should have increased the number of monthly convictions for poaching, particularly as many people in his new range subsisted heavily on wild animals. There was, however, no rise in the number of convictions for poaching.

If one accepts the conventional view that poaching for food is less important than commercial poaching, then Adamson concentrated heavily on the former and almost ignored the latter: the reverse of what he should have done. According to him the majority he arrested were Dorobo and Turkana people.

His use of the word 'Dorobo' is revealing. The term derives from two Maa (the language of the Masai) words, *'l'torro'* and *'bo'*, meaning those with bees in their stock enclosure and applied specifically to honey hunters and hunters and gatherers generally. Swahili caravaneers modified them to Wandorobo as a catch-all term for hunting people. White folk guided into East Africa by the Swahili, picked up the name, often dropped the prefix Wa and, like George, spoke of the Dorobo tribe with impressive but spurious authority. To this day most whites, even those born in Kenya, do not realise that there are many distinct hunting and gathering tribes – Ogiek, Il Tikiri, Hadza, Wata, Boni, among them – each with their own specialities, characteristics and sometimes languages, but all of whom are lumped as 'Dorobo'. Yet there never was a tribe of that name.

Within the Northern Game Reserve there were at least three hunter gatherer Dorobo tribes: the Mukogodo who spoke an eastern Cushitic language (Yaaku); the Mumuuyot, and the Il Tikiri who spoke Plains Nilote languages akin to Maa. They were all small groups and it is easy to see how they were overlooked and ignored by the general run of both black and white people. However anyone whose business was to stop natives hunting should have been aware of all this as the knowledge was central to his work.

George did know that there were several thousand Dorobo men, women and children in his area, who did not cultivate, owned few domestic animals, and quite openly depended on hunting and bee-keeping. Their way of life was recognised but disapproved of by the

---

[21] in litt. Adamson G., 1946 to the Game Warden.

George Adamson's range at its greatest just after the Second World War.

Administration who wanted them to settle and be 'civilised'. To this end an agricultural project was established for them at Wamba on the western slopes of the Mathews Range where these hunters could be converted into farmers.

George supported the programme to the extent that on more than one occasion he arbitrarily and quite illegally seized (kidnapped) Dorobo women and children, sending them under guard to the settlement. He hoped that their men folk would come looking for them and settle where they found them. In 1947 the project was abandoned because, hardly surprisingly, its supposed beneficiaries kept running away. The issue is mentioned here because despite the shallowness of his knowledge of the people, his efforts to settle them proved that he at least knew they depended on hunting. It is in the face of these facts that George's penchant for romanticizing stands out. In an excerpt from his monthly report for January 1944 he wrote[22] –

---

[22] Adamson G., 1944. *Monthly Report for January 1944 to the Game Warden.*

> *"Most of the damage [i.e. poaching] is done by a gang of five men, three of whom have already served sentences for poaching, including the ring-leader. The only way to stop him is to give him a job."*

In the following April report he reported –

> *"I sent some of my men into the Mukogodo country and they succeeded in arresting four more of the worst poachers. Three were convicted and received heavy sentences. There are now only three real professionals left at large and the position ought to remain satisfactory for some time, particularly as the two cleaverest [sic] and boldest of the gang have now got jobs as Game Scouts."*

Here we have George the raconteur iterating the romantic belief that poachers readily turn gamekeepers. It also supported a legend in the making of old George Adamson being 'fully in control of the NFD'. The tale might have suited rural Lincolnshire, but was at odds with the reality of northern Kenya where several thousand people depended in large degree on hunting.

* * * *

Where the Turkana people were concerned (and they were his favourites) he showed greater interest and knowledge. Writing to the District Commissioner Isiolo on July 21st 1943 about his worry that the Turkana were poaching more than ever before and –

> *"... will continue to do so, however rigorously the game laws are enforced. Fines and imprisonment have no effect, simply because the majority of the men are destitute and in order to live and support their large families are compelled to either steal stock, or poach game.*
>
> *"Periodical round-ups of vagrant Turkana and their return to [Turkanaland] mearly aggrivates [sic] the problem and causes a great deal of hardship ..."*

George's concern for the Turkana illustrated awareness that the game laws conflicted fundamentally with the people's way of life. He knew that the Turkana (and Dorobo) poached of necessity, that the necessity related directly to their numbers and that it was an activity regularly undertaken by many Turkana and the majority of 'Dorobo' men.

This contrasts with the correspondence to his headquarters creating an image in which all or most illegal hunting was done by a few gangs of 'professionals'. It is also a comment on his superiors that they accepted what he wrote.

Adamson's treatment of those who hunted for meat and out of necessity contrasts starkly with his approach to commercial poaching. Within a week of commencing his first safari as a warden George received information that certain Somali villages were buying much rhino horn (recorded in his first monthly report). Yet, red-hot and as an eager new boy out to impress his superiors, he never mentioned these Somali horn buyers again and the issue was not followed up despite a continuous flow of evidence.

The Administrative Officer in charge of the NFD – Gerald Reece – wrote to his DCs in Garissa and Isiolo[23] drawing their attention to an extract from a Kenya Police report dated August 1942 –

> "Captain Corfe, Assistant Political Officer, Kismayu estimated that about 3,000 elephant were shot in the Garissa District and the Somali District adjoining Garissa from March 1941 to March 1942. The most profitable market for poachers now is for giraffe hides and leopard skins.
>
> "As regards leopard skins, the Political Officer, Lugh (in Jubaland), told me recently that the trade in them was officially encouraged in Somalia."

A Turkana from Kenya's north-west: Adamson's favourite tribe (Photo Peter Davey).

George Adamson received a copy of this letter and he made several trips across the border after which he wrote to Archie Ritchie, the Game Warden on 11th November 1942 from Moyale –

> "Within the last twelve months 389 leopard skins have been sold in the Lugh market place alone ... many from Kenya ... Prices up to 60/- [shillings] were paid. Traps were sold in the market. Also a substantial leopard skin trade into Abyssinian Moyale flourishes."

In his monthly report for November 1942 –

> "... Small parties of Bon Marihan poachers are said to cross over the border between Mandera and El Wak and poach leopard and giraffe, only a few are armed with rifles."

In January 1943 –

> "I came across many places where leopard traps had been set along paths ... gather the Boni carry on considerable trade in giraffe skins and leopard skins with Somalis ... the leopard traps are probably supplied by Somalis from Jubaland."

---

23 Reece G., 1942. *Administration Letter GA/22/2/1/1159 of 16.10.1942.*

In July 1945's report George mentioned Kamba/Somali/Arab leopard skin rackets. His November 1946 report said –

"... a Somali trader with a leopard skin ... got 6 months in jail."

It also recorded that the informer critical to this conviction was subsequently murdered. While George clearly knew of widespread Somali trafficking in game products, for reasons best known to himself, he did little about it.

George also showed a similar reluctance to prosecute white poachers. During the war ranchers were allowed to shoot game on their land and to sell it to butchers in Nanyuki to feed Italian prisoners of war and military labour. It was good money for anyone with a rifle and the necessary transport. Inevitably game on the ranches was reduced, leading some entrepreneurs to look for it elsewhere in the Northern Reserve where they were not allowed to hunt. George knew who was involved, yet he never got them to court. Writing to the Game Warden on the 20th April 1944, he asked if 'Mac' (MacArthur: who had been a policeman before joining the Game Department) would prosecute a case involving white men –

"I am rather afraid of making a hash of the case, as most of my experience has to do with native poachers, the majority of whom one can usually persuade to plead guilty."

This reluctance to prosecute people who were likely to put up a fight either in the field or subsequently in court was most emphatically not because Adamson was a coward. His death while charging to rescue a party under attack by Somali bandits illustrated a brave man. More likely, he just did not enjoy such work.

Evidence of this comes from an incident, which occurred in 1944. Until 1945 game scouts unaccompanied by a ranger or policeman had no powers of arrest (many did not even have identifying uniforms or badges). George disregarded the laws and used his men to bring in poachers with neither a ranger nor policeman present. Eventually a Somali who knew his law picked this up and his defence was that he had been arrested unlawfully. The DC hearing the case agreed, the prisoner was freed and the two game scouts were themselves prosecuted. It also resulted in amending the law to give game scouts powers of arrest.

* * * *

Occasionally George was caught up in the irony of it all. Where in 1938 he had arrested wretched Africans for possessing an ostrich feather or two, in 1939 in northern Laikipia he killed over 1,000 zebra and oryx to aid the war effort: leaving them to rot where they fell so cattle could have the grass the game would have eaten.

From the outbreak of the war George's personal file contained letters asking to be released to join the army. He felt uncomfortable seeing his contemporaries in uniform while he remained a civilian. The Government was in no hurry to let him go. The Italians were expected to invade Kenya from the north, which made the local tribes apprehensive. George's ability to get around the District on foot and his familiarity with the people was seen as a calming influence, but it also confirmed that those in authority knew what he did best: wander about remote places on safari, preferring to be under canvas than in a house. He was ordered to spend the early months of 1940 showing the flag among the tribesmen: a rôle he detested.

In July 1940, to his gratification, George Adamson was called up and, as was the case with so many from the wildlife and hunting field, posted to military intelligence. Keen to get into action he wangled a posting to Grant's Irregulars – a force of Somalis under Hugh Grant (a former soldier turned DC and temporarily returned soldiering). Their hopes of being an independent fighting force were dashed when their unit was used to provide guides and scouts for regular army units. They saw some action as the Italians were driven out of southern Somalia but reaching the port of Kismayu, Grant's Irregulars were withdrawn to administer captured territory. This was not what George wanted and on his first leave back to Kenya, he wriggled his way out of military service and back into being a game ranger. Patriotism had distinct limits!

A letter to the Game Warden of 16th April 1946 illustrates another difficulty game rangers experienced: meeting day to day expenses out of their own pockets, then reclaiming them from Government at the end of each month. Thus George wrote –

> *"I am finding it increasingly difficult to meet all expenses on Government work out of my own pocket . One way out is to get a larger imprest, which I am not very keen on, or for the Department to pay wages and lorry expenses directly and not via my pocket. Sorry to have so many grouses but I am getting really worried about money."*

Without clerical staff, George wrote in longhand in a duplicate book, which doubled as a filing system. No other department proceeded on so *ad hoc* a basis. It would never have been possible if the Department had not shared in the romantic delusion of what a game ranger or warden was.

In recent times people hark back nostalgically to the good old days when George Adamson ran the NFD single-handed. Truthfully, his presence probably made no difference to wildlife in that region. George Adamson *was* the quintessential game warden not for his achievements, but for the inconsequential rôle he played.

He spent the last decades of his life collecting captive lions and releasing them into the wild, claiming to conserve them. And while many did not see it as such, his apparent sincerity hushed criticisms. His reputation for high principle was tarnished when he secretly killed at least one truly wild competitor in a national park to make space for his personal lions[24]. Coming after convicting more than 400 people who killed animals principally to eat, such contempt for the law purely to please himself, Adamson should have suffered the maximum penalty possible. It was not the only poaching in a national park of which he was guilty. He was caught red-handed by warden Myles Turner, shooting antelope for his lions in the Serengeti National Park, despite having been told he was not to do so. It exposed his lion keeping for the extravagant personal indulgence it was.

I write critically and the image I have painted is at odds with the received picture of George Adamson. He never was a great conservationist and deserves no place in history as

---

[24] House A., 1993. *The Great Safari: the Lives of George and Joy Adamson.* Harvill/Collins, London.

such. He created the myth about himself within the Game Department through his reports and writing. Outside the Department others added to it. His delightful nature predisposed people to believe the best of him. It was further enhanced in counterpoint by his wife's abrasive, publicity-seeking and neurotic personality. From the purely conservation standpoint, his own records prove an ineffectual career. Though this is indisputable, it is neither what the public wished to hear, nor makes him a lesser man.

In truth George was an adventurer romantic. His long foot safaris to remote places with a few local tribesmen as companions caught his colleagues' interest in the same way as Thesiger's walks among the Arabs. There was a great deal of impractical romanticism in the image of a game ranger and he fitted the bill perfectly. His subsequent involvement with lions was folly in the grandest sense. He wasted time magnificently but entertained the world. We would be the poorer had he not done so.

Archie Ritchie, the most influential individual in East Africa's conservation history.

CHAPTER SIX

AND THERE WERE OTHERS

When one uses individual cases to make a point, as with George Adamson in the previous chapter, those not mentioned tend to acquire an unnatural obscurity. Many others contributed to Kenya's conservation image and 'shape'. Early in the century Blayney Percival stood out, mainly because he was on his own for so long and because he subsequently wrote books on his experiences[25] Yet he was primarily a naturalist and while his observations are historically valuable, he had small reputation as a law-enforcer or policy maker.

Captain Archie Ritchie was Kenya's Game Warden from 1923 until 1949. He had served in the 1914-1918 war first in the French Foreign Legion, then in the Grenadier Guards when he was decorated for gallantry. While he had a degree in biology, he was not particularly scientific in his work. His attitude reflected gamekeeping in the sense that it was practised on English estates. Thus in 1926 he approved Lord Ramsden introducing red deer to Kenya. The first batch were acclimatised at Kipipiri and then released to breed and populate the moorlands of the Aberdares (years before it was made a park). They died out quickly. Apparently an essential stimulus in their reproductive biology is variable day length and the more or less constant day length on the equator did not suit them. The last surviving red deer may have been seen in the late 1930s on the heights of Eburu, an extinct volcano in the Rift Valley well south of where Ramsden released his founding stock. The observer, a Masai herdsman, reported seeing a buck 'with horns like trees', which fits deer antlers.

Ramsden's experiment excited Ritchie to propose introducing Indian Black Buck onto the Athi and Kapiti (Kaputei) Plains south of Nairobi. Superficially trivial, this hang-over attitude from the Victorian era is what led to New Zealand being colonised by not only the British people, but British red deer, blackbirds, thrushes, chaffinches, goldfinches, starlings, Indian mynahs, rabbits, ferrets, brush-tailed possums from Australia and many other species at no little cost to the native flora and fauna.

Red deer, some of which were introduced to Kenya in the 1920s, but needing variation in the length of daylight to trigger their reproductive cycles, they failed to breed and died out.
(Painting of the North African race, Jonathan Kingdon).

---

[25] Percival B., 1924. *A Game Ranger's Notebook*. Nisbet & Co., London, &
Percival B., 1928. *A Game Ranger on Safari*. Nisbet & Co., London.

Ritchie staunchly advocated national parks and put his views strongly to the Kenya Land Commission in 1933. In this he was but one among many who from the turn of the century understood that big game and humans do *not* get along, and in the long run the animals would only survive if special areas free of humans were set aside for them. The list of people advocating national parks included many senior administrators, hunters, and naturalists such as V. G. L. van Someren. By the 1930s rising luminaries like Dr Louis Leakey lent their voices to the cause and the issue arose whenever the Government formally discussed game policy. Nevertheless, by virtue of both character and his official position as the Game Warden, Ritchie probably had a greater influence than all other individuals in committing the Government to setting up national parks. As a midwife for national parks his influence was reflected in the following extract from his Annual Report for 1933[26] –

> *"In October 1933, there took place in London an event of first class importance to game preservation – The International Game Conference. The main objects of the Conference may be epitomised as follows. To secure the setting aside, in every territory in Africa where there is considerable indigenous fauna and flora, of areas which shall in perpetuity, or such approach to it as is humanly possible, harbour in a natural and primitive state examples of the wild life of the territory; such areas to be known as National Parks or Strict Natural Reserves. To secure the universal institution throughout Africa of as adequate a degree of general game preservation as local conditions will warrant or allow. To facilitate the co-operation between territories to the above ends, as also to secure such uniformity as is possible with regard to game legislation, to prevent the extermination of certain rare or threatened animals.*
>
> *"My reason for mentioning the conference in this chapter is that we may claim without immodesty to have taken no small part in the preparation of the convention which was placed before the conference and formed there the basis of discussion and agreement. The convention was indeed modeled, in its exposition of the theory and practice of game preservation, very largely on existent Kenya conception."*

Large, bluff and forceful, Archie Ritchie was a Kenya 'character'. One did not expect the Game Department to typify the civil service, but his style made the point. Irate farmers whose crops had been despoiled by Ritchie's animals were mollified when he pulled open a drawer in his vast desk, produced glasses, gin and Angostura bitters and insisted that discussion be over a drink. Would that civil servants generally were as civilised: it would make for a happier world. Archie Ritchie was a gentleman of the old school who believed that it was the Game Department's duty to ensure that animals were hunted sportingly and deplored any changes. Thus he wrote[27] about the growing use of motor cars –

---

[26] Ritchie A., 1933. *Annual Report of the Kenya Game Department.* Govt. Printer, Nairobi.
[27] Ritchie A., 1928. *Annual Report of the Kenya Game Department.* Govt. Printer, Nairobi.

"A safari of months with its blankets and guides and preparations now set out for a whirling week-end with a few puncture patches, while a bag once valued for the hardships, disappointments – even dangers – which had gone to obtain it, now counts its achievements in terms of broken springs. The romance has gone, and with it, for the most part, all the old canons of good hunting."

Given that it was the general style to write anecdotally, Ritchie nonetheless took this to the same level as George Adamson and also enjoyed telling stories. In his reports data are hidden in verbiage and appear inconsistently. Seemingly his Department's duty was to 'protect the game' by making sure that white men hunted legally and sportingly, black men did not hunt at all and 'bad' animals were controlled. He appreciated that as land was developed it would progressively be denied to big game. Yet by the same presumption, though he never wrote it in so many words, the Game Department's duties would gradually dissolve away. Most likely, this will have seemed so distant a prospect that it did not warrant considering. It is only in the last quarter of this century that people have grasped the very immediate costs of human increase.

Using game commercially troubled Ritchie. In his Annual Report for 1924[28] he felt the fact that zebra hides fetched shs 4.00 each was beneficial in that it solved the problem of too many zebras. In 1926[29] he reported on the danger of allowing game to be killed for profit.

In 1927[30] he says 'happily' zebras were no longer a pest because they had been shot out. In 1934[31] damage to agriculture by game would be resolved if the –

"killing of game could in some way be put on an economic basis: if, in other words, the killing of zebra and the rest could be made to pay."

Yet he never did allow anyone to develop commercial wildlife uses. It figured when he had to resolve damage to agriculture, but was never taken to the level of sustainable commercial offtake for its own sake.

In Kenya little was done to enforce the game laws where Africans were concerned until Ritchie hired a Mr C. G. MacArthur, who had previously been a detective in the Police Criminal Investigation Department in late 1926 and, *inter alia*, was a champion heavy-weight boxer (it was he who George Adamson called for to prosecute white men). Mac, as he was widely known, was not a writer and left few records. He retired at the end of the Second World War, became a Director of the hunting firm Safariland Limited, established Mac's Inn at Mtito Andei and died c.1950 leaving little material evidence of his activities. Such as remains indicates that he was an outstanding field ranger whose reputation lingers on in the folklore of Ukambani and the Coast Province hinterlands, in Ritchie's occasional written comments and a paper in Kenya's National Archives that he wrote on the Wasanye (Waliangulu)[32]. MacArthur seemed to have a single goal: to jail as many poachers as he could.

---

[28] Ritchie A., 1924. *Annual Report of the Kenya Game Department*. Govt. Printer, Nairobi.
[29] Ritchie A., 1926. *Annual Report of the Kenya Game Department*. Govt. Printer, Nairobi.
[30] Ritchie A., 1927. *Annual Report of the Kenya Game Department*. Govt. Printer, Nairobi.
[31] Ritchie A., 1934. *Annual Report of the Kenya Game Department*. Govt. Printer, Nairobi.
[32] MacArthur C.G. Typescript, Kenya National Archives.

MacArthur, policeman, boxer and most successful anti-poaching warden Africa has seen.

In the Annual Report for 1927[33] Ritchie wrote –

"I am glad to be able to record that, before the end of the year, had been laid the essential foundations for a vigorous campaign in the country of the Wasanya and other Wanyika tribes in the Kipini Mombasa hinterland; and that such campaign has since been attended with the utmost success, much ivory being recovered and a vital blow being dealt to the main stem of the ivory ramifications throughout the area . ."

In 1928[34] he wrote of MacArthur's work at the coast –

"He not only obtained valuable information about game distribution but recovered for government over 100 elephant tusks and obtained some 75 convictions for the illegal killing of elephant and other game by the various Wanyika people."

The term Wanyika was a catch-all Kiswahili phrase that the British adopted for Giriama, Chonyi, Rabai, Ribe, Duruma, Digo and Pokomo tribes of the Coast Province. In 1930[35] MacArthur was making his presence felt. His work was the first time that anyone in the Game Department had come to terms with *any* aspect of native hunting. In that year he secured 495 convictions which resulted in fines of £3,268 (or more than £326,800 in 2003 values) and jail terms totalling 71 years (averaging just over two months per conviction). One case in Malindi concerned the capture of 218 rhino horns weighing 434 kg (955 lb) that resulted in the Arab concerned being fined £560 (more than £56,000 in 2003 values). An additional two cases concerning 34 and 38 rhino horns respectively were prosecuted in Nairobi. MacArthur said that the Wasanye, Wakamba and Giriama hunted chiefly for rhino owing to the high value of the horn, but also significantly for elephants. He had evidence that Somalis were trafficking both ivory and rhino horn on a large scale from Kenya to Italian Somaliland and thence to Zanzibar which, at the time, was the major entrepôt for both commodities. In 1929 and 1930[36] Zanzibar imported rhino horn of respectively 1,375 kg and 1,222 kg from Somaliland (representing approximately 625 and 555 rhino).

In 1931 MacArthur's convictions were 477 (in one year more than Adamson had arrested in eleven) and Ritchie wrote[37] –

---

[33] Ritchie A., 1927. *Annual Report of the Kenya Game Department.* Govt. Printer, Nairobi.
[34] Ritchie A., 1928. *Annual Report of the Kenya Game Department.* Govt. Printer, Nairobi.
[35] Ritchie A ., 1931. *Annual Report of the Kenya Game Department.* Govt. Printer, Nairobi.
[36] Anon, 1929 & 1930. *Blue Books for 1929 and 1930 for the Protectorate of Zanzibar.* Government Printer, Zanzibar.
[37] Ritchie A., 1931. *Annual Report of the Kenya Game Department.* Govt. Printer, Nairobi.

> "The outstanding event of the year was MacArthur's dramatic campaign in Wagalla country. These favoured folk had previously enjoyed complete immunity from interference in their ivory operations; and boasted, with some reason, that they could not be divested of their freedom from restraint. MacArthur, whose plans had been laid thoroughly and well beforehand, gave them a rude shock. He arrived at Kipini in mid-April with – I quote his words to me – '90 prisoners and witnesses. Some 70 Wagalla and Wapokomo, 7 Wagiriama, and 2 Wakamba were convicted. Total fines amounted to about £1,400. In my presence about £600 was paid in cash'. In addition to the fines, much ivory and some rhino horn were recovered."

Ritchie also reported that Italian Somaliland was an open door for the illegal trophy traffic. In 1932, '33 and '34[38] MacArthur kept up his pressure and convicted 492, 504 and 751 people respectively for poaching offences. He also trekked through country that no white man is known to have crossed before. Therein, of course lay the key to much of MacArthur's success. He was operating in a huge tract of land that was virtually unknown to white men and in which the natives openly went about their hunting. His statistics led Ritchie to write –

> "We must admit, however, that we have not succeeded in perhaps the most important aspect of our work – the suppression of rhino poaching and dealings in illicit rhino horn ..."

Convictions in 1935, '36 and '37[39] were 590, 500 and 457 respectively. MacArthur was not of course on his own; he had a staff of fifteen Game Scouts. Nonetheless the total of 4,341 convictions, the great majority of which were secured by him and his men, in a space of ten years may be an all-Africa and perhaps a world record. As already indicated, arrests were easy to make because the people made little attempt to hide what they were doing. On top of this MacArthur used tactics that through lack of probity eroded Kamba belief in white men keeping their word. One occasion is long remembered with wry humour in eastern Kitui. MacArthur declared that if those in unlawful possession of tusks or rhino horn surrendered them to him they would not be prosecuted. Thinking that it would be a good way to get him to leave the district, many Kamba brought some trophies to the 'amnesty' meeting, whereupon MacArthur went back on his word, seized them, and prosecuted them to a man. The hiatus caused by the Second World War saw the end of MacArthur's service with the Department.

---

[38] Ritchie A., 1935. *Annual Reports of the Kenya Game Department for 1932/34*. Govt. Printer, Nairobi.

[39] Ritchie A., 1935, 1936, 1937. *Annual Reports of the Kenya Game Department*. Govt. Printer, Nairobi.

* * * *

Mervyn Cowie, brought up in Kenya, an accountant but a wildlife enthusiast, believed in National Parks. Without them he foresaw wildlife perpetually giving way to expanding human development. It has become part of received conservation history that he, almost alone, was responsible for the formation of Kenya's national parks. It is a false perception as others had called for them while Mervyn was still a child. When Ritchie was making his case for parks to the Kenya Land Commission in 1933, Cowie was only 25 and not long articled. To give Mervyn his due, he never claimed sole responsibility for the evolution of the Kenya parks. He came to prominence in wildlife matters when, under a pseudonym he wrote to a local paper advocating wildlife extermination. As hoped, it stirred outrage and, though Government had already decided in principle that parks would be a good thing, Cowie's publicity hastened translating intention into action. He became the first Director of the Royal National Parks of Kenya, running the organisation until 1965.

Cowie, the first Director of Kenya's national parks, and a clever and far sighted strategist.

Mervyn Cowie was not particularly knowledgeable about natural history. His forte was administration and as a political tactician. This strength was reflected in his rapid rise to the rank of Colonel during the war. Mervyn was far-sighted and, unusual among conservationists, foresaw that there would come a point when part of the conservation estate might have to be surrendered in the face of human increase and gave the matter much thought. It was his misfortune that the political tides swirling about the demise of Empire eclipsed this forward thinking being applied and robbed him of more enduring kudos in the conservation record than that he is currently accorded.

Mervyn's foresight gave his overview of conservation an edge of pessimism. While population increase was one aspect, he was also perturbed by the degree to which western support for conservation was emotional rather than logical. Though he rode the emotional horse when it suited, he knew it was capricious. Discussing this, I asked him if he had noticed the cyclical nature of conservation emotions, with issues coming up in regular sequence? Outbursts over poaching, for example, came up in the 1930s, 1950s and 1970s. And in some synchrony they were separated by spells in which the public talked about wildlife use. He had, and felt that in 1979 things really had not advanced much since his boyhood. The issue perplexed him and with a twinkle he said, "Ian, it so puzzled me that I even went to an astrologer about it!"

* * * *

Evelyn Temple-Boreham was born on the 6th of March 1913. He grew up loving the bush, open-air life and hunting. Aged eighteen he became a laboratory assistant in Kenya's Veterinary Department, but soon changed to being a Livestock Officer because the work was out of doors and took him to wild places in the 'Native Reserves' where game was abundant. In 1939 just before war broke out, he transferred to the Game Department. He enrolled in the Kenya Regiment, was seconded to the King's African Rifles with whom he served in Ethiopia and Burma, where he won a Military Cross and rose to the rank of Major. Returning from the war, it was then his good fortune to be posted to Narok District – perhaps the best wildlife area in Kenya because it contained the northernmost part of the 'Serengeti Ecosystem'.

Lyn (Evelyn) Temple-Boreham, an admired game warden very well liked by the Masai of Narok District.

Lyn or 'TB,' as he was known, was a larger than life character who made Narok his personal fiefdom in a manner unmatched by any other game ranger. So much was this the case that even the Chief Game Warden let TB select his stand-ins to run the Narok station when he went on leave. Only a small minority of the Department ever qualified for the honour. He got on with the Masai exceptionally well and, always being in the field and not being transferred away from the District as other government officers were, they knew him better than any other official. Indeed in their eyes he was more than just a game official: for them he was Government's representative even above the DC and consulted him on issues far outside the conservation ambit. Such was their liking and respect that years before his service was due to end, the Masai gave TB some land to live on when he eventually did retire from the Department. Therein lay his strength as a game ranger: communication with the local people, interest in their lives and knowledge of their cattle created trust that served him well.

TB turned a blind eye to Africans hunting for food. Only if they were blatant or took elephant, rhino and leopard did he arrest them. Yet where white men were concerned, he enforced the laws rigorously. Using his good relations with the Masai he developed so effective an intelligence network that hunting parties could do little that went unobserved and unreported to him.

TB's lasting memorial is the Masai Mara Game Reserve. Knowing that the Masai had a deep aversion to surrendering more land for animals (Nairobi National Park and Amboseli in Kenya, Serengeti and Tarangire National Parks, the Ngorongoro Conservation Area and Mkomazi Game Reserves in Tanganyika had all been carved out of Masai land), he got them to create their own locally controlled reserve in 1962.

TB acted as Chief Game Warden on several occasions when Willie Hale and Ian Grimwood were on leave and was offered the post when Willie retired. He declined it, however, preferring to stay in *his* Narok. He died of a heart attack aged 56 in 1969.

* * * *

Rodney Thomas Elliott was born into a military family in Bodmin, England, on the 14th June 1921. Commissioned a Second Lieutenant in the Duke of Cornwall's Light Infantry (DCLI), he volunteered and was accepted into the Commandos ending as a subaltern in 6 Commando. His active service was confined to Algeria and after six months of more or less continuous action, principally reconnoitring behind enemy lines, he tripped a German Teller mine, was blown up and rendered unfit for further service with the Commandos.

At the war's end the Second Battalion of the DCLI was posted as part of the British Army of Occupation on the Rhine. Here Major Elliott got his first taste for gamekeeping. Germans were not allowed firearms and their foresters had no means of controlling roebuck and wild boar. Appealing to the British Army for help, Rodney was appointed 'wildlife officer' for the entire Corps. Foremost among his problems was poaching by members of the British forces.

Five fruitless and very cold nights in ambush for poachers were rewarded on the sixth when a vehicle loaded with gunners appeared, spotlighting and shooting anything that was held in the beam – be it deer, fox, hare, rabbit or boar. Rodney stopped the vehicle whose occupants were all from the nearby RAF station. A lance-corporal was driving and next to him was a man who appeared to be in charge, but whose heavy jacket concealed all rank badges, just as Rodney's overcoat hid his. The budding gamekeeper delivered a blistering rebuke to the vehicle's occupants and ordered them back to their base. A day or so later the commander of the RAF base (a wartime hero), telephoned Elliott, revealed his rank and confessed it had been he who had received the tirade. He invited the young major to the base for a drink, ending the incident on a friendly note.

Eventually Elliott decided that a life in the colonies offered better prospects than Britain, left the army, and sailed for Kenya as a soldier-settler. Once there, his experience in Germany stood him in good stead and in 1948 he was selected as a game control officer based at Nanyuki on a salary of £350 a year. Among his early recollections was being shocked by the disorder in the Game Department headquarters' store. Valuable rifles stacked higgledy-piggledy and general untidiness suggested an inefficiency that was later both proved correct and had dire consequences.

Elliott spent the best part of the next two years shooting buffalo and other game on the farmlands around the base of Mount Kenya and the northern Aberdare Mountains. He proved himself in Hale's eyes and was promoted to be a game ranger on the Department's permanent staff in 1950. On leave at the end of his first tour of duty, Elliott was in Caithness for a grouse shoot when a cable arrived requesting his immediate return to Kenya. The Mau Mau rebellion had broken out and all men with military experience were wanted back in the country as quickly as possible. His reply –

"Delighted to return immediately provided His Majesty's Government refunds my share of the grouse shoot costs,"

was followed by an instruction to return as soon as the shoot was finished. Back in Nanyuki he and his game scouts were enrolled in the Kenya Police Reserve and were available for combat on call by the local police. There was no shortage of action. The Mau

Mau gangs were new, had yet to become bush-wise or to learn that it was not tactically sound to stand and fight the better armed government security forces. Elliott's small team of scouts ran up a big score of successful actions in counterpoint to the British troops' general failures. Asked to explain the reason for such a difference in results, he said that the key was in tracking and what is best summarised as 'bush sense' which the British troops notoriously lacked. In short order he was commanded to form a tracker school for them. He ran this until Mau Mau was breaking in mid-1954, before returning to Game Department duties.

Maralal became Elliott's station and the area with which his name was associated for the rest of his service with the Game Department. His correspondence gives an accurate picture of his work in 1955. On 7th May he wrote to Willie Hale about game matters on the 95,000-acre Laikipia Ranch managed by a Mr Southey. He

Rodney Thomas Elliott (The Major) who, it was said, would prosecute his grandmother if she broke a game law.

said that Southey had a reasonable attitude towards game, accepting its presence. He shot buffalo to keep their numbers down as he feared they were a source of cattle diseases and to provide meat for his labour. His offtake was around one a week. Elliott recommended that Southey be allowed to sell forty buffalo hides a year.

On 14th May he wrote about Inspector G. E. Touche of the Kenya Police Reserve who had pleaded guilty to shooting game from a motor vehicle and for which he had been fined shs 300, given a month in which to pay or, in default, serve one month in jail. His licence was automatically revoked and he was barred from taking out another for three years. At the time that he committed the crime he had been with two other white police inspectors.

Several days later Elliott reported charging a Samburu – Lebesoi Lelisimon – for spearing a buffalo for which the man had been fined shs 200. The same letter recorded that he was investigating farm guards who had allegedly killed a giraffe, a rhino and a female eland. On 23rd May he wrote that a Mr Burton had been fined shs 150 or three months in jail for shooting a female eland without a licence and arrested a police constable for unspecified reasons.

On 6th June Elliott caught a Mr G. Fox of the Desert Locust Control Organisation for shooting an eland unlawfully. The accused said that it had been an accident: he had been shooting at a Grant's gazelle but hit the eland by mistake. This assumption of his gullibility so incensed Elliott that he had charged the man for a second offence: shooting at an oryx while within 200 yards of his vehicle. Fox caved in, admitted the offences and asked if the matter could be resolved by 'summary punishment'.

On 10th August he wrote to Gilbert Sauvage, a trapper, asking if he could stock two dams near Maralal with hippo. He said that if the Game Warden agreed, Sauvage could catch hippos where they were unwanted and in lieu of a fee take some of the hippos caught while releasing the others into the dams.

A month later three Turkana hunters were convicted and given eighteen month sentences for setting snares, possessing arrow poison and causing unnecessary suffering. The relative severity of their sentences arose from them having earlier been convicted of killing a giraffe.

This was followed by an angry letter concerning a newly fledged professional hunter – R. W. Ryans – who on his first safari for Ker & Downey Limited wounded and lost a rhino. Instead of following up the animal himself, he had left two trackers to do so while he went back to camp with his client. The incident happened at 10.30 in the morning and Elliott arrived in Ryans' camp at 14.30. Such sudden appearances at awkward moments were characteristic. The hunter should never have returned to camp, but kept on the animal's tracks until they were irretrievably lost or night fell. Caught with his feet up and a beer in hand left him with no excuses and by the time Elliott was through with him he certainly thought that his professional licence would be revoked.

Three days later a letter to headquarters notified his superiors that a well-known professional hunter, had been fined shs 2,000 for taking a greater kudu illegally. This automatically meant that the man lost his licence to take out safari parties. He also asked Willie Hale to notify the Tanganyika Game Department of the conviction to prevent him hunting in that country. In the same letter he reported a Samburu had been fined shs 500 for killing a cheetah.

On 21st September professional hunter John Cook reported wounding and losing a rhino. After Cook described the event Elliott was satisfied that he had taken reasonable steps to rectify the mistake and that no blame attached to the hunters. He then recommended the party's members be issued special licences for an ostrich and a Grevy's zebra. Later when he found out that the rhino had been a female with a large calf Elliott regretted accepting Cook's story.

Later on the same day he wrote to headquarters stating that professional hunting safaris had wounded four big game animals in the preceding four months and wanted Willie Hale to register strong displeasure to the hunters.

The following month he wrote to W. Thoms Esq., allowing him to keep a cheetah as a pet, if he could tame it and that he could keep a stock-killing leopard's skin. Elliott said that he would personally arrange for the necessary permits to be issued.

Four weeks later four American missionaries, M. Donohew, J.M. Retherford, F. Reeve and a Mr Yutzy, were taken to court for killing a hartebeest unlawfully. On 4th January 1956 Elliott wrote his first admirably brief (all of two pages) Annual Report with an additional page of statistics covering the first seven months in his new post. His staff comprised a driver and twelve game scouts, all of whom lived under canvas. Four elephants had been shot on control, three buffalo and a larger number of zebra. They had monitored twelve safari parties that had hunted in the area and shot six elephants, eight rhinos, fifteen buffaloes and seven leopards. Seven white men and thirteen Africans had been taken to court. The game on the Leroghi Plateau and the El Barta plains had been counted for the first time.

The attention Elliott paid to hunting safaris earned him a reputation of being a 'bit of a bastard' and for enforcing the game laws more fiercely than other rangers. Yet he shared Lyn Temple-Boreham's perception that the game laws were unrealistic, and usually turned a blind eye towards Africans taking animals for meat and traditional uses. Whereas it was technically illegal for the Samburu to possess ivory, he took no action against the many who openly wore ivory ornaments. As long as they didn't kill elephants, rhinos or leopards commercially he was not concerned. And when they were troubled by stock raiding predators or a truculent elephant or buffalo, he encouraged them to dispose of it themselves, thus both saving the Department the bother of attending to it and giving the local young men the fun of a spear hunt. He had to be informed of such actions as soon as practically possible.

Some poaching was commercial. His neighbour – George Adamson – brought one case to Elliott's attention. George had an informer who was bringing in a rather regular supply of rhino horns, complete with details of how he had recovered them from poachers – as a freelance game scout, so it seemed. All the horns reportedly came from Elliott's Maralal range and not George's Isiolo range. Investigating, it transpired that George's informer had set up a gang of young Samburu to spear rhino in Elliott's area. He then took the horns across to George's range for the Government reward, concocting tales about wresting them from poachers. The entrepreneur was arrested and jailed. Elliott's early 1950 records thus confirmed that Game Department staff occasionally poached. The most embarrassing of his cases arose while standing in for Lyn Temple-Boreham in 1956 when he caught some of 'TB's' most trusted Game Scouts shooting elephant for their ivory.

Resilient, able to live off poor grazing, the common or Burchell's zebra (or quagga) holds out against humans usurping of wild range longer than most other wild species. (Photo Peter Davey)

Elliott was ahead of his contemporaries as an early advocate of channelling money from wildlife away from central government coffers into those of local authorities or into landowners' pockets. Further, believing that landowners should benefit from game on their properties, he was among the earliest to advocate that they be given game quotas from which hides and skins could be sold to bring in revenue. Further, in as far as possible, he put the allocation of such quotas in the hands of his honorary wardens.

He even induced Willie Hale's successor, Major Ian Grimwood, to recommend that 50% of the value of found ivory and tusks from elephants shot on control should go the local District Council[40]. Elliott saw all too clearly that the strongest incentive for anyone, black or white, to welcome wild animals on their land was if they generated revenue to at least offset the costs of their presence or, better, return a profit. In the 1950s and early 1960s, Elliott's reputation revolved about his law enforcement activities. Retrospectively, however, it was his steps to make game profitable for landowners that set him ahead of most colleagues.

---

[40] Chief Game Warden to Rodney Elliott. Letter GA/3/3-1/4 of 11th January 1961.

As with George Adamson there has been a tendency to mourn the passing of the 'good old colonial days' when conservation was successful in Kenya. Yet it never was successful. Other than in the national parks, it never worked for the simple yet profound reason that there never was a viable policy. Some wardens, like TB and Rodney Elliott, recognised the rôle that hunting played in African lives and deliberately did little to stop it except where ivory, rhino horn and leopard skins were involved. Yet others, like Adamson, tried to enforce the inappropriate laws. For the most part they were ineffective because they did not try very hard and, in any case, they were too thin on the ground to have much impact. The exception, MacArthur, certainly made the Kamba aware of the laws, yet even his staggering record of convictions never stopped them hunting: it merely made them cautious. Elliott's endeavour to allow landowners to profit from wild animals on their land was a rational break with the past. A minority in the Game Department recognised this and by the late 1950s were thinking about change along radical lines.

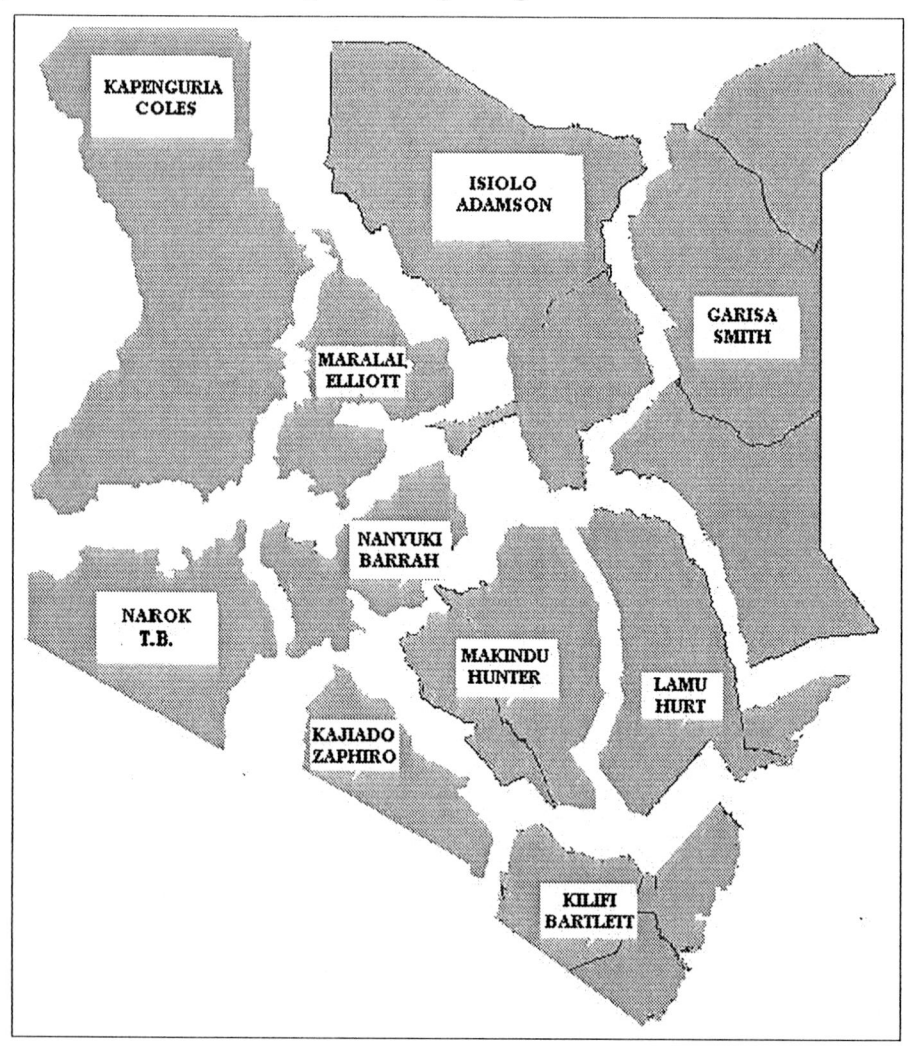

In 1956 Kenya was split into ten game ranges.

CHAPTER SEVEN

A BAND OF GENTLEMEN

The previous chapters provide some background to conserving in Kenya. They are the backdrop that gives perspective for the personal narrative to which I now return. I wrote elsewhere[41] that by being in the right place at the right time I became a game ranger, transferring from temporary status in the Administration into permanent service in Kenya's Game Department. The official ground for my selection was assumption that my recent experience following Mau Mau gangs would make me good at chasing poachers. This was augmented by a basic knowledge of natural history.

* * * *

In mid-1956 the Kenya Game Department was a band of gentlemen. It had twelve white officers, the Chief Game Warden's Secretary Mary Henderson, two Goan clerks – Michael and Eddie – and about 200 African game scouts. As already told, two new Ranger posts were added to the establishment and filled by David Brown (D.B.) and myself, bringing the establishment to fourteen officers (excluding the smaller Fisheries Section). As a hangover from a distant past, we referred to ourselves as game rangers. Yet as the law designated the head of the Department the Chief Game Warden it followed that we should more properly call ourselves game wardens. In early 1958 our chief – Willie Hale – decreed that we drop the use of ranger and call ourselves wardens.

The Game Warden (Headquarters) acted as Hale's deputy, and was Col Neil Sandeman, previously an officer in the Sudan Defence Force. Kenya was split into ten 'ranges': Kapenguria, Maralal, Isiolo, Garissa, Lamu, Narok, Nanyuki, Kajiado, Makindu and Kilifi.

In 1956 the lynchpins of Kenya's conservation system were two legal Acts: the Wild Animals Protection Ordinance (WAPO) No. 18 of 1951 under which the Game Department functioned, and the Royal National Parks of Kenya Ordinance No. 215 of 1948 under which the country's national parks were managed. The Game Department was a government department; its head was the Chief Game Warden, beneath him were Senior Game Wardens, Game Wardens and a subordinate staff of Game Scouts ranked as Sergeants, Corporals and Scouts. Between Game Wardens and the subordinate were Game Control Officers who tended to be employed on a temporary basis. The national parks came under a Board of Trustees, who appointed a Director and staff of Wardens and subordinate African Rangers to run them. The national parks were therefore independent from the Game Department and quasi-governmental in status.

WAPO gave game wardens wide powers. Section 41 permitted any competent officer to enter upon any land for the purpose of carrying out the provisions of the Ordinance. Section 42 extended that power to inspecting and searching premises or tents or baggage or vehicles and seizing any item which appeared to have been obtained in contravention of the Ordinance or used in committing an offence against it. Section 43 gave competent

---

41 Parker I.S.C & Amin M., 1983. *Ivory Crisis*. Chatto & Windus, London.

officers the power to erect temporary barriers across roads and public places without defining barriers. Section 44 gave them the power to arrest anyone 'reasonably suspected' of having committed an offence or who, if left free, they believed would cause an unreasonable delay in being made answerable to justice. We game wardens needed no search warrants or warrants to effect an arrest. To top this off, Section 46 allowed us to exercise all the powers of a public prosecutor appointed under any ordinance for the time being in force. The Chief Game Warden could also appoint honorary game wardens for renewable five year terms and could delegate any or all of his powers to other wardens or to honorary game wardens. Technically the Chief Game Warden could substitute anyone for himself without reference to higher authority. At his sole discretion, he could confer on any untrained, inexperienced member of the public, the legal power to enter and search any premises or properties without warrants, erect barriers on highways, arrest and charge any individual with offences under the game laws and then prosecute them.

With hindsight these sweeping powers were astonishing. Imagine the situations that could arise if the heads of police forces or Chief Justices were permitted to enrol just anyone as either honorary policemen or honorary judges and delegate their duties to untrained amateurs. Yet when the powers accorded wardens by the law were taken into account, there can have been few, if any, parallels to them elsewhere in modern British or British-based law.

One accepted that the Chief Game Warden would be a man of probity, yet there was also a general acceptance that game and park wardens were in some way special. As with religion's priests, padres, ministers and imams, the western public accorded them an exalted *moral* status. Priests and game wardens were traditionally paid very small salaries: they were expected to be dedicated and have vocations for their respective callings. What they did seemed spiritually *good* – whatever that may mean. And just as such senses are acquired more through cultural osmosis than any training, so everyone was expected to understand the need for gamekeeping. It called for no special knowledge. This is certainly still widely believed in the modern world by the innumerable amateurs involved in conservation. The crucial requirement is that one should *believe*. The fact that, in due course, the game laws would be read grammatically without this western cultural background or concern for their spirit was inconceivable in 1956. Yet, as will be shown later, this was to happen in Kenya.

All that, however, is running ahead of my tale. I read the game laws as part of my 'training'. The other part of my training was to read the Departmental files, particularly the Annual and Monthly reports. I have always been grateful for the ten days I spent doing this as it gave me insight into the Department's workings that I could never have otherwise gained. George Adamson's files made the best reading for he was a born storyteller. He had just acquired Elsa and there was still no inkling that the affair between her and her owners would become of worldwide interest. At that early stage Willie Hale's attitude towards Elsa was one of amused tolerance, but even when the Adamsons had only had her for five months or so, he was already aware they were putting the lion cub ahead of duty and that it would lead to trouble.

Michael de Souza, our senior Goan clerk, then went through the prevailing accounting procedures, and I was 'trained'. My equipment was one ·475 double-barrelled rifle, a small tent, camp chair, table, bed and basin, some cooking utensils and my brand new Land Rover. In those days game wardens *had* to provide their own transport and it was a condition of my employment that I bought a vehicle which Government advanced me the money to buy. Just as with George Adamson, Rodney Elliott and all my new warden colleagues, my office consisted of a foolscap-sized triplicate book in which I would write all my official correspondence: the original copy going to the addressee, a duplicate to anyone else I felt should have a copy and the triplicate remaining in the book as my own filed copy. There were no typewriters or clerical staff to help out with office work. I had a Local Purchase Order (LPO) book with which I could buy fuel, and for convenience I paid for many things that could not be covered by an LPO out of my own pocket. For such payments Government would reimburse me later.

The issue of the day was anti-poaching. The Governor, Sir Evelyn Baring, was an enthusiastic conservationist. Reacting to reports of extensive poaching in and to the east of the Tsavo National Parks, he had ordered the Game Department, National Parks, Administration, Judiciary and Police countrywide to crack down on poaching. Greatest attention was focused on Tsavo East National Park where the warden, David Sheldrick, had drawn public attention to the widespread illegal hunting then going on there, principally by Wata and Kamba hunters. I was ordered to the Makindu range to the west of Tsavo and told, quite simply, catch poachers.

The regular ranger in charge of Makindu, J. A. Hunter (known simply as 'J.A.') was away on leave. No-one at headquarters knew where his game scouts were; I would have to go and find them and take them in hand. The District Officer (DO) Makueni, in which Makindu lay, might know where some of them were. There was no accommodation available in Makueni, so I would have to organise whatever I could on arrival, or live in my tent.

I found five game scouts at Makueni. The day after arriving in early October 1956, I took these Scouts, drove out into the bush and raided the first sixteen huts I came to and arrested sixteen men for possessing antelope skins. That this indicated game must have played an important part in their economies went over my head. The prisoners were tried that same afternoon and each sentenced to six months in jail plus a shs 200 fine and I thought that I had made a fine start. Others did not think so.

Unknown to me, David Sheldrick and his assistant warden, Billy Woodley, under whom I had served in the Kenya Regiment, were setting up a special paramilitary mobile anti-poaching unit of platoon strength to take on the poachers. No-one had told me that the National Park anti-poaching team under Woodley had been authorised to operate beyond the park borders and had spent weeks collecting information and planning a raid in the same area to which the Game Department had sent me. When he arrived with a friend (David McCabe) a day or so after my raid, they found the community alerted and Bill's plan spoiled. Unfairly, I thought, I was criticised for this.

Smarting, I went to Voi to meet David Sheldrick and demanded that he liaise more closely before barging into other people's areas. I was a cocky little twerp and the meeting was not auspicious. However, as I was responsible for anti-poaching in the Makindu Range I should have been consulted before Woodley started operations and demanded this be done in future. Later we laughed at the episode, but at the time David and Bill felt that I needed taking down a peg.

In fact the plan whereby National Parks, Game Department and Police were to work together emerged from a meeting chaired by the Commissioner of Police on 10th October

1956[42] which was *after* I arrested the sixteen poachers, so Sheldrick, Woodley and McCabe had jumped the gun. At that meeting it was agreed there should be two more mobile forces modelled on that which Sheldrick had already established. The three would operate throughout Coast Province under Sheldrick's control and as part of Sir Evelyn Baring's anti-poaching programme. I was placed under Sheldrick's command. It was the first time Game Department and National Parks had co-operated in this manner. Sadly it was to be the only time.

The Governor's personal interest focused government attention on poaching in a manner never seen before. Ordering all arms of Government to help, he told the police to go out of their way to bring poachers to book and stressed that magistrates should take cases seriously and hand down deterrent sentences. The Kenya Police Airwing placed an aircraft at Sheldrick's and Woodley's disposal. A chief inspector of police, Alan Child, was attached to the anti-poaching force in Voi, a small railway town on the Nairobi/Mombasa road, to handle all prosecutions and to liaise generally between the National Parks and Police. A senior policeman, Superintendent Rassie Potgieter from the Criminal Investigation Department, was detailed specifically to go after the illegal ivory and rhino horn buyers. It was the Empire's gamekeeping high point and also its last hurrah – though this would not be apparent for years to come.

The difference between Game Department and National Parks was made clear when J.A. returned from leave. In Game Department eyes he was a legendary figure who had shot one thousand rhino between 1946 and 1948, clearing the way for settlement at Makueni. In fact he did not personally shoot the thousand rhinos: he killed many, but his game scouts had shot a substantial part of the total. Further, not all were killed in the Makueni area, though I doubt that such minutae are of much consequence. Prior to joining the Department he had been a professional hunter, was an outstanding shot, a great elephant man, a poacher and a ladykiller. By October 1956 only the reputation was left, though the twinkle when he looked upon a lass revealed an active imagination. Soft-spoken in his native burr, he seemed dreamy and unable to comprehend Sheldrick's and Woodley's military approach. And he couldn't for the life of him see why the natives should not be left alone: they hunted certainly, but they always had. J.A.'s sympathies were with the poachers and all this talk about stopping poaching seemed against the natural order of things. In the national park officers' eyes J.A. was all but useless.

As a certain amount of red tape had to be attended to before recruiting new game scouts for the two Game Department Field Forces could start, I was at David Sheldrick's disposal and attached to the Park Field Force. Sent on patrol to the north-west of the Tsavo Park, we found no evidence of poachers but came across the skeleton of a large bull elephant that had died months previously. Had I been on my own, I would not have found its tusks because they were nowhere near the skeleton. The park rangers, being experienced, looked for them in widening circles, saying that other elephants and not people may have removed the ivory. Sure enough, they found both tusks widely separated from one another and several hundred metres from the skeleton. Each weighed around 80lbs (35 kg) and showed no tooth marks or other evidence that they had been moved to where they were found by

---

[42] Police Minutes of a Meeting held at Colony Police Headquarters at 10 a.m. on Wednesday 10th October 1956, to Discuss Methods of Combating Game Poaching.

predators like hyaenas. Subsequently I saw abundant evidence that living elephants are peculiarly attracted to dead elephants' tusks, picking up, fondling and carrying them – often for hundreds of metres. It is difficult not to be subjective about why elephants do this and why they select them up in preference to a skeleton's more abundant bones. Seemingly they have an aesthetic sense about ivory.

Another of our patrol's tasks was knocking down all beehives we came across. There were many – probably hundreds – of them as every baobab tree had at least one and often several hanging from its branches. Knocking them down was ostensibly to stop their erstwhile owners returning to the park illegally to harvest honey. It was also removing evidence that Kamba had once lived there: the Park was then eight years old.

I then joined Bill Woodley at Mutha, a small trading centre a few miles outside the

Simba Singi Nzenga: an exceptionally accomplished leopard trapper.
(Photo Bill Woodley)

park where he had a series of raids planned. For several nights in succession we visited the homes of poachers on the wanted list, arresting and interrogating them, then rushing off to follow up any lead the interrogations produced. Success depended on speed and the pace was exhausting as the work went on literally for twenty-four hours of the day. Five days later we got a lead on the whereabouts of 'Simba' Singi Nzenga a particularly wanted leopard trapper and Bill sent me to pick him up. He was, so our informants said, posing as a railway labourer at a small railway station called Kyulu in Tsavo West National Park, a mere hundred yards or so off the main road that bisects Tsavo.

I was to drive with four rangers the 100 odd miles to Kyulu, arrive there after mid-night, arrest Simba and take him to the Park headquarters at Voi some seventy-five miles further on. Arriving on schedule we went to the only house we could see. It was occupied by the African stationmaster who, hardly surprisingly, was not amused. To Africans, then and now, game wardens rank next to tax collectors in the degree to which they are disliked. Despite being told to keep his voice down, the stationmaster spoke out loudly, denied knowledge of Simba, declined to show us where the labourers' quarters were, questioned our right to be on railway property and asked whether we had search warrants. His tactic was obvious: by making a hullabaloo he was warning others of our presence. After four nights without proper sleep, I was in no mood to be mucked about.

A cuff to the side of his head from the ranger corporal and a jab to the ribs from me changed his mind and he showed us to the labourers' quarters. Simba was long gone. The Stationmaster's ruse had worked. Yet he was not finished with us. While we were searching for Singi he had telephoned the railway police at Voi, lodging charges of assault against us. The police wanted to know who we were. This was the final straw. If he wanted trouble he would get his money's worth. To press charges of assault he should have a doctor's report on the evidence. The sooner this was obtained, the better. It passed through my mind that, if given time, it was not past so determined a 'bush lawyer' to 'arrange' a couple of abrasions. To cut a long story short he was driven seventy five miles to Voi, presented to a very sleepy government doctor, examined and declared free of physical evidence of assault and dumped at Voi police station to make his charges and find his own way back to Kyulu.

Highhanded and ill-considered, my action certainly neutralised charges of assault, but it had also brought the railway to a temporary standstill. The stationmaster's presence at Kyulu was necessary and his removal without any arrangements for his absence being taken care of stirred a real hornets' nest. Sheldrick was asked to explain why his anti-poaching operation had disrupted the line and I was banished on patrol to the furthest corner of Tsavo East until all the dust settled.

Shortly after this I had an altercation with David's Somali sergeant-major Mohamed which, coming as it did, on top of the railway stoppage and my initial poor start, strained David's patience too far. He sacked the Sergeant Major and requested that I be transferred away from the anti-poaching operation. I think David over-reacted. It did me no harm and probably made me a shade less cocky.

The anti-poaching operation ran for two years: 1955-7. In this period 429 African poachers were caught and jailed, ivory and rhino horn worth £25,000 was recovered (worth more than £1 million today) and paid for the operation. Poaching and trespass in Tsavo East Park was greatly reduced until 1971 when it reappeared; a period of fourteen years. Within three months of starting work Superintendent Rassie Potgieter had recovered 4,069 lbs (1,846 kg) of ivory, 35lbs (16 kg) of rhino horn and prosecuted 115 Africans, eleven Arabs, five Indians and one Somali. All the Africans were convicted, as were seven Arabs, two Indians and the Somali. Thereafter his results tailed off and Rassie was relieved of his duties and retired from the police force rather mysteriously amid a welter of rumours that he had in some way been compromised by the buyers. That they offered huge bribes was proven on at least one occasion when another policeman – Johnnie Potgieter (no relation to Rassie) – apprehended an Indian with a carload of tusks. On that occasion he turned down the offer of a house and a large sum of money.

Rassie Potgieter's initial results were striking evidence of the relative ease with which the illegal buyers were brought to book. By the same token they were not very cautious about the way they went about getting ivory, illustrating their contempt for Game Department efficiency in running them down.

Yet the hero of the anti-poaching operation was of course the man who ordered that it be undertaken – the Governor, Sir Evelyn Baring. This has been overlooked ever since. Instead, perhaps a little unfairly, the spotlight fell on the partnership of Sheldrick and Woodley, who though crucial, were nonetheless cogs in Baring's machine.

In the annals of African conservation the Sheldrick/Woodley partnership was unique and made points that are obvious enough with hindsight, but not all of which were appreciated at the time. The first was that poachers are most easily caught, not in the field, but in the precincts of their homes when they are not on guard. Getting evidence to convict them calls for solid detective work: the painstaking, unspectacular, boring, evidence collection that underlies all good police routines. There never was any reason for thinking otherwise. To control poaching, as with any illicit activity, a permanent informer network of people who know their local patches, and a team of detectives attuned to pick up all evidence are required. Sheldrick and Woodley put such a system in place, showing that poaching in the short term can be controlled. In the 1980s, a zoologist, Richard Bell was to repeat this in Zambia and again make the point that routine police work where the poachers lived was what was called for. Field patrolling should be a secondary support to a good intelligence system but, alone, is rather inefficient, regardless of how well armed and equipped the law enforcers may be.

The degree to which the 1956-7 anti-poaching programme was successful owed a great deal to the speed with which it was implemented. The rural hunting communities literally did not know what had hit them. Neither knowing the law nor their rights and, being poor, unable to hire defence lawyers, those arrested were easy to convict in court. As with Mau Mau, the keys to success were informers through whom basic evidence was accumulated. This was not difficult in communities that held hunting no crime, but a laudable skill and the centuries-old tradition which fed them.

Sir Evelyn Baring, the man who in 1956 and 1957 demanded that all arms of the Kenya Government tackle poaching seriously, and as Governor, was in a position to make this happen. (Photo Camerapix)

To a degree, the Wata were themselves the primary reason for the anti-poaching campaign's successes. To them elephant hunting was the most admirable of activities. Brought up to honour the successful hunter and to speak out about his achievements, they were an ingenuous and friendly people who found it culturally difficult to lie about or deny hunting successes. Thus when asked about an incident they tended not only to acknowledge it, but give their interrogators even more information. Consequently virtually all the Wata men who lived in the coastal hinterlands were sent to jail for hunting. The whole business of catching and imprisoning them was sordid.

The Kamba whose lands were to the west of Tsavo did not have the Wata total commitment to elephants. With a more cynical attitude towards Government and white men generally, they were less willing to divulge their hunting records. Consequently, the anti-poaching successes achieved against the Wata were not duplicated against the Kamba. While many were jailed and the anti-poaching made them cautious about hunting in Tsavo East, it did not stop them altogether as it had the Wata. In Tsavo West Park they kept on poaching very much as they had before the campaign.

Having virtually stopped poaching in Tsavo East, both David and Bill felt that it was not their duty to sustain the anti-poaching work outside their park and control of the operation was handed over to the Game Department. The police withdrew their aircraft. The two

Cape buffalo, the truly wild cattle of Africa, dangerous and exciting to hunt.
(Photo Peter Davey)

Game Department anti-poaching platoons were separated and sent elsewhere without an overall controller or the organisational and logistical base that Sheldrick had provided. No game wardens liked anti-poaching and both units gently fell apart with their scouts and wardens filtering off into more routine Game Department duties until with no overt decision to scrap them, the units dissolved.

* * * *

For three months after leaving the anti-poaching operation I shot buffalo at Nanyuki. These animals were abundant and disliked by the farmers whose wheat they grazed. Obviously I redeemed myself for I was then posted to Kilifi on the coast. That it was a senior warden's post was no reflection of any particular merit on my part. It happened because my predecessor resigned at short notice and there was no-one else available to take over the post. Ironically, at the time, Kilifi was the range in which the anti-poaching activity was still intense and I had to re-establish close links with both Sheldrick and Woodley, though no longer as their subordinate. From thence forward, however, we became firm friends.

Government officers were charged rent for the accommodation government provided and deducted it from our monthly pay cheques. Kilifi was a senior station and I was the Department's most junior officer. My salary of under £50 a month was inadequate to carry rent for a senior officer's quarters as well as paying back the loan for my Land Rover and I was left with £12.50 a month on which to live. Even in those inexpensive days, that was too little. Had I not been bailed out more or less constantly by my fiancée, Christine Mowat, I could never have survived as a game warden.

Many other wardens found themselves in the same position and the man I replaced at Kilifi, Fred Bartlett, had resigned only because he could not educate his children on a game warden's salary. Badly paid even by civil service standards, we were paradoxically supposed to look after thousands of elephants and rhinos whose tusks and horns were worth huge sums of money. The Kilifi range was roughly a triangle whose apex was the Tana River delta and base the Kenya/Tanganyika border from the foot of Mount Kilimanjaro to the Indian Ocean with an area of approximately 12,000 square miles (>30,000 square kilometres). It contained perhaps 25,000 elephants and maybe 5,000 black rhino. The standing crop of ivory will have been of the order of 250,000 kilos worth around the same value in Sterling pounds (which in 2003 terms would be worth over $100 million). The standing crop of rhino horn will have weighed around 10,000 kilos and was worth around £50,000 (worth more than $20 million in today's terms). I had nine game scouts to stop people from illegally taking this ivory or rhino horn. With the same game scouts I was also supposed to prevent these animals and others from damaging people and their crops.

My Kilifi range was not unique in this sense. Indeed it was one of the smaller ranges. Nevertheless there was so much to do that we wardens could pick the work we fancied and be fully occupied to the exclusion of what did not interest us. My particular interest as a warden was the Wata – the same elephant hunters who had been the most prominent poachers in the Tsavo East National Park. Most of the tribe lived in my Kilifi range. Throughout 1956 and 1957 they were being thrown in jail for following their traditions and even that early in my career, it was obvious that this was no solution to the problem. Over the next eight years my life was dominated by the fortunes of this small tribe.

Both game and park wardens annually took elephants on licence: some for both the thrill of the hunt (like Bill Woodley) and the rewards, others merely for the ivory to sell. This was Woodley's last elephant taken on licence in 1961. The man on the left is the Kamba Elui Nthenge. In the centre is Boru (Magonzi) Debasa, Elui and Boru were both ace bow hunters and great friends of Bill. The man on the right is a lesser Watic hunter called Wario.

CHAPTER EIGHT

NATURAL GENTLEFOLK

When Sir Evelyn Baring initiated the anti-poaching campaign of 1956 and 1957, we wardens knew nothing about the ivory trade, its history or how it had driven African history. We knew nothing about the poachers, their backgrounds and why they hunted. That changed as the programme unfolded and we perceived that poaching in eastern Kenya would never be resolved unless this knowledge was acquired. Those we were calling poachers only warranted this term because a law created in ignorance of their way of life said that they were. It was so unfair that it undermined our own self-confidence, particularly in relation to the justice and fair play we believed was the essence of being British. In this chapter I summarise thoughts on the Wata accumulated over the years, the better that what we did to them can be seen in perspective.

* * * *

In the prologue I sum up hunting experiences with the Wata. The incident described could have happened in 1956 or before Christ was born 2,000 years ago. The Wata sang for and with me when I shot an elephant with them in 1958 near Kilifi. In the following years I hunted with them many times. They spoke of elephants as people with names for males, females, sizes, temperaments and features. The better I knew them, the greater the sense that their elephant culture was so ancient that its origins lay far beyond the horizons of their own folklore. The more I knew of them the greater my respect and liking for them. I know they never trusted me completely while I was an officer of a Government of whose objectives they were never sure, but we were friends. They trusted Christine, my wife, far more. And to our daughter Sue they gave their own name of Safo, traditionally bequeathed to the eldest daughter of a successful elephant hunter and by which they still know her. Of all people I have known, the Wata were the most kindly and courteous: they were natural gentlefolk. And if I have an ineradicable regret, it was that I ever jailed one for hunting.

Just who were these Wata? This name – which they use for themselves – is not widely used by others white or black. To Sheldrick and Woodley they were Waliangulu. On the coast itself they were more widely known as Wasanye. In total, in the 1950s, they probably numbered not more than five thousand – men, women and children. The majority lived in Kenya's coastal Kilifi District, but some dwelt in the contiguous Tana River, Taita and Kwale Districts – all of which were within the Kilifi Game Department range. They spoke Oromo, the Galla spoken from Ethiopia in the north to the Tana River delta in the south where the Cushitic Orma pastoralists live. Socially they are divided into clans along the same lines as the Orma.

At the outset, I thought the Wata were hunters and gatherers in the conventional mould. That is they were the original inhabitants of the region and had stuck to ancient ways, never adopting livestock or agriculture – rather like the popular (but erroneous) image of the

A typical Wata homestead.

Kalahari Bushmen. In keeping with that image, pastoralists and farmers looked down on them as primitive. They had been written about by Charles New[43] and Arthur Neumann[44] in the 19th century and a distinguished early administrator, C. W. Hobley in 1912[45]. A. Werner also recorded interviewing them in 1913[46]. These authorities and the administrators supported this view of backwardness. Yet the more I learned, the less it fitted.

As a rule, hunters and gatherers tend to generalise, taking and eating all sorts of animals. If they tend to hunt more of one species than others, it is most likely because of its greater availability. Thus some Bushmen around the Kalahari catch and eat a lot of springhares not because they particularly like eating springhares, but because these animals are abundant and easily caught. Elephant were abundant in Wata lands, but so were many other mammals large and small. Thus no shortage of other game had made them specialise on elephant. Their mighty bows and robust arrows (displayed in the photograph on page 18 were designed specifically for elephants: for penetration, and not distant, accurate archery. Elephants and ivory were at the core of Wata culture. A boy became a man by taking an elephant. Ivory was the main item of bride price. Such specialisation seemed odd among hunters and gatherers.

It might be argued that as an elephant provides a lot of meat, specialising on them was a matter of economics as happened along the edges of the glaciers in ice-age Europe. The circumstances were different, however: people near the ice caps not only had to store large

---

[43] New C., 1874. *Life Wanderings and Labours in Eastern Africa.* Hodder & Stoughton, London.
[44] Neumann A. H., 1898. *Elephant-hunting in East Equatorial Africa.* Rowland Ward, London.
[45] Hobley C.W., 1912. *The Wa-Langulu or Ariangulu of the Taru Desert.* #9 in the Anthropological Journal *Man.*
[46] Werner A., 1913. *A Few Notes on the Wasanye.* #107 in the Anthropological Journal *Man.*

Traditionally, upon killing an elephant the hunter's family and friends would camp close to the carcass, from which the flesh was removed and split into thin ribbons that were festooned on nearby bushes where they rapidly dried in the sun and could be taken home and stored. Little was wasted.

quantities of food for long periods to survive the cold, but the cold itself was the means for preserving it. Africa with abundant fauna available year round did not make storage an imperative, while its warm climate and abundant pests made long storage difficult. Africa's selective pressures will have been the opposite to those prevailing in ice-age Europe: economy would have leaned towards smaller animals. Further, if there was a fundamental advantage in specialising upon elephants for food, it would surely have occurred widely across the continent wherever there were hunters and elephants. Yet the only hunters to have done so were the Wata and the closely related Boni who live north of the Tana River in the northern coastal hinterlands of Kenya and southernmost Somalia.

Another oddity was the name by which Wata are widely known on the Kenya coast – Wasanye (aka Wasania, Wasanya). It stems from a Bantu term for working iron or smithing. In Giriama (the Giriama are also close neighbours to many Wata) and the coastal Mijikenda Bantu of whom they are part, smithing is *usanye* and a blacksmith is *musanya*. It is Mijikenda oral history that they were taught to work iron by the Wata hunters their forebears found in the region when they first arrived. Because of Wata iron working skills they have been called Wasanye – those who smith – ever since.

The fact that agriculturists and herders learned iron working technology from 'primitive' hunters reverses the normal course of events. Usually new technologies evolve in the 'higher' cultures and pass to the 'lower' cultures. This, too, suggested that the Wata were not 'normal' hunters and gatherers. It also raised the question of how they acquired iron working technology (as distinct from mining and smelting – which they did not do) in the first place?

The East African coast has a long history of contact with the 'civilised' world. The discovery of Mycænean pottery on the Somali coast by Neville Chittick[47] suggests Mediterranean trade with the East African coast 1,000 years or more before Christ. The earliest surviving record of circumnavigating Africa was the expedition sent by Pharoah Necho II in 611 BC and Hanno established West African settlements in c. 570 BC to acquire West African gold.

The *Periplus of the Erythræan Sea* (a treatise written by an anonymous Greek geographer in Alexandria about 60 AD) set me thinking. It described maritime commerce through the Red Sea reporting that ivory was exported from there in great quantities. As used, the term Red Sea applied to somewhat more than the area we know by it today, and included the East African coast as well. The nearest sources of abundant elephant close to the Red Sea in this wide sense would be the Ethiopian highlands and the East African coastline.

Who produced this ivory? Here the *Periplus* drops a gem: it said that the elephant hunters used bows so powerful that they had to be drawn by more than one man. On the face of it ridiculous, this makes perfectly good sense to anyone who has watched someone untrained in the technique of trying to draw a one hundred and seventy pound Wata longbow. As with the Welsh and English medieval longbows, drawing a Wata bow needs both special technique and training. Recall that the writer of the *Periplus* got his information second, third and fourth hand from returning matelots. I can imagine some sailor in a far off port being challenged to draw a great elephant bow – as I have seen happen in modern times – and be totally defeated. Later, recalling the incident among his mates, he might well say that the bow was so strong that it *could* only be drawn by two men. And by the time that the story had passed through several other mouths and reached Alexandria, what had originally been uttered to stress how difficult the great bows were to draw had become fact. Familiarity with the Wata elephant bows, gives what sounded like fanciful speculation the ring of truth and hints of a connection between, if not the Wata, at least their technology and ancient commerce.

The next hint also came from the *Periplus* as it mentions iron ingots being exported to Africa. Here we have a critical connection with the big bows. Their huge power was needed to drive an arrow deep into a very big body. Harnessing such power needed sharp arrowheads robust enough to withstand the high impact forces. Arrowheads made from fine flint or obsidian could have sufficed, but if they were generally used they would be found today because they do not rust away like iron. All over highland Kenya one finds obsidian points and blades. Yet such artefacts are absent from or exceedingly rare in coastal Kenya. If they are absent in the areas where the big bows were in use, then their arrows most likely had metal blades which, rusting away, did not become artefacts for later archaeologists to find.

Thus from the *Periplus* we have ground for thinking that around 2,000 years ago ivory could well have been coming from the East African coast. Further, elephants were being taken with bows so powerful that the associated arrow-heads had to be made of stone (for which there is no evidence) or iron. That iron could have been used is suggested by the export of ingots to the region from the 'civilised' world. To top it off, those who carried

---

[47] Chittick N., 1976. Pers comm.

the iron returned with ivory in exchange. That the iron would have had to be to smithed into arrow-heads and blades and tools makes the drift of my thinking obvious. We have an hypothesis to suggest how elephant hunters acquired smithing skills and the name – Wasanye. It brings together the Wata, smithing, big bows and ivory production.

Other evidence supports the budding hypothesis. The ancients' demand for ivory was seemingly insatiable. The Bible makes the point in numerous references to it. In the Song of Solomon –

> *"My beloved ... his body is as ivory ... Oh Prince's Daughter ... thy neck is the tower of ivory"*

it is clear how even in those distant times, people were turned on by the commodity. King Ahab's palace contained so much that it was called the Ivory House. The Mediterranean civilisations from Egyptians, Minoan, through the Greeks and Romans had an enormous demand for ivory. So, too, did those in the Middle East and through to the Indus basin and beyond. The Greeks developed chryselephantine art in which a sculpture of wood or other material was overlaid with a layer of ivory which in turn, was decorated with gold and gold leaf. Phidias's statues of Zeus in the temple of Olympia and Athena in the Parthenon were made in this way and were over forty feet high (>12 metres). A 30 to 40 kilo tusk can produce three square metres of veneers three millimetres thick and I calculate that each of these statues must have needed 720 tusks averaging 35 kilos in weight (i.e. 25 tonnes of ivory). Yet these were exceptional works made at the height of Greece's prosperity, and it is difficult to estimate what the country's annual imports were. Indeed it is not until the Roman Empire was at its zenith that we get some indication of volume. It seems ivory was used in all walks of life. Quoting from Wilson and Ayerst[48]–

> *"Rome – powerful, acquisitive, Philistine Rome – consumed ivory in great quantities, as she consumed everything which symbolised wealth and majesty. Republican and imperial officials used ivory for their staffs of office, insignia and curule chairs. Votive offerings of tusks adorned the temples. Cicero spoke of 'houses of marble that glitter with ivory and gold' and from the writers of every Roman century we learn that the substance was used for luxury items from furniture inlay to book-covers, from bird cages to brooches and from combs to scabbards. Seneca owned 500 tripod tables with ivory legs while some of the emperors displayed profligacy in their use of ivory that showed a complete lack of understanding of the difficulties of supply ...*
>
> *"Caligula, not content with making his horse a consul, gave the creature an ivory stable."*

Such profligate use of ivory compared with Indian and Chinese consumption in the fifteenth and sixteenth centuries[49] and suggests that Roman consumption will not have been less than 200 tonnes a year. Tenuous as this figure may be, it can be translated back

---

[48] Wilson D. & Ayerst P., 1976. *White Gold: The Story of African Ivory*. Heinemann, London.
[49] Parker I.S.C., 1979. *The Ivory Trade*. Report to the United States Congress. Typescript.

into tusks and elephants. If the average tusk weighed 10 kg (an arbitrary guess not too far out of keeping with modern production), consumption would have been around 20,000 tusks a year. Even if consumption was half this, it would still involve several thousand elephants annually. Where would they have come from?

By 500 BC the Asian elephants in Syria and Asia Minor were extinct[50] so it is reasonable to assume that Africa was the main source. Yet where in the continent will the ivory have come from? Acquiring it in any quantity placed a specific demand on the acquirer: ability to transport it. Getting it out of the continental interior called for, above all else, manpower. Where tsetse flies occurred – which was usually where dense elephant populations were – draft animals could not be used. Rivers were few, ruling out water transport in most places. Portage by men was the only practical method of extraction. Not only would this have called for many men as porters (200 tonnes of ivory would break down into 7,400 x 35 kilo loads), but many more to defend the carriers from those who would steal the tusks and yet more to carry the food necessary for the carriers and their guards. The civilisations of Meroe and Aksum might have had the organisational ability to do so and extract ivory from the areas under their control, but overall, the ancients did not have the manpower to extract ivory from the continental interior. The most consistently exploitable elephants will have been those within reach of mariners and where no long distance land portage was necessary.

By the time of Christ all of Africa's shoreline could be reached by mariners. On economic grounds the elephants most heavily exploited are likely to have been those closest to the Mediterranean, Middle East and Indus Valley civilisations. Conventionally the Romans are thought to have acquired their ivory mainly from North Africa and particularly the Magreb. Indeed elephant extinction in North Africa is attributed to their demand for it. Yet I doubt that North Africa's elephants were ever in a position to have provided 200 tonnes of ivory annually over several centuries.

If one draws a straight line from Cape Verde in the west to Cape Gardafui in the east, all Africa to the north of it and within one hundred miles of the sea is arid or semi-arid and in historical times unlikely to have had the vegetation to support abundant elephants. Dense elephant populations within 100 miles of the sea will only have occurred south of this line. And the most accessible to maritime traders from the whole arc of Mediterranean, Middle East and Indus Valley civilisations will have been those in southern Somalia and coastal Kenya. Is it coincidence that the continent's only elephant cultures – the Wata and the Boni – are in southern Somalia and coastal Kenya?

My hypothesis: the Wata specialisation with elephants is the outcome of living in that area of Africa that (a) could sustain high elephant populations, (b) was accessible to maritime traders and (c) was the nearest to Mediterranean, Middle Eastern and the Indus civilisations. If I am right, then the Wata were indeed no ordinary hunters and gatherers. Their elephant culture was more an ivory culture of some antiquity. Producing a valuable commodity, they will have had the means to become 'civilised'.

---

[50] Scullard H.H., 1974. *The Elephant in the Greek and Roman World.* Thames & Hudson, London.

Africa outlined: dark shading indicates areas with rainfall of more than 800mm annually where elephants could be dense. Light shading is the area of between 200 and 800 mm rainfall annually where elephant existed, but in light densities only. No shading indicates less than 200 mm rainfall a year where there were no, or only very few elephants. Hardly coincidentally, the Wata (and Boni) elephant cultures occur where high elephant density touched the shortest sea routes to the Mediterranean, Persian Gulf and Asia.

In the years intervening between Roman times and the present, the very incomplete written record indicates ivory kept flowing away from the Kenya coastlands to lands beyond the seas. In the eleventh century the Arab traveller al'Masudi[51] told of it going to China, though he makes no mention of who the producers were. The Portuguese, who arrived on the coast in the late fifteenth century do not record Wata but they do refer to the 'Mozungullos'. They also comment on Monhicas and Mossegejus using the

---

[51] Davidson B., 1966. *Africa – History of a Continent.* Spring Books, London.

prefix *mo-* where the Swahili and later writers used the Bantu *wa-*. The Monhicas are clearly the Wanyika – a catch-all term for the Mijikenda tribes still in use – and the Mossegejus were the Wasegeju who today live along the south Kenya and northern Tanzanian coasts. The Mozungullos are thus Wazungullos and I am not alone in taking this as a corruption of Waliangulu (or *vice versa*). Gray[52], who looked into the matter wrote –

> *"Mozungullos is clearly a Portuguese corruption of the Kiswahili word 'Wa-Langulo'. The tribe in question is evidently the Wa-Sania,... The neighbouring tribes call the Wa-Sania the Wa-Langulo ... "*

That the 'Mozungullos' to whom the Portuguese refer were more than hunters and gatherers is apparent from the raids they made on Mombasa in 1612, 1614 and 1625. In the last they murdered the Portuguese Governor. And Gray quotes Pedro Baretto de Rezende writing in 1634 that the Mozungullos had neither law, nor king, ... that they did not exceed three or four thousand in number ... that they fought with poisoned arrows ... and that they were regarded as vassals of the Kings of Mombasa and Mellinde (Malindi). Clearly they were a political force to be reckoned with. That, in turn called for relative wealth which they could have had if they were ivory producers. It would have given them both the means and incentives to become more organised and cohesive than true hunters and gatherers. While they do not say that these people were elephant hunters, they record that their source of ivory was the 'kaffres' – i.e. the Mozungullos and other native people.

In the same era that the Portuguese arrived on the Kenya coast, the Orma pastoralists arrived in the coastal hinterlands from the north. Trading with the coastal towns as they did, these Orma would have appreciated ivory's value. Militarily organised and powerful, it would have been consistent with their traditions to have subjugated the Mozungullos – aka Waliangulu – to gain control of their ivory production. If the Waliangulu were reduced to helots, it would explain their adoption of both Orma customs and Oromo as their language. Hobley[53] was told that the Wata were formerly serfs of the Orma and that they had to present one tusk from every elephant killed to the Orma chief, which indicates that they were at some time subjugated. Prins[54], another anthropologist, also refers to the Boni and Sanye as 'two helot hunter tribes'.

When I suggested to some of the old Wata I knew that they had once been Orma slaves, they put it to me slightly differently. They said that a Watic (single Wata male) would form a friendship with an Orma man to whom he would both give and sell ivory and rhino horn in return for sheep, goats and occasionally cattle. The initial friendship would be reinforced by this trading. There was intermarriage with Orma taking Wata women as wives and concubines. The genetic drift was redressed somewhat by the occasional Orma man turning hunter and marrying into the Wata and from more casual liaisons. Hence the appearance among some Wata of straight Cushitic hair. While the essence of the

---

52 Gray J., 1947. *Rezende's Description of East Africa in 1634*. Tangayika Notes & Records 23.
53 Hobley C.W., 1912. *Ibid.*
54 Prins A.H.J., 1952. *The Coastal Tribes of the North-Eastern Bantu*. International African Affairs Institute, London.

relationship between Orma and Wata was friendship, invariably the hunter was the junior friend. They saw themselves as younger brothers who, because of this status, should show some obedience to their elders. The suggestion that the Mozungullos = Waliangulu = Wata were subjugated by the Orma thus has some substance. Ivory was at the core of this relationship which was more formal and organised than that which existed elsewhere between hunters and gatherers and either pastoralists or agricultural neighbours.

A final indication that the Wata were different from conventional hunters and gatherers was that they themselves occasionally bought slaves with their ivory. When David Sheldrick assembled Wata family trees he learned that Arabicho, son of Boru from Garbete, was only a Watic through adoption. He had, in fact, been captured by Swahili slavers as a teenager in Malawi, in the nineteenth century. Sold in Zanzibar, he had been bought by an Arab on the Kenya coast. In turn, he was exchanged for tusks and became the property of one Boru Badiva of Garbete in the hinterland near Debaso hill just outside what became the Tsavo Park. In due course Boru adopted and named him Arabicho ('Arrbicho', with the *r* rolled). From then on the Wata treated him as one of their own and he settled into elephant hunting as though born to it. He was an old man when his sons Boru and Chobe came to Sheldrick's attention in 1956. Acquiring slaves was, again, not what one would expect of true hunters and gatherers.

* * * *

Thus I tentatively suggest that the Wata were not 'primitive' hunters, but the modern representatives of an ivory producing culture that had existed in the same place in some form or other for perhaps 2,000 years and more. Their fortunes may have waxed and at one time given them the military capacity to sack a port city such as Mombasa, and at others waned to subservience under the Orma. In 1956 neither I nor my colleagues knew anything of the foregoing. We knew virtually nothing about Vasco da Gama or how the twin keys of ivory and gold had underwritten so much of Africa's history. I recount it at this point to place our gamekeeping endeavours in perspective. Had we had the evidence now available, perhaps history would have been different. Emphatically, this is not a criticism, but an observation. Any suggestion that we *should* have known more is nullified by the fact that Africans *should* have told us more. Our ignorance was at least in large part the product of their own silence over their history. Were they even aware of it? Most of Africa's history is lost before it is written.

David Sheldrick: Warden of Kenya's Tsavo East National Park from the late 1940s to the mid 1970s.

CHAPTER NINE

THE PARTNERS

Woodley and Sheldrick are names almost synonymous with Kenya's Tsavo National Park. Both were instrumental in evolving policy towards the Wata, and as close friends influenced me greatly in my formative years as a game warden. They stand out in the annals of East African conservation, so I record a little about them here.

* * * *

When proclaimed in 1948, the Tsavo National Park was originally a single 20,657 square kilometre (8,069 square mile) block bisected by the main Mombasa/Nairobi road and railway. The first warden was Ron (Steve) Stevens, a somewhat humourless chemist who had seen wartime military service, had no experience of either bush life or wild animals and who did not last. He was aided by two junior assistant wardens: Frank William (Bill or Billy) de Medewe Woodley, born in Nairobi on 3rd March 1929; and Peter Robin Jenkins born in Gilgil in the Rift Valley on 26th June 1930.

Myth has it that the Tsavo Park was a pristine wilderness, a term which connotes that it was empty of people and their influences and the domain of raw, man-free nature. Reality is that humans had used the area for many thousands of years, the evidence being in the stone artefacts that lie scattered across its surface. Orma Galla grave cairns confirm their presence in what is now the park for at least several centuries past. When the first white men entered the land, the various place names they learned all revealed not only human presence, but spheres of influence. In the north-west the names are all Kamba, and Kamba lived, hunted, grazed cattle and gathered honey there. In the south there were Masai. In the centre were the Taita who hunted and grazed the land around their highlands. Indeed the name Tsavo itself derives from a German mispronunciation of the Dabida (one of the Taita languages) 'Chafo' for what is now the Tsavo river. All through the eastern half, the names are Oromo and bequeathed by the Orma or Wata who speak it and who were present when white men came. The greater reality is that there are few habitats in Africa that have not been inhabited and influenced one way or another by people. This should not be surprising as they have lived in Africa for more than a million years. What is odd is that anyone should think otherwise.

How Tsavo came to be a national park is instructive. By the late 1920s the depth of the white man's ignorance of land tenure among Kenya's more than 40 tribes was becoming apparent, together with its political ramifications. In 1932 a commission of three men – Judge M. Carter, F. O'B. Wilson and R. W. Hemsted – was established as the Kenya Land Commission to determine rights of tenure. After interviewing people of all races the length and breadth of the country, it published its findings in 1933. The Kenya Land Commission Report[55] (also known as the Carter Report) is a seminal work on Kenya's history and, minor detail excepted, has been generally accepted by Kenya's indigenes and aliens alike as historically sound.

---

55 The Kenya Land Commission Report 1933. Government Printer, Nairobi.

The Commission examined both Kamba and Orma claims to what were then referred to as the 'eastern grazing lands'. It concluded that as the Kamba had already been awarded sufficient land for their exclusive use, they should not also own those parts of the eastern grazing lands that by tradition they used nomadically. And the Commission felt that as Orma density over the area they inhabited was less than one person per square mile, they were too few to own any of so vast a tract in perpetuity. Ironically the Commission did not appreciate that low human and stock densities were ecological imperatives for sustained use of such dry lands. Nonetheless the Commission recognised that as both Kamba and Orma used the area traditionally, it recommended their rights to go on doing so should be safeguarded.

Yet when the Tsavo Park was proclaimed, Kamba and Orma traditions as well as the Kenya Land Commission's recommendations were ignored. A world war had been fought and perhaps much had been forgotten in the interim. The only white official who had ever spent any length of time in the Tsavo area was the now retired game ranger – C. G. MacArthur – and it was he who actually recommended the park borders. The Government had no alternative or corroborating sources of information in its ranks, forgot what the Carter Commission had said about safeguarding Kamba and Orma rights, and took what he said as fact. In the words of Woodley's biographer, Dennis Holman[56] referring to the Tsavo boundaries –

> "... *You merely have to look at the map to see how arbitrarily and thoughtlessly most of the lines were drawn."*

Wrongly, MacArthur said the land within his boundaries was unoccupied.

In 1981 I asked Mervyn Cowie, the National Parks of Kenya's first Director, why Tsavo was so large? His reply reflected the strategist that he was. Tsavo came about for no better reason, said Mervyn, than at the time no Government Department wanted the area. Even the nascent national park authority did not really want this dry, flat, homogeneous *Commiphora* woodland, not known for abundant wildlife. Its ambition, following Ritchie's recommendation to the Land Commission and going back to the turn of the twentieth century, was to have what had originally been Kenya's Northern Game Reserve. This had all the dry country elements contained in Tsavo, together with botanical endemics that Tsavo lacked, but in addition had mountains, montane forests and scenic magnificence. The parks were denied this area because the Government would neither evict the pastoral residents from it, nor subordinate their interests to conservation. Mervyn therefore took the biggest tract of land available, not because it would make a good national park, but because no-one in government objected to him having it. Particularly germane, he saw it as a tactical acquisition which could be traded off, at least in part, for smaller, more desirable conservation areas at a later date. Such foresight was typical of him and had the political tides of Kenya's independence not swept him from office, it may have borne fruit.

Technically in law, Tsavo is still a single unit, but in 1949 it was divided into two administrative entities – Tsavo East and Tsavo West – for administrative purposes. Since then they have generally been regarded as two parks. Tsavo East was all of the original

---

[56] Holman D., 1978. *Elephants at Sundown*. W. H. Allen, London.

park east of the Mombasa-Nairobi railway and Tsavo West all to the west. Tsavo East was placed under warden Major David Sheldrick with Woodley as his assistant.

Sheldrick was born in Egypt in 1919, son of Captain Alfred Sheldrick and his wife Avis. The family came to Kenya in the 1920s to farm near Nyeri. In the prevailing Imperial custom of being sent 'home' to be educated, David went to Canford, doing well academically and shining as a boxer. His interests, however, lay in the outdoors and natural history and on leaving school he came back to Kenya just before the start of the Second World War. All career plans were shelved with the onset of war and, called to the colours, he trained with the Kenya Regiment before being seconded to the King's African Rifles (KAR) with whom he served in Ethiopia, India and Burma.

By inclination David was a military man, saw things in an orderly manner, declared an objective and formed a plan to achieve it in sequential steps. A good organiser, but a martinet, subordinates thought him a good leader. To his peers he was reserved and somewhat cold, but this was a façade hiding an emotional personality with a strong sense of humour which only became apparent on deeper acquaintance.

* * * *

It was immediately clear to Sheldrick and Woodley, that Tsavo East was far from empty. Kamba were living in the north-west, grazing their cattle and collecting honey on a large scale. Wata were resident along both the Voi and Sabaki rivers and Orma were resident on the Tiva and Sabaki rivers together with cattle in their thousands. Some Wata were re-settled at Mangea Hill in Malindi sub-district, but the Kamba and Orma were simply told to get out without compensation. Terminating their residence was relatively simple. In later years Woodley confessed to misgivings over what had been done, but at the time assumed that his superiors knew what they were doing.

The greater problem Sheldrick and Woodley faced was illegal hunting. Both Wata and Kamba were taking elephants and rhino on a substantial scale. Far from being unable to take big game with their primitive weapons as was widely assumed, using their powerful longbows whose arrows carried *Acokanthera* poison, they killed elephants and rhinos easily. Ambushing animals where they came to drink, the bowmen collected and sold ivory on a scale that seemed prodigious by prevailing standards. Woodley had first reported this prior to Sheldrick's employment as a warden and had been so disbelieved by his superiors as to be within an ace of being dismissed for lying. Not until Sheldrick threw his weight behind the evidence was the stigma removed from Bill's record.

The system of law enforcement initially adopted by David Sheldrick and the national parks generally in Kenya was copied from South Africa's Kruger National Park. A network of ranger posts at strategic points across the park was established. Up to six African park rangers occupied a post and patrolled the country around it regularly. It was assumed that by knowing their areas well, they would quickly spot and stop illegal activity. Between 1949 and 1952 the staff was trained and ranger posts were gradually set up. Yet the system did not reduce poaching. Initially thinking that failure rested on the quality of rangers involved, Sheldrick tried to improve their training, motivation and morale. Not till the 1950s did he realise that the system itself was inappropriate.

Men from Sheldrick's mobile Field Force, a platoon-sized unit that could be moved to wherever required. (Photo Alan Root)

Isolated rangers were difficult to supervise. Aggressive poachers in big gangs could intimidate them into turning a blind eye toward illegal activities. Operating from fixed bases made them easy to observe and for poachers to keep clear of areas being patrolled. Once thinking along these lines, David abolished isolated ranger posts and consolidated his men into a centrally based platoon or 'field force' organised on military lines into three sections. This broke the ease with which rangers could be approached by poachers to 'cut deals'. Their presence in an area became unpredictable. Able to operate as a platoon by combining sections, the rangers had sufficient strength to counter any physical threat that poaching gangs could pose. And perhaps more important, the rangers were easier to supervise.

By nature easy-going and happy-go-lucky with a ready laugh, Woodley was deeply influenced by Sheldrick's care for detail and obsession that if something was to be done, it be done perfectly. Even more, he came to value careful planning, of catering for detail and the care that should be given to logistics, though until his dying day he never changed an almost African disregard for punctuality. The relationship between the two men was in this initial phase very much one of mentor and pupil. They caught some of the poachers who were investing Tsavo East and themselves became known to the poachers. From the outset both liked the Wata and the more they learned, the greater became their sympathy for them.

The names the Africans gave the two wardens reflected something of their perceptions. David's nickname 'Saa Nane' derived from his system of starting the day early so that everyone could knock-off work at 2 p.m.: an appropriate schedule for hot places like Tsavo (Saa Nane – the eighth hour after dawn – is the Swahili for two o'clock). The choice of this impersonal name matched his rather austere and distant demeanour. Woodley's was the opposite: he was just 'Bilu', reflecting his warmer personality. On the one hand this was the local pronunciation of Bill, on the other *bilu* in Oromo means 'knife'. While in later years there may have been some subtle interplay between the two meanings, at the outset Bilu meant no more than Bill.

Bill's park service was interrupted in 1952 by six months compulsory national service training in the Kenya Regiment. No sooner was this completed than the Mau Mau insurrection broke out and he was conscripted for two years' military duty. In fact he served three years in the Kenya Regiment, rose from private to Captain and was awarded the Military Cross. Bill's Regiment service proved the value of good intelligence and gave experience in managing informers and spies to gather it. Perhaps the most potent of all tactics in this respect was the skill of 'turning' new-caught prisoners to work against their erstwhile comrades. There was of course nothing novel in this and it must have been central to good intelligence for as long as men have fought one another. Bill worked with a very able pseudo-gang operator, Francis Erskine. As is perhaps natural, there is a tendency to recall the lighter moments of otherwise tense circumstances. Such was the case of 'Flasher' Leete.

Sergeant Dennis Leete was with one of Erskine's pseudo-gangs when they contacted a big Mau Mau group at night. As was the practice, the white men were blacked with stove polish, but even so they tended to stay at the back and let their tamed Kikuyu do the talking. The contact was protracted and, at the back of the crowd, Leete was overcome with the need to pee. Turning away from the others he unthinkingly flourished the one unblacked bit of his anatomy. Had any Mau Mau been close by, it would certainly have given the game away and, back at camp after the operation was over, Leete was the only man in the Regiment's history to be charged for 'flashing'.

In 1955 Bill Woodley left the Kenya Regiment, returning to Tsavo as assistant warden both decorated and with the confidence that military command bestows. From then on the relationship between Woodley and Sheldrick ceased to be mentor and pupil and became much more a partnership.

* * * *

Spearheading Sir Evelyn Baring's 1956-7 anti-poaching operation put a spotlight on Woodley and Sheldrick and for both it was the high point of their careers. Yet their greatest achievement in this endeavour has been virtually overlooked throughout African conservation. The first step towards this was taken when David realised that concentrating his men into a field force was a more efficient way of using his manpower than dispersing it in small static bases – the ranger posts. In turn, this brought home the great difficulty of intercepting poachers in a huge area with very limited forces. No matter how high the morale and motivation of the law enforcers, the initiative is with the poachers. While not in the field actually poaching, they are cryptically hidden within the general population. Like guerrillas, they stay concealed until, to best suit themselves, they decide when, where and how they enter the field to hunt (or raid). Until that moment, the law enforcers can only watch, wait and guess. In a place like Tsavo and no matter how good the available intelligence on poacher activity, there must always be a very large element of luck in intercepting poachers on one of their forays.

Appreciating this and greatly stimulated by its application in the anti-Mau Mau campaign, Sheldrick and Woodley switched their effort away from field patrolling to

Billy Woodley, originally David Sheldrick's subordinate, went on to surpass his mentor in length of service and overall contribution to East African conservation

(Photo Phil Snyder)

gathering intelligence within the communities of which the poachers were part. They provided meat as food for families. Eating is not an activity that is easily concealed and, inviting others to share food is common to most of the world's cultures and particularly in Africa. An informer within a community soon hears of who is eating game meat and where to ask questions. Women and children are innocent sources of information. The secret of the anti-poaching campaign's success lay not in patrolling the parks, but by placing spies in the communities of which the poachers were part.

This is, of course, the essence of all good detective work: collecting information, bit by laborious bit. Meticulously Sheldrick and Woodley did this, compiling dossiers on individuals in all the communities that lived around the Tsavo park. Once a basic net of evidence had developed, the field force would swoop in the early hours of the morning. Houses were surrounded, suspects rounded up and arrested. Interrogated immediately and out of the hearing of others, they were confronted with facts. Presented in such a way that it seemed as if these had been provided by their friends and neighbours, many retaliated by giving information on these friends. The net of information rapidly developed into a more solid fabric. It was unpleasant work, perfused by the interrogators' lies and innuendoes sown skilfully to make the victim think that he had been informed on by others. It was also very effective. Surprisingly, however, the technique never caught on and both in Kenya and elsewhere across Africa, conservationists still regard field patrolling as the method of choice to stop poaching.

CHAPTER TEN

A NEW APPROACH

Both David Sheldrick and Bill Woodley felt that though the anti-poaching campaign achieved successes, law enforcement of this type and on such a scale was not the long term solution and both predicted that the problem would persist. Both felt a strong empathy with the 'poachers' and understood the central rôle that hunting played in their cultures. As no other wardens or rangers before them, they felt that game laws which did not allow Africans to hunt lawfully were unrealistic and that something should be done about it. However their partnership was nearing its end.

Perhaps this was inevitable as Bill's rising experience and maturity warranted promotion and a park command of his own. Yet personal matters accelerated the process. David's first marriage had ended. Spending so much time chasing poachers in the field left Bill's wife Daphne in David's company more than it was reasonable to expect a normal marriage to endure. Inevitably David and Daphne were drawn to one another, leaving Bill on the sidelines. After Bill was posted to become Warden of Kenya's Mountain Parks based at Nyeri, Daphne swapped husbands. In the civil service an officer taking over his subordinate's wife would not have been tolerated. In the more relaxed national park system, however, it was overlooked. Not least among Cowie's reasons for tolerating the situation was that the two men remained friends, stressing how remarkable a partnership theirs had been.

* * * *

During the anti-poaching campaign David had discovered a latent flair for attracting public attention. With Hollywood looks he could have been his own publicity agent. Yet while liking praise well enough, he did not like displaying in public. Instead he approached figures in the public eye and fed them the information he wanted publicised. His greatest coup in this respect had been drawing the Governor's attention to the poaching then plaguing Tsavo, for it was this that had stimulated Sir Evelyn to launch the anti-poaching campaign.

Initially David's most effective ally was Noel Simon, an ex-Fleet Air Arm pilot turned Kenya farmer who was also a wildlife enthusiast. It was he who initially raised an outcry about poaching that was supported by journalist Anthony Cullen of the *Kenya Weekly News*. Now David turned to Simon and discussed the problem of legitimate African access to wildlife. When Noel produced information on how native peoples in northern Canada were allowed to hunt, both felt that a similar provision could work in Kenya. Their thinking caught my interest and in July 1957 the three of us with Bill Woodley proposed that the Wata be allowed to hunt elephants lawfully as a long-term solution to poaching. We rationalised that if the people could legally use game they would stop incontinent poaching. This thinking was summed in a seminal paper by Noel Simon[57]. In an addendum to Simon's paper David Sheldrick warned –

> *"There is no doubt that if the Waliangulu were to think that this scheme was for the benefit of Europeans or other tribesmen it would fail ..."*

The concept entered Government's files through the Game Policy Committee which tentatively supported the idea. Willie Hale, as a member of that Committee, was party to that position and therefore receptive to my suggestion, supported by Sheldrick and Simon, that I should take the next step of formally studying the proposal. Willie agreed and in early October 1957 I was relieved of my duties as Game Warden Kilifi to concentrate on the 'Waliangulu Scheme'. Bill Woodley and I made several field trips to meet and discuss the idea with the Wata. I submitted my recommendations in January 1958. Before I had even reported, however, Willie reversed his previous support and summed his views[58] thus –

> *"Frankly, I do not think this scheme will work. I consider that the correct solution is to attempt to convert the Waliangulu to Agriculture and get them settled in places such as Hola."*

Blunt, as always, Neil Sandeman, the Department's Deputy Head, grunted what many game rangers felt –

> *"Our job is to look after animals; not bloody people!"*

I was told to resume duties as Game Warden Kilifi. This I did and had I been a dutiful subordinate, that would have been the end of the proposal. I was not, however, a dutiful subordinate. Determined not to let the Game Department kill off the proposal I circulated it to the Provincial Commissioner Coast Province, all the District Commissioners in Coast Province and many District Officers as well, to a man the Administrators supported the new approach. As Government's most influential arm this backing carried immense weight. The Provincial Commissioner, Desmond O'Hagan summed their reaction in a letter[59] to the Permanent Secretary, Ministry of Forest Development, Game & Fisheries –

> *"My first reactions are that a scheme of this nature is well worth a trial... because it does make a constructive attempt to solve the problems of the Waliangulu hunters."*

The East African Professional Hunters' Association, on behalf of its members, objected vehemently to my plan, holding that the project would affect their businesses ruinously, be cover for poaching and that no Africans should ever be allowed to hunt. Both Game Department and the Ministry of Forest Development, Game & Fisheries endorsed this

---

[57] Simon N.M., 1st August 1957. *Future Conservation Trends.* A paper submitted to the Kenya Government Game Policy Committee.
[58] Letter GA 3/5/6/38 of 30th October 1957 from Chief Game Warden to Game Warden Kilifi.
[59] Letter ADM. 1/20/15 of 5th March 1958 from Provincial Commissioner, Coast to the Permanent Secretary, Ministry of Forest Development, Game & Fisheries, Nairobi.

reactionary outlook. On 9th May the Permanent Secretary in the Ministry in a letter[60] to the PC Coast rejected the proposed Waliangulu Scheme. Yet such objections merely strengthened the Administration's support for it.

Outside Government Noel Simon had also pushed the proposal hard, presenting it first at the East Africa Fauna Conference held in April 1958, then at the 10th Assembly of the International Union for the Conservation of Nature & Natural Resources (IUCN) in Athens where a resolution drafted by Dr Frank Fraser-Darling was passed –

> *"This Assembly of the IUCN notes with approval the pilot Game Management Scheme being planned for the Waliangulu tribe in Kenya, whereby the economic welfare and cultural patterns of the tribe will be sustained, and records its full support of this attempt to show that protein production is possible by game management from areas of low or non-existent agricultural value. The scheme provides an opportunity of obtaining valuable scientific and sociological data from this form of land use."*

The outcome of Noel's work was the Nuffield Foundation in Britain offering a grant of £10,000 (equivalent value in 2003 = £470,000) to get the Scheme started. Faced with massive support from the Administration and this international endorsement, the Ministry of Forest Development, Game & Fisheries caved in.

In late 1958 Willie Hale retired and for the next year the Department was without a head, though Neil Sandeman stood in on all important issues. The Permanent Secretary, John Webster, wrote me a personal letter[61] on 29th April 1959 in which he said –

> *"You will be pleased to hear that the Government have now approved the Galana River Game Management Scheme... I would like to congratulate you on all that you have done in the preparation of this project, and to wish you the best of luck in carrying it out."*

This courteous commendation from the top man in our Ministry was much appreciated because by civil service norms I had been disobedient, had fought my seniors in the Department and Ministry, ruffling their feathers severely. However, without the support of Sheldrick, Woodley, Simon and the Administration, the project would have got nowhere. Ironically the independence that had originally made me a nuisance to David, was the factor that forced the Waliangulu Scheme (aka the Galana Game Management Scheme) into being.

Having been coerced into accepting the scheme in principle, my Department and Ministry now tried strangling it with red tape. All trophies – meat, hides and horns from any animal taken – could be used and sold with the money going back to the project directly. All, that is, except the ivory that was central to the project's finances. All revenue from ivory, from whatever source, had to go to the Treasury. No reasons for this were ever given. By law, ivory was a game trophy the same as any other. There was nothing under

---

[60] Letter 22/4/1/8 of 9th May 1958 from Permanent Secretary, Ministry of Forest Development, Game & Fisheries to Provincial Commissioner, Coast.
[61] Letter 22/4/1/8 of 29th April 1959 from John Webster.

the Wild Animals' Protection Ordinance that set it aside. And it was within the Chief Game Warden's powers to waive Government's right to any trophies.

None of us understood the Treasury's obdurate refusal to discuss the issue. Even the Administrators were baffled. The term that cropped up time and again was that the Treasury could not 'hypothecate' ivory to the scheme. Hypothecate means to pledge or to mortgage. Hypothec, as a noun is a right established by law. Yet no law created an hypothec whereby ivory was exclusively Treasury property. We brought up the issue with the Governor, Sir Evelyn Baring, when he attended a meeting to discuss the scheme on 4th September 1959 in Malindi – just before he nearly drowned saving an Asian girl from being pulled out to sea by an undertow. Against the combined pleading by Provincial Commissioner, the District Commissioner the District Officer Malindi and myself, Sir Evelyn was adamant. The minutes of that meeting read –

> *"His Excellency agreed that the Treasury could not hypothecate the ivory as it would be a dangerous precedent and might lead to all African District Councils claiming the ivory in their areas."*

Why that should be *bad* remained obscure. All present except His Excellency felt the opposite: that it would be *good* if local people benefited directly from the ivory produced in their areas. A precedent already existed whereby the Royal National Parks of Kenya owned the ivory from the Parks and National Reserves. However Sir Evelyn diplomatically weaselled his way round our problem by commanding the Treasury to give the project an annual grant equivalent in value to the ivory it produced. Cumbersome as this was, it was the best we could hope for. What I did not appreciate until years later was that in Africa ivory has always been a coinage in its own right. Quite literally a country's elephants were a mint and the arguments against letting us have it directly were the same as those for keeping national mints under state control.

At the end of 1959 I left Kilifi to go on brief leave, get married, and then take up the new post of Warden in charge of the Galana Game Management Scheme.

* * * *

Summing my first four years as a Game Warden, I joined a band of gentlemen most of whom became firm friends; for all that I espoused ideas outside the Department's traditions. Whereas I felt that Africans should be allowed to hunt, not through any liberal ideology but from common sense, Willie Hale summed the counter view succinctly but crassly in his 1953 Annual Report[62] in a section on poaching –

> *"Every African is a poacher... In Kenya one recognises no customary hunting rights, though certain sub-tribes, such as the Tharaka in Kitui, the Teita in Voi and the Wasanye in the Coast Province, not to mention the ubiquitous forest-dwelling Dorobo, are nearly all full time hunters."*

---

[62] Hale W., 1954. *Annual Report for the Kenya Game Department for 1953*. Govt. Printer, Nairobi.

His solution to this –

"Poaching must be stopped and, therefore, poachers must be prosecuted."

When it came to 'rights' Africans had as much theoretical right to wild animals as anyone else. Where they depended on hunting partly or wholly, that right was surely greater than any immigrants' to hunt for fun? The game laws were deficient in this respect and needed to be changed. In the long run the future of conservation depended more on African acceptance than anyone else's and this would be withheld while they were not allowed to hunt and were thrown in jail for doing so. Ironically, today the catchphrase of current Kenya Wildlife Service policy is 'community conservation' and all the emphasis is on local involvement as though it was a new idea. There is now no recall that in the late 1950s the logical outcome of the great anti-poaching campaign was a 'community conservation' endeavour orders of magnitude greater than anything that has happened since.

In those early days I did not appreciate that there was rather more than conservation behind the game laws' racial bias. White men's security was a potent element and it is no coincidence that the three countries in Anglophone Africa with sizeable white immigrant communities – Kenya, Southern Rhodesia and South Africa (including what is now Namibia) – were also the *only* three where natives were disbarred from hunting. In all three it was primarily a security and not conservation policy to keep firearms out of their hands. In those countries without white immigrants, natives *were* allowed to hunt. In Uganda the majority of elephants shot on licence were taken by African licence holders. In 1940 Tanganyika introduced a game licence especially for the natives: it cost five shillings and permitted them a range of animals including buffalo, but not elephant or rhino. Further, there was no attempt to make them use modern firearms: the old muzzle-loading muskets, with which the country was awash from the old slaving days a century earlier, were quite acceptable[63]. Clearly expediency and not principle underwrote British Africa's game laws. Only if the Tanganyikans wanted valuable species like elephant and rhino did they have to pay the same as anyone else.

Against such a background my views on allowing natives to hunt were labelled variously as ultra-liberal, socialist, communist and some of the professional hunters swore that I was a kaffir-lover – the ultimate derogatory epithet in the racial lexicon of the times. If those of us who advocated letting the Wata hunt – Bill Woodley, Noel Simon, David Sheldrick and myself – had political leanings, they were decidedly to the right. Yet, regardless of our political orientations, our experiences in the anti-poaching campaign had led us to suggest a new and radically different approach to the colonial game policy that prevailed until 1957.

* * * *

I ran the Galana Game Management Scheme from early 1960 until mid-1964, assisted for most of this time by Tony Seth-Smith – a friend from childhood's bird and butterfly

---

[63] see, for example, Caldwell K., 1948. *Report of a Faunal Survey in Eastern & Central Africa.* Occasional Paper 8, Society for the Preservation of the Fauna of the Empire, London.

collecting days. Looking back, those years were idyllic. Both of us were newly married. Our wives took to the bush readily, and both had produced daughters within a few months of one another. The nearest 'civilisation' was either Malindi, eighty miles away on the coast, or Voi, seventy-five miles to the west across Tsavo East National Park. The Sheldricks with whom we were in daily radio contact on the National Park network, acted as our base in Voi. Isolated in an area where no white people had ever lived before, the 7,680 square kilometres (3,000 square miles) designated the Scheme area was without development of any sort. Everything we did there was new to us and we had the satisfaction, albeit it in small measure, that Victorian explorers of yesteryear must have savoured. With hindsight in this shrinking world, our view excluded the perfectly obvious evidence that Africans had been there all along, but that is how things were then.

The Wata with whom we worked were as decent a band as one could wish to have met anywhere. Our office involvements were minimal and we had a wonderful open-air life. Best of all, we believed in what we were doing: that we could make wildlife pay, give the Wata a way to continue their long involvement with elephants and give conservation new direction by becoming a means of using land rather than being a barrier to such use. Wholly involved with what we were doing and within ourselves, we paid little attention to what was happening in the world about us.

The basic idea behind the project was that the Wata should be allowed to hunt traditionally, but with a ceiling on the number of elephants they might take and our guidance in selling the produce. That was crystal clear in the plan submitted in early 1958. In the process of getting the plan accepted by Government it became severely distorted. That is not surprising for several reasons, not least of which was my callow understanding of civil services. The Wata were never free to pursue their way of life: the Scheme benefited them by only providing employment. A few selected men got to hunt with rifles. As David Sheldrick so perspicaciously observed, if the Wata did not believe that it was truly for them the Scheme would fail. When Government refused to let it market its own ivory, that made them certain that it was not for their benefit.

Somehow we got something very basic badly wrong. How could we as very young white newcomers teach the Wata to live off elephants? With an elephant-dependent culture of great antiquity, surely they should have been the teachers? There was much so anachronistic about the programme that it was bound to fail. Yet at the time we could not see this. We were running the first project of its sort, the first 'modern' game management scheme in Africa and in relative terms still probably the biggest. We were young, enthusiastic and there was immense goodwill between the Wata and ourselves. At the time I certainly thought we were on the right track. Now, by the same token, *mea culpa*, we were not. All remaining intact is that our intentions were right, notwithstanding that such intentions may line the route to hell!

* * * *

Kenya's independence on 12th December 1963 was traumatic. The greatest shock was our position vis-à-vis the United Kingdom. The British Government offered its colonial civil servants the option of serving on under the new Government or taking immediate retirement and compensation for lost earnings in the form of a lump sum – the 'golden

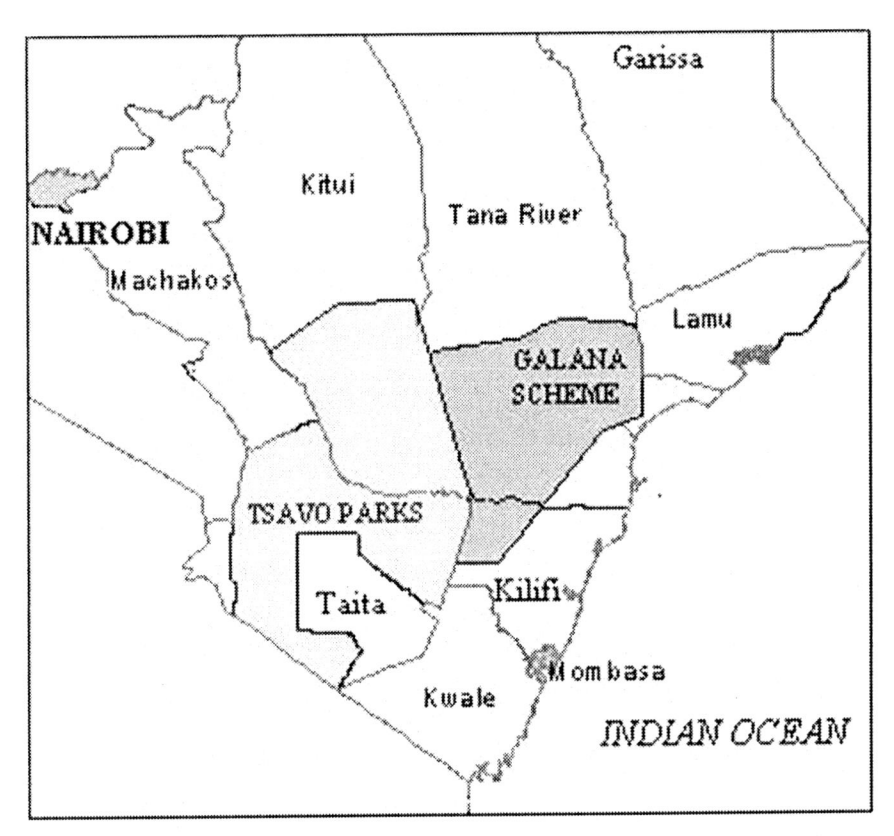

The combined Galana Scheme and Tsavo National Park totaled 28,600 km2.

bowler' – that would help them set up in new careers. Excluded were locally recruited civil servants. For all that we were as 'British' as any that came from Britain, we were not entitled to golden bowlers. Further such pension as I might be due on retirement at some distant date would not be payable or guaranteed by the British Government, but by the incoming, untested, Kenya Government.

If there had been one thing I had never questioned, it was that the British would do right by us. Whether signed on in the Colonial Service from Britain or Kenya, all of us civil servants worked for the same employer – Her Majesty. We were of the same nationality and whether we signed on in Kenya or Timbuktu should have had no bearing on our terms of service, but it did. While we received the same basic salaries, the 'Poms' from Britain received additional inducement to serve abroad that raised what they received by at least a third over the basic level. Ironically many of them would have taken a lesser salary to work outside Britain! This policy had twin roots.

The first was ethnic. Britain refused to accept that its civil servants in Kenya were split three ways into those who were black and ethnically Africans, brown and ethnically Asian and whites who were ethnically European. The Africans were indivisibly of, for and from Africa. They looked upon the Government and their loyalty to Her Majesty the Queen as impositions and on us whites as aliens. It is an inarguably correct outlook. The Asians – in keeping with their traditions – tried not to take sides, but leaned towards what was personally and pragmatically advantageous. Their Asian enthicity was never doubted. The whites were equally obviously new to Africa and in all other aspects European. They saw themselves as indivisibly British and their loyalty to the Crown something as much part of them as their names and faces. Britain never asked locally recruited Britons whether we wished to opt for retirement or to become Kenyans; we were simply told we didn't qualify for golden bowlers.

Britain argued that to have acknowledged ethnic reality would have been 'racialist'. From this position to have offered us local whites the option of golden bowlers would have been unfair. If we were offered them, why shouldn't all members of Kenya's civil service be offered them? For a civil service whose pay scales had until very recently indeed entrenched ethnicity in its pay scales (Europeans were paid more than Africans for the same job) this switch in position was one of convenience rather than principle. Yet it did raise a point. Let us accept that we were colourless civil servants. Where we were concerned Independence constituted a change of employer from Her White Majesty in London, to the Black Government in Nairobi. All us civil servants should have been asked whether we accepted the change of employer. The principle involved had nothing to do with one's place of birth or ethnic status. In this case, however, Britain did make a very obvious ethnic distinction. There was no way that it would compensate black Africans if they did not wish to accept their new black employer. There were not many of us locally employed whites in the civil service and the easiest way out was to deny ethnic reality. We had to lump it. The term 'Perfidious Albion' fitted rather well! Britain's attitude towards Australia and New Zealand when it came to joining the European Union was very much in the same vein.

The second root goes back to the American Revolution when London learned that it could not count on 'colonial' civil servant loyalty. For this reason the officer commanding the Kenya Regiment and those in the KAR had to be serving officers of the British Army. The exception to this rule that proved its point was of course Rhodesia where, as happened in America, when the chips were down, colonial loyalties lay differently to London's ideas.

Britain treated British colonials as a second class citizenry. Despite the assurances that Kenyatta gave Kenya's white community, independence brought a sense that we would be inferior in our adopted land. There are those who feel that we deserved no less, having lorded it over the indigenous inhabitants, treating them as lesser beings and one can understand if not approve of the attitude. Yet our suspicions have been justified. Criticism of government by whites has invariably been rebutted in racial terms and most recently the politician, Richard Leakey, was told that regardless of Kenya citizenship, no white should ever aspire to lead the country.

It was in a negative frame of mind that in 1964, I reviewed the Game Management Scheme's progress. The project needed further capital, which Government said it could not provide. I thus recommended that provided Wata interests could be looked after, the project should be opened to the private sector. I did not appreciate then that the Scheme had already failed in its principle objective. If, as I believe, the Wata represented an elephant hunting culture that spanned the past two millennia, who was I or any other alien to show them how to live off elephants? The only rôle that Tony Seth-Smith and I should have played was marketing Wata hunting produce and ensuring that they did not exceed such quotas as may have been allocated to them. Other than for this, the Wata should have been left to themselves. Such clarity took many years to evolve.

Suffice it that Ian Grimwood, the Chief Game Warden who replaced Willie Hale, authorised me to invite private investors to take over running the project. Having done so successfully I was mortified when he changed his mind and tried to move the goal posts. Having originally said that they would have a very broad degree of autonomy, he reversed this and insisted on extremely tight Game Department control. By now feeling that there was little future for me in government, Grimwood's revocation of agreements already made with the private investors was the last straw. I resigned in 1964.

Wanting to stay in Kenya and in conservation, with my wife and four friends we established a wildlife research and management consultancy – Wildlife Services Limited. This had phases of profitability, success and dismal failure. We closed the Company in 1976 after running it for twelve years. Since then I have continued as an occasional consultant and, while natural history has been an abiding interest, involved myself in other fields as well.

Elephants were so widespread and abundant in Kenya in the 1950s that it was thought their numbers must be increasing. Little did we realize that they had been in decline since the start of the colonial era.

CHAPTER ELEVEN

ROTTEN APPLES IN THE BARREL

Earlier I made the observation that while game wardens were among the most poorly paid of civil servants, ironically they were responsible for a vast treasury of ready currency in the form of ivory and rhino horn. I have also pointed out that there was a prevailing mystique about wardens as people of such moral standing that despite the temptation their circumstances created, they would not succumb. Like priests, to do so would be contrary to their vocation. This public esteem was rather pleasant, but it was not, regrettably, lived up to by all to whom it was accorded. Just as there have always been naughty priests so, too, were there bad gamekeepers.

* * * *

That at least some of the Department's subordinate staff were corrupt was borne out by Rodney Elliott's arrest of several of TB's scouts for poaching elephants. Some of George Adamson's men, too, were taking rhino as apparent to a colleague, Peter Saw[64] –

> "Things in Isiolo are pretty chaotic. The Scouts at Archer's Post, Liman, Komalirr etc. are implicated in the killing of a rhino near Shaab."

Yet it would be wrong to imply that corruption was monopolised by the subordinate scouts. When one's job involved shooting elephants on control, there was always the temptation to shoot the biggest and then select the best of the season's crop for one's licence. This was not permitted and if a warden did it and was found out, he would have been in serious trouble. I knew two wardens who took control tusks on licence. One was given the option to resign quietly, which he did and went on to become a very well known national park warden in a neighbouring country. The other was only found out after he had already resigned. Not that he attempted to hide the fact: his station ivory register showed a neat line through an entry with the words 'taken on licence' written against it.

Temptation was added to by the fact that poaching has never been considered real crime in either black or white societies. The English folk song, *The Lincolnshire Poacher*, sums the common attitude in which, at worst, the poacher is a loveable rascal. It would have been exceptional if no game rangers had succumbed to yet greater temptation and I had not been long at Kilifi when I came across evidence that one had.

I was in the northern end of my range at Karawa near the Tana River in the months when elephants came into the area in their thousands. Driving along the main road north, which was straight for miles on end, ten minutes did not pass without seeing a herd crossing it ahead of us. The scout sitting beside me reminisced about the elephants one of my

---

[64] Saw P. Letter dated 7th June 1961 to Rodney Elliott.

predecessors had shot along this stretch of road in the late 1940s, with the unspoken implication of 'Why didn't I do likewise?' Keeping responses neutral I asked why elephants had been shot here, miles from the nearest agriculture.

"Because they did not pay taxes," he had laughed.

This, apparently, had been the euphemistic justification for taking elephant illegally. Gradually the story came out that he had shot at least many tens, and possibly hundreds, seasonally when they were crossing the road near Karawa. A check in the District Officer Malindi's ivory register (that being the nearest administrative centre), the Kilifi District Register, the Ivory Room registers and the station monthly and annual reports, showed no evidence of such numbers of tusks having been handed in. Seemingly the game ranger had run a tidy little sideline in ivory.

Coincidentally, in 1949 Bill Woodley got a whiff of this. He had been on the track of a large pair of tusks a poacher had hidden in the area where this game warden had been 'tax collecting'. His entry into the area had resulted in the warden writing a furious letter to headquarters demanding that the young assistant park warden be disciplined for operating outside the national park. That Woodley was also an honorary game warden and entitled to do so was not taken into consideration. The fury was inexplicable, other than as an attempt to prevent anyone else coming across awkward evidence. More than seven years had passed when I took up the trail, by which time the man was dead and further investigation pointless.

Many years later when collecting historical information from the Wata, by which time we knew one another well and neither side had an axe to grind, I threw a four day party for old acquaintances on the banks of the Sabaki. Supplying beer *ad lib.* and several sheep which were slaughtered and roasted as we talked, I learned much that was new to me. One tale concerned Maitha, an old Giriama, who was spoken of with awe. As a hunter he ranked with the very best which, coming from the Wata and concerning a man from another tribe, was extraordinary. I was intrigued because, familiar with the great hunters' names, I had never heard it before. Later, when I checked with Bill Woodley, neither had he. To cut a long story short, not only was Maitha an exceptional hunter, but also well-known as a particularly effective *mchawi* (witch doctor). No-one had informed on him during the anti-poaching campaign as the consequences would have been altogether too horrible to contemplate.

Maitha hunted a great deal with the Wata, but had two great hunting buddies – Wambua Makula, a Kamba and Kathuo Kagala, a Giriama who had married a Wata girl and whom the Wata regarded as one of their own. And it was listening to Kathuo and Boru Debasa, a Watic ace, that I learned of another white man's involvement. The talk was in Giriama which I do not speak, but follow roughly. Boru was asking Kathuo about the rhino horns that he and Maitha had delivered to a white man called Posita. Curious, I interrupted and the language switched to Kiswahili.

Who was this Posita? I knew him, they said, of course I knew him: yet to our mutual exasperation the name Posita was beyond me. They described where he had lived in Voi, but it wasn't until Boru Debassa mimicked the man, tilting his head to one side slightly that I immediately recognised who Posita was. It was an honorary game warden – Bob Foster. The Bantu interchangeability of *P* and *F* and separating consonants with a vowel made Foster quite naturally 'Posita'.

* * * *

Bob Foster had worked for Voi Sisal Estates. As an honorary game warden he shot many elephants on control for damaging the sisal. He was also reputed to have shot more elephants with 100 lb tusks than anyone in the past century, or possibly ever. For a while he had also been a Game Department elephant control officer. A quiet, soft-spoken Scot who had served as an aircraftman in the Royal Flying Corps in the First World War, no-one that I know ever thought of Bob as a poacher or illicit dealer – except possibly Bill Woodley. In Tsavo's early days he had been curious about how far into what became the park Bob Foster had shot elephants 'on control': some over 25 miles (40 km) from the sisal. Yet Bill's curiosity had never got beyond a question mark in the top shelves of his mind.

In the professional hunting fraternity Bob Foster was greatly respected. Yet, from what I heard on the banks of the Sabaki, over a period of several years he had bought many rhino horns from Maitha and Kathuo. That the tale Kathuo told was the truth I have no doubt at all. In truth, it did not make me think any the less of Bob Foster. Had I discovered his racket while it was going on and while I was still a warden, I would without question have arrested and prosecuted him. Yet given the value of rhino horn, I was more surprised by how few people had taken it. Maybe more had but we never found out?

Did my Wata friends know of other *Wazungus* (white folk) who had been poachers and or trophy buyers? Only one other case had been recalled and that had involved Galogalo Kafonde, another ace of whom I have written elsewhere[65], in one of his earliest brushes with white men and their laws. It was before the Second World War and Galogalo had found a huge pair of tusks, the biggest he had ever seen. He found them near Sala hill, close by the Sabaki river on the eastern edge of what became the Tsavo National Park. The elephant had been dead some months.

One William Harvey, a butcher from Mombasa, who was hunting for elephant seventy miles away on the Voi river had heard about the tusks from his Wata trackers and secretly bought them. Galogalo said that they had never been weighed, but were over seven *mikono* long. After the tusks were delivered to Harvey, he had shot a hartebeest, daubing its blood on their butt ends to make them look freshly killed, and returned to Mombasa saying he had taken the elephant on his licence.

They had been too big: word of tusks *that* size went about town fast and attracted traders, hunters and ivory aficionados like flies to a cowpat. A photograph of them appeared in the local paper and they went on display in Harvey's butchery. Among those who came to see them was MacArthur. The blood had not fooled him and he had seized the ivory. As Harvey never let on that he had bought the tusks, and steadfastly claimed that he had found them while hunting, he was never prosecuted. The loss of his tusks and licence was deemed punishment enough.

Hardly surprising, given the size of the tusks, I had already heard this story years earlier from Bill Woodley. He, in turn, had got it from Peter Jenkins, who in turn had got it from MacArthur's own mouth. Like me, Bill had also heard of the tusks during his

---

[65] in Parker & Amin. *Ibid.*

anti-poaching years. What impressed him then was Galogalo's ingenuity: to avoid carrying his finds to his home a dozen miles or so from Sala, he had lashed them to logs and floated them downstream.

There is yet more to this tale. In 1936 the Swedish professional hunter Bror von Blixen had taken an American – Winston Guest – and his sister to hunt elephants, on the Yatta plateau, upstream from Sala hill. Guest's passion was to get the biggest elephant on record. With the help of Beryl Markham, the aviatrix[66], spotting for them from the air, they had come upon two monsters. Winston had dropped the bigger animal with one shot and the party then chased the second animal which Winston's sister killed. With tusks weighing 130 lbs apiece, this was a fine trophy. All were agreed, however, that it was small compared to that which first fell to Winston's bullet. Alas, when they returned to the site of the first shot, the elephant had gone. Obviously only stunned, it had recovered and fled toward Sala. They followed it for two days before losing its spoor.

Winston Guest never forgot 'his' elephant and when news filtered back in late 1939 of MacArthur confiscating Harvey's tusks which had come from a dead elephant, he sent an employee from America to claim them on his behalf. The Game Department had not been convinced that Harvey's tusks were likely to have come from Guest's lost elephant. The Voi river where Harvey said he had found them was fifty miles or so from Sala Hill where Guest and von Blixen lost their victim's tracks. The tusks were then sold to the American Museum of Natural History. How much more bitter would Winston Guest's gall have been, had he known that the tusks were found, not on the Voi river but at Sala within a very short distance, possibly less than a mile, of where he and Bror Blixen had lost the trail.

The tusks the Game Department sold to the American Museum of Natural History in New York weighed 189 and 178 lbs respectively[67]. They would have dried out by about five per cent by the time they were weighed so that, fresh, the heavier of the two may have weighed 198 lbs and the lighter 187 lbs. As such they are the biggest tusks ever known to have come out of Kenya.

So much for the tale of William Harvey's (or Galogalo's, or Winston Guest's, or MacArthur's) tusks. It hardly qualified Harvey as an ivory poacher or trafficker. With some certainty in the last decades of the colonial era, I do not think that there were any white ivory racketeers of consequence. I know of a number of fiddles concerning elephants taken on licence, but none were commercial enterprises of substance. Another Mombasa butcher whom I had had cause to investigate in 1957 – Ginger Bell – later said that he had bought and sold ivory on some scale and that he once had to dump a substantial amount in Old Mombasa harbour when he thought the law was closing in on him. It might have been true, but I took it as romancing by an old man who cultivated an image of being a bit of a rascal. I think he knew (as I did) that scuba divers had found ivory at the bottom of the Old Port. It was not whole tusks but butt-ends – which didn't fit with his story.

---

[66] Markham B., 1942. *West with the Night*. Houghton Miflin, New York.
[67] Ritchie A., 1937. *Annual Report of the Kenya Game Department*. Govt. Printer, Nairobi.

The greatest tusks to come out of Kenya in living memory.

* * * *

From personal experience I knew that all was not above board in the Game Department. I have recounted elsewhere[68] how I stumbled across an ivory racket involving the headquarters' clerical staff, the warden in charge of the Ivory Room and the second in command of the police Criminal Investigation Department in Mombasa. I was too dumb to recognise the evidence I was seeing: as new as I was it was inconceivable to me that members of the Department might be stealing tusks.

Those involved did not appreciate that my silence arose from stupidity and had to assume I kept quiet about the evidence I had seen for reasons best known to myself. I had to be neutralized so that any accusations I might later make would be discounted. To impugn my reputation they accused me first of shooting a rhino, then killing six elephants and finally of stealing a pair of control tusks and taking them on my licence.

Each instance made me angry, but in my paranoia I took it to be a person or persons trying to block the Waliangulu Scheme which the Department so disliked. On the first two occasions the charges were easily disproved and I let the issues rest because I was damned if anyone would get a rise out of me that might compromise the Scheme. The CID was called in on the third instance and, once the investigating officer was satisfied that the accusation was trumped up, he asked what I had seen that warranted such persistence in 'setting me up'. The Officer, Ron Winterburn, did not for a moment believe that it was a plot to stop the impending game management project. From professional experience, he thought I knew something which threatened someone who was bent on neutralizing the threat.

Winterburn's surmise was correct. Until he cleared my name I had been unaware that he was already investigating the Game Department. This process had been set in motion by my colleague David Brown. Standing in for Neil Sandeman when the latter went on leave, D.B. had been struck by how easy it would be to fiddle the records and steal trophies: so easy that he made a few checks. Having just sold some trophies but leaving the issue of appropriate receipts and handling the money to Michael de Souza, he surreptitiously checked to see if the money had been properly accounted for. There was no record of the sales he had made and Michael had pocketed the money.

In due course Winterburn exposed de Souza's long-standing ivory thefts and in 1960 Michael and his junior Eddie were arrested and tried for stealing Government trophies. Eddie went to jail for nine months. Michael was given three years on probation, despite the evidence that he had taken around $74,000 worth of government ivory (valued at $3.5 million in 2003). The magistrate was lenient because he felt that more senior men were culpable (and he must have been alluding to Willie Hale and Neil Sandeman), albeit only through negligence. He referred to the Department as 'Gilbertian' and stated that its management verged on a public scandal[69].

Ron Winterburn's case against de Souza tripped a switch in my memory and I had indeed seen evidence of what he was up to, and that this was the root of the accusations against me. However, it also meant that de Souza was in cahoots with the game warden

---

[68] in Parker & Amin. *Ibid.*
[69] reported in *The East African Standard,* 1960.

then in charge of the Ivory Room – an ex-policeman – and the head of the CID in Mombasa, for it was they who attempted to frame me. Not until years later did I get corroborating evidence of the degree to which members of the CID in Mombasa were implicated in ivory racketeering from Mike Webley, a fellow game warden who, in the 1950s was a CID policeman and had investigated the matter. In the 1950s, no less than three superintendents connected to the Mombasa CID retired and disappeared in a hurry over matters to do with ivory. Yet, despite knowing that both game wardens and senior colonial policemen had been involved in illegal ivory, I still felt that, certain individuals excepted, the Department *was* a band of gentlemen.

In the wake of de Souza's conviction, Ian Grimwood, the new Chief Game Warden, should have introduced more stringent accounting procedures for trophies – particularly the more valuable elephant tusks, rhino horns and leopard skins. Yet he did not, and the management remained as muddled as it had always been.

Take the trade in game trophies for example. Theoretically no-one could possess a game trophy – that is any 'part of a game animal whether added to or changed by the work of man or not which is in such form as to be recognisable as a durable part of a game animal' – without a permit (Section 33 of WAPO). If one killed an animal on licence, that licence was an acceptable permit. However, if one acquired a trophy through gift or purchase, the person making the gift or sale had to get a 'sale permit' for it from the Department (Section 34 of WAPO). The sale permit endorsed in ink by the seller then stayed with the trophy as evidence of its lawful origin. If, at a later date, the owner of the trophy wished to take or send it out of Kenya, he or she had to surrender the appropriate sale permit for an 'export permit' from the Department (Section 37 of WAPO). On top of all this, no-one could sell game trophies commercially without a Dealer's Permit under Section 35 of WAPO. In theory, therefore, the demands of the London Conference in 1900 and the Brussels Conference of 1933 that trade be controlled were well taken care of. Dealers could only operate with Game Department sanction; the legitimacy of any game trophy was proven through the licence on which an animal was taken or a sale permit; no trophy could be exported without an export permit.

In practice the system was a shambles. The law made no allowance for the fact that what started out as a single trophy – a zebra hide for example – could become several hundred trophies when converted into watch straps, each of which called for its own sale permit. Once turned into watch straps it was difficult if not impossible to prove from which particular hide individual items had originated, compromising the purpose of the permits.

By the late 1950s tourists were the main market for game trophies and the demand for sale permits for items made from game animals was running at between 65,000 and 75,000 a year. In turn this called for a similar number of export permits. This volume was beyond the Game Department's clerical capacity to issue, so it took a truly Gilbertian step and placed the onus for issuing export permits on the dealers who sold the trophies. Export permit books were handed over to the very people that, above all others, the permits were supposed to control. In effect the dealers became self-licensing elements within the official licensing system. The headquarters staff were supposed to randomly check the permit books to ensure that they were not used to cover illicit trophies, but this was little more than face-saving formality.

* * * *

In sum, not only did the Game Department present unique opportunities for theft with small chance of getting caught, but some members of it seized those opportunities. From the very outset it seemed that those responsible for framing both policy and laws lacked basic law enforcement savvy. Nothing brought this out more bizarrely than when the dealers in trophies were entrusted with issuing permits to the customers who bought game trophies. The probity of the colonial civil service notwithstanding, it was amazing that there were not more rotten apples in the Game Department barrel.

CHAPTER TWELVE

ON TUSKS

At Mutha in Kamba country, Bill Woodley had glanced at a tusk and said, "Nice cow ivory," and I learned that male and female tusks of the same weight have different shapes. The male will be thicker at the base and taper pronouncedly all the way from base to tip. The female's will have so slight a taper that for most of its length it will appear cylindrical and only come to a sharp point through wear and use in the last fifth or so of its length. While the biggest female tusk known weighed over 70 lbs (32 kg), any over 20 lbs are unusually large and over 30 lbs exceptional. Male tusks grow much bigger, the largest recorded being 228½ lbs (104 kg) and those over 40 lbs are common. Up to this lower weight the conical shape that distinguishes male from female tusks is obvious. Thereafter, as they get bigger, the taper become less pronounced: in hunters' jargon, with a lack of taper a tusk was 'carrying its weight'.

Female tusks were slender and cylindrical: the pair shewn here being typical of an East African savanna elephant over forty years old.

At any given age, a male elephant's tusks were much bigger than a female's and tended to have a more pronounced taper between lip and point.

When judging a tusk's size, the aficionado first looked to see if it carried its weight or not. There are other differences between the sexes. The ratio of a tusk's length inside the elephant's skull relative to its overall length, for example. With smaller skulls that proportion is usually less with females than with males.

Like most game wardens and some of the park wardens including Bill Woodley and Peter Jenkins, I took out licences to hunt elephant. Getting a decent pair of tusks was an easy and legitimate way to increase a paltry salary. The average elephant taken on licence in Kenya in those days had tusks that weighed over 70 lbs each, that is it yielded 150 lbs of ivory which sold for £1 per pound weight. The licence cost £75 so one could double one's money. An individual was allowed two licences a year, the second costing £100. Thus taking only average elephants, for an annual outlay of £175 one could get a return of £300, doubling what I had to live on. The Kilifi range was prime elephant country and although I had to borrow money to buy my licences, the average tusk weight I took on licence was 103 lbs (i. e. 206 lbs per elephant). Recall that, having deducted rent, tax, loan repayment, water, electricity bills, etc. I was trying to live off £12.50 a month (£150 a year). In such circumstances the income from an elephant licence was critical.

Few commodities have been the subject of more romantic speculation than ivory. It has a strange fascination for humans, though why this should be has yet to be explained. In part it must be ivory's tactile qualities. This was obvious when watching traders handling it. Millionaires who normally employed others to carry out mundane physical tasks would take off their jackets and move tusks about a storeroom floor. This was particularly the case with big tusks which they seemed compelled to touch and feel.

Listening to a wealthy European trader's wife wonder rather critically why her husband bothered with tusks, I had teased her, "Christine, if you want to understand, watch William touch a tusk; watch his hands – the sensuality – it's almost as though he touches a woman!"

Instead of the defensive protest I had expected, William had agreed enthusiastically, "Yes! Yes, that is it! That is it! You have to *feel* beautiful ivory to understand."

His response went to the very core of the commodity's attraction. Wherever it has occurred man has acquired ivory and considered it valuable regardless of sources. These span the planet: ivory is produced by narwhals and walruses in the Arctic, sperm whale in the oceans, hippo and many pigs as well as elephant in Africa, Asia and Siberia's tundra where a vast store of mammoth ivory lies frozen.

From the turn of the century the Ivory Rooms in the ports of Mombasa and Dar-es-Salaam auctioned thousands of tusks annually. Until 1956 both had been run by their countries' Customs & Excise Departments. Ivory was revenue and as the Government's principal revenue collecting arm this had been wholly appropriate. Other than for ivory acquired in the course of their duties, Game Departments had nothing to do with how the Ivory Rooms disposed of government ivory. The colonial ivory collecting system pivoted around the Administration's district revenue officers. Each district headquarters had an ivory register. Tusks found by the local people from natural mortality were brought in and surrendered in return for a reward which was of the order of ten per cent of prevailing ivory prices. The reward was referred to as a 'portage' fee because governments did not admit that they bought ivory. The semantic difference may have been apparent to civil servants steeped in cryptic grammar, but where the public was concerned, governments *did* buy ivory and did so very cheaply.

Annually, thousands of tusks were auctioned on the Ivory Room floor, yet no one appreciated the vast amount of information they contained on the state of elephants both generally and locally that went unrecorded. (Photo Camerapix)

Tusks were also collected in the district headquarters from seizures by game wardens or police and animals shot on 'control'. At regular intervals every district sent its ivory to the Ivory Rooms where it was sold and the ivory bullion was converted into the national currency. The Kenya Game Department had no clearly defined procedures for its wardens to dispose of ivory or rhino horn. There were official ivory register books, but I was never issued with one and in my tenure as Game Warden Kilifi, the station never had one. I delivered such tusks as came into my hands from control work, natural mortality or seizures, directly to the Mombasa Ivory Room. This seemed sensible as Mombasa was a mere forty-five miles from Kilifi. The only mishap I ever had was on rounding a bend in Mombasa's then Salim Road somewhat faster than I should have, a large seventy five pound tusk was launched out of the back of my Land Rover. Like a torpedo from a destroyer it flew across the road and down the far pavement. Like a great white sabre spinning about its point of balance, it sent a party of Arabs leaping in a wild sword dance, narrowly missing shins and ankles. Fortunately no-one was hurt and after an embarrassed recovery, I left the street a-jabber with amazement.

Other more distant game wardens either delivered tusks and rhino horns to their district revenue officers, or to Game Department headquarters: the choice was up to them. To say that the system was lax would be understatement. Be that as it may, the Mombasa Ivory Room received most of its tusks directly from the administrative districts. Consequently between 1900 and 1956, the Game Department never kept the country's primary ivory records. In addition to handling Kenya Government and National Parks ivory, the Ivory Room also sold all of the Uganda Government's and National Parks' tusks. Less regularly, it also sold large numbers of tusks from the Belgian Congo and other countries. The Kenya and Tanganyika Governments charged a commission of between two and three per cent for such services. Their ivory rooms held two auctions each a year so timed that an auction occurred in either Mombasa or Dar-es-Salaam in each quarter of the year.

In 1956 both Kenya and Tanganyika decided that their Ivory Rooms should more properly be under their respective Game Departments. In that tusks came from elephants this was superficially logical. Yet the issue was never carefully thought through. The control of state ivory sales, in which accurate accounting was at a premium, was taken from organisations, Customs & Excise Departments, designed specifically to collect revenue and handed to two Departments which were the least accountable or numerate of all government organs. Until 1956 the Game Departments only had theoretical control of the ivory resource in the field. From that date on, however, they were given real control of its disposal. *De facto*, the Game Departments acquired the immense commercial power of an ivory monopoly. At the time this was not appreciated, though in due course the consequences would become all too obvious.

In those early days I went often to the Ivory Room, and observing, talking to the staff and the traders, picked up information on tusks. These were classified according to an Indian system that dated back to the eighteenth century when Mumbai (Bombay) was the premier market for East Africa's ivory. *Vilaiti* was all tusks over forty pounds and exclusively male. *Cutchi* was all male tusks between twenty and forty pounds. *Fankda* were male tusks between ten and twenty pounds. *Calasia* were all female tusks heavier than ten pounds. *Maksub* and *Dandia* were tusks of both sexes respectively between five and ten pounds and less than five pounds. Defective tusks were *Chinai*.

On arrival, tusks were broken into lots of like weights and shapes and I was soon struck by consistent regional differences. The most notable were those from westernmost Uganda and Zaire. They were, on the whole, much straighter and more slender. Except for the points where wear and use exposed the underlying ivory, most of these western tusks between lip and tip were either black or a lovely deep, polished mahogany, presumably stained by plant saps from lush tropical vegetation. Their underlying colour which could be seen in the points where all staining had been worn off, was more yellow than white. Where polished by wear, they also had a gloss that was unusual on tusks from elsewhere. The buyers called these western tusks *gandai* or 'hard' (later I learned that the Chinese referred to them as 'yellow' and other tusks as 'white' or 'soft') and they were sold separately.

Other than in the lightest two weight categories, male and female tusks were separated and the distinction between them obviously well appreciated in trade. When the regional and sex differences were taken into account, it was patently obvious that ivory displayed consistent local characteristics that permitted one to identify area of origin. Not only did tusk shape relate to places of origin, but the ratios between males and females told a great deal about what was happening to the elephants in any area. Taken over all, Uganda produced far more female tusks than Kenya – presumably because there was less discrimination in favour of shooting males. Within Kenya, ivory from the National Parks that derived principally from natural mortality differed in its composition of males and female tusks to that from agricultural districts, which in turn was not the same as that from pastoral areas. I realized that collected tusks provided a vast source of information on both elephants and people. What I learned and saw in the Mombasa Ivory Room stood me in good stead.

* * * *

An early Wildlife Services consultancy for the Kenya Game Department's Fauna Research Unit involved extracting and placing all faunal records from Departmental files, in a card index system. Ironically we were the last people to go through these records as shortly afterwards they were destroyed. The destruction took place in two waves. The first was undertaken by the white staff before the first African Chief Game Warden took office when it was felt that certain issues might be racially sensitive and bias the incoming African officers against their white predecessors. The second was after the first African had taken over and (presumably) felt it appropriate to be rid of the 'colonial' past. I am sorry that I did not steal the files and save sixty years of records. While certain administrative, police and political files would inevitably have been extracted from the system in circumstances of changeover – as happened wherever independence was granted in the Empire – I know of no other case where the records were so wilfully obliterated as in the Kenya Game Department. Sadly, and for all that the yield of faunal records from sixty-four years of gamekeeping was rather sparse, our card index system has also vanished.

By 1964 an obsession with counting animals had set in. Every Game Department and National Park system wanted to know how many animals they had. The logic was seductively simple: before you manage you have to know what there is to be managed and to know what there is, it must be counted. It is the prerequisite for monitoring the health of populations. No-one questioned such apparent common sense – which was fine by us who wanted to make money counting animals. That is not to say we cynically thought otherwise for, in the 1960s we believed the logic. Yet across Britain and Europe game populations – birds, mammals and fish – had been managed for several centuries without counting them. In the USA deer populations had been 'managed' by adjusting annual offtakes up or down on the basis of the average age from the previous year's offtake. A rise in the average age of stags killed = enough stags are surviving to raise the average age = therefore offtake can be raised. Conversely a drop in average age of the annual kill = not enough stags are surviving = decrease offtake. Simplistic, functional, but counts were not necessary. Obviously sometimes knowing numbers is useful, but the *necessity* of general counting was and still is grossly overstated. Nevertheless, we learned to fly, got an aeroplane, and were hired to count animals.

Some idea of our involvement with research comes from the number of papers we published in what was then the East African Wildlife Journal between 1964 and 1974. The Journal published 105 full papers of which members of our company produced 10, together with the raw data for a further six authored by others which made us responsible for 15% of the published material in the region's scientific wildlife journal. We also published in other international scientific journals.

Our largest and most profitable contract was culling 4,000 hippo and 2,000 elephants in Uganda's Murchison Falls National Park. The need for this work arose from the policy pursued by the Trustees of the Uganda National Parks. They held that the purpose of their parks was to preserve as great a diversity of indigenous plant and animal species as possible. Anything that increased diversity should be encouraged. *Vice versa*, anything which reduced diversity should be controlled. Hippo had reduced the diversity of other

We learned to fly, got an aeroplane and were hired to count animals.

grazing species within their ranges. Similarly, elephants had widely eliminated forest and woodland in favour of grassland and, in the process removed woodland and forest faunas from the park. The Trustees decreed that both species should be reduced in number to facilitate a general recovery in diversity.

To do this work we set up a Uganda-based company, Game Management (Uganda) Limited, and brought in two extra partners – Tony Seth-Smith and Mike Rowbotham. The most satisfying aspect of this project (in addition to good profits) was the collaboration with Dr Richard (Dick) Laws (later succeeding Sir Vivian Fuchs as Director of the British Antarctic Survey, then becoming Master of Churchill College, Cambridge) collecting and analysing the scientific data from animals culled. It marked the start of a close friendship between our families that persists into the present. In short order, elephant social organisation, longevity, growth, nutrition, reproduction, and population dynamics were described with confidence for the first time.

This does not mean that certain aspects of their biology were not understood before our work. The Wata perceptions, for example, were centuries if not millennia old. Other scientists had tackled aspects of elephant biology. Yet this was the first time it had ever been approached quite so holistically, on such a large scale and in such depth. We summarised the findings in a book[70] of which the reviewer in *Nature* wrote –

---

[70] Laws R.M., Parker I.S.C. & Johnstone R.C.B., 1975. *Elephants & Their Habitats*. Oxford University Press, Oxford.

"I think it is safe to say that there is no other mammal population in which growth, nutrition, habits, reproduction, age structure and behaviour has been so thoroughly studied in relation to climate, flora and fauna and land use of its habitat."

Later behavioural studies confirmed Laws' findings and though excellent in themselves, few acknowledged their debt to Dick's accurate description of elephant biology on which they rested. Looking back over the past thirty years there have been only two discoveries: elephant communication with very low frequency sound waves, and the phenomenon of musth behaviour in the male that were not described or anticipated from the work in Uganda. We have one great regret: that DNA analysis was not available in the 1960s. With it we could have vastly enhanced the culling programmes' output of knowledge.

The research dispersed much of the romance that had previously enveloped elephants. Nothing lost its mystique more than ivory. We now knew that size was basically the product of age; the older the elephant, the longer and heavier its tusks. Both tusk length and, even better, circumference at the lip were good ageing criteria: just as horse-traders down the years have aged horses accurately by looking at their teeth, we could now do the same with elephants. One did not get big ivory on Mount Kenya and the Aberdares because few elephant lived long enough to produce it. Surrounded by high densities of people and intensively farmed land, an elephant which ventured outside the forests' cover was soon shot. It was as simple as that.

Knowing the relationship of tusk size to age, being able to separate males' from females' tusks at a glance and recognising areas of origin from tusk shape and wear, it was obvious that if these data were taken from tusks in trade, one could accurately assess the status of a country's elephants. It was knowledge that I used extensively in the future. Unfortunately, however, it only came to light as the great public ivory auctions where it could best have been applied were on the verge of disappearing. In 1967 Uganda decided to hold its own auctions and dispense with the service provided by the Kenya Government in Mombasa. Kenya's ivory was soon being sold by private arrangement and Zaire, as the Belgian Congo had become, ceased being a legitimate ivory exporter after its tumultuous descent into 'freedom' in 1960.

All the elephants' tusks and hippo tushes from animals cropped were ours to sell, and now as a bulk producer I saw the ivory trade from a new angle. As the expected volume of ivory from the 2,000 elephants exceeded twenty-five tons, we received friendly approaches from many buyers. My initial response had been that we would keep in touch, but make no deals until we had a substantial volume of ivory in hand. That was in early 1965. Wanting to have the hippo reduction programme running smoothly before starting on elephants, we did not plan on having tusks to sell until the year's end.

Thus I was surprised in June 1965 when several Asian buyers in Kampala, Uganda's commercial capital, reproached me for not having asked them for quotes for our first ivory. They took much convincing that we had not yet produced any elephant tusks. Large quantities had been on sale in Kampala and they naturally assumed it must have come from our elephant reduction contract. Assured otherwise, they soon identified its source: the Uganda Army. In 1965 Zaire was chaotic and parts of the north-east still in rebel hands. At some time in late 1964 or early '65 the Uganda Army under General

Amin had crossed the border and established themselves at Aru in north-eastern Zaire on the pretext of preventing the chaos there from spilling into Uganda. While there the Ugandan Army acquired a lot of ivory and, so it was said, a substantial amount of gold. All was brought back to Uganda and sold. It was this that had initially been thought to have come from us. To my knowledge, it was the first large ivory heist in post-colonial Africa and a sign of things to come.

When our Uganda work commenced in 1965, the three East African countries of Kenya, Uganda and Tanzania still had their colonial infrastructure in place and functioning. The East African Community still ran Railways, Harbours, East African Airways, the Customs and Excise authorities, the Income Tax Department and many research bodies, as efficient federal services. The currency was still the colonial East African shilling and residence in one country gave one automatic access to the other two without need for immigration formalities. Both Uganda and Tanzania were nonetheless very much commercial appendages to Kenya. Even in Uganda, by far in a way the wealthiest of the three East African territories, nine out of ten items – whether they were spare parts or items for our camp commissariat – had to be ordered from Kenya. Consequently, though going through the public relations formula of registering a Uganda Company, we ran the show from Kenya.

Our principal bank account was not in Uganda, but at Molo in the Kenya highlands, for it was there that Mike Rowbotham farmed within easy reach of a bank. We were taking over £2,000 a week from the sale of hippo carcasses – a lot of cash in those days. Our flight route from Murchison to Nairobi took us over Mike's farm. To bank the take, which we did *en route* to Nairobi once or twice a month, we buzzed his house to alert him, whereon he bade all but members of his family stay well away, then placed wife and children around the lawn, no cricket captain ever placing his fielders more carefully. When they were ready, we made a low pass and dropped the money.

The Ugandans resented our dependence on Kenya, while we did not trust Uganda. It had an air of instability that soon manifest itself. In May 1966 the Prime Minister Milton Obote set the army upon the President – the *Kabaka* (King) of the Baganda tribe – setting in motion events that culminated in General Idi Amin seizing power in 1971. My introduction to soldiers out of control occurred at Gulu airport. Gulu was the regional administrative headquarters and I had gone there on some errand. Leaving my aircraft I walked to a public telephone to call for a lift into town. A hard blow in the back was the first I knew of the Army's presence. Collecting myself I turned to see a great, wild-eyed sergeant with flaring nostrils and an FN .762 rifle pointed at my midriff. At that instant there was as much in common between us as there would have been had I faced a thing from outer space.

There was no reason for the assault other than it was, quite literally, a form of drunkenness emanating, not from alcohol or a drug, but unbridled power and a feature that African soldiery has since displayed all over the continent in the following three decades. What was I doing there he wanted to know. Refusing to answer the question, I demanded to see his commanding officer. Eventually a young lieutenant turned up and, having conceded that I was in the country legally, entitled to use the airport, the telephone, and had no liability whatever to report to the Army, he agreed that I had been assaulted gratuitously. "We live in the dangerous times," was the only solace that he offered. I got off lightly. Others stopped at roadblocks were severely beaten up. On another occasion a Ugandan Army aircraft spotted one of our field camps in the park and in the early hours of the morning a company

From the air and when it was bunched together in open grassland, we could see this herd of over 2,000 elephants as a brown smudge in the landscape at a distance of 56 kilometres (35 miles).

of soldiers surrounded it. Fortunately they had had to come through a regular Park entrance where the gate rangers told them who we were. It was not long before direct flights from Kenya were banned and entry to Uganda had to be through the main international airport. The three states produced their own currencies and regional disintegration was under way.

In late 1967 or early 1968 there was an event in Uganda which never was widely known and which seems to have vanished from the records. An amnesty was declared for all illegally held ivory: tusks handed in to Government would be paid for without questions asked. The arrangement was supposed to last for six months but I recall it being extended once and possibly more times. Although I tried to find out how much ivory was surrendered, and heard that eighty tons had come in, Uganda's Chief Game Warden at the time – a veterinarian called Lawrence Tennant – had been reluctant to discuss the amnesty or even say whose idea it had been. The whole business was kept remarkably secret and out of the public's ken.

In 1970 one of my partners, Alistair Graham, and I were contracted to count elephants in the Murchison Falls National Park and the contiguous Gulu Elephant Sanctuary which lay along the right bank of the Albert Nile. It was an area we knew well and, like the Murchison Park itself, was declared a game reserve early in colonial times. This status was given, not for conservation, but as a legal device to bar human residence in areas whose tsetse flies transmitted human sleeping sickness. It was not the only part of east or central Africa where a game reserve was declared to break human contact with infected tsetse. The elephant sanctuary was a forgotten corner of Uganda whose elephant population nearly rivalled that in the Murchison Park itself.

Alistair and I found elephant hunting on an enormous scale taking place. There were so many elephant skeletons along the shores of the Albert Nile that we called the stretch ten miles or so north from the Pakwach Bridge 'the Ivory Coast'. Deeper in the Sanctuary we found camps in which there were piles of smoke-dried meat between twenty and forty paces long, ten paces wide and at least eight to ten feet in height. There were vehicle tracks all over the ground where there had been none before. There will have been a big yield of ivory, but it seemed that the elephants were being taken primarily for their meat. The operation had been under way for some time – from the carcass remains, possibly as much as two years – and presumably there was some link between it and the ivory amnesty.

We never learned who was responsible for the massive hunting in the Gulu Elephant Sanctuary. From our own experience we knew demand for smoked elephant meat was not great in Uganda. The traders who had bought it from us disposed of most of it in Zaire, bartering it for coffee, cotton and that country's general produce. Presumably the meat coming out of the Elephant Sanctuary would have been used for the same purpose. One way or another, I think we had witnessed the second great ivory heist of the post independence era. It, too, preceded Idi Amin seizing power.

* * * *

Back in Kenya the Chief Game Warden Ian Grimwood had also been facing a crisis. In 1964 he received orders – ostensibly from President Kenyatta – that certain Freedom Fighters should be allowed to sell game trophies they acquired and hid during their Mau Mau days in the forests. The original command implied that the amount would not be substantial. In 1964 the active forest war had been over for nearly a decade and while buried ivory could still be in good condition after such a period, rhino horn and leopard skins would not. That both horn and leopard skins figured in the trophies at all made it certain that they were not relics of the forest campaign but of more recent origin. Grimwood was apprehensive that the special permission to 'ex-Mau Mau' would develop into something greater.

Among those allowed to bring in trophies was 'Field Marshal Muthoni Kirima', a woman. Her story is told in a book by David Njagi[71] where she claimed to have lived in the forests, presumably of the Aberdares, from 1952 until 1963 (eleven years) when Kenya became independent. This sounds improbable. One or two isolated groups did stay in the forests after the Emergency's active phase ended in 1956[72], but it is unlikely that so high-ranking and well-known a person as she claimed to be would not have appeared in the official records somewhere – even if merely to mention that she had disappeared. No-one I know could recall her name. The title 'Field Marshal' was self-given and arose from her recollection that –

> *"Kimathi* [the Mau Mau leader Dedan Kimathi] *said that anyone who will fight to the end of the war will be called Field Marshal."*

---

[71] Njagi D., 1993. *The Last Mau Mau Field Marshals.* Published by Ngwataniro Self Help Group & David Njagi. Meru, Kenya.

[72] Franklin D., 1996. *A Pied Cloak: Memoirs of a Colonial Police (Special Branch) Officer.* Janus Publishing Co., London.

I suspect, as do some ex-members of Mau Mau I have consulted, that Muthoni's Mau Mau status was partially if not wholly bogus and her permission to collect trophies had strong post-independence political origins. Regardless of Muthoni's claims on behalf of Mau Mau, she was, without question, given permission to gather ivory –

> "*We finally met Kenyatta and we told him our problems and made a request; we wanted a permit to collect old ivory lying in the forests of Kenya. He agreed.*"[73]

Any suggestion that this would be confined to the forests in which the Mau Mau had operated was quickly abandoned. By her own admission –

> "*I travelled to many parts of Kenya collecting the ivory.*"[74]

Ian Grimwood foresaw that allowing ivory to be collected by such as Muthoni would be uncontrollable. Unable to refuse the presidential edict, he nonetheless initially tried to direct events by insisting that all the trophies brought in should be surrendered to Government. Upon surrender, the collector would be paid shs 12.00 per kilo and the trophies would then be disposed of by Government in the normal way through the Ivory Room. This price was raised to shs 22.00 per kilo after Muthoni again successfully sought the President's intercession on the matter. Grimwood's very obvious dislike for the whole arrangement and reluctance to surrender control did not sit well with his superiors. As his days were clearly numbered, he resigned in 1965.

Herein lay the reason why Ian Grimwood went back on his word over the future of the Galana Scheme. It was when Kenyatta was ordering that Muthoni be allowed to collect tusks. He could not give a group of white investors freedom to manage game on the one hand, while at the same time trying to tie down African entrepreneurs as tightly as possible. My Galana plan became an early political casualty and while Ian's decision is now understandable, at the time it made no sense because I was ignorant of Kenyatta's ivory edict.

* * * *

Ivory had involved me one way or another every since I joined the Game Department. We recovered it from poachers, picked it up from natural mortality, acquired it for government from the animals we shot on control or for ourselves from elephants taken on licence. The elephant culling in Uganda produced both greater scientific knowledge of tusks than had previously prevailed and a greater quantity of ivory to sell than any private individual had legally produced since pre-game law days. This led to greater contact with ivory traders and to an apparent paradox when they asked me to analyse the trade for them, as will be explained later. Ivory was leading in many directions and into disparate fields.

Ivory from natural mortality has always been a very substantial component of ivory traded: yet consistently overlooked or discounted by those analyzing the trade.

CHAPTER THIRTEEN

THE BATTLE OF WOUNDED FOOT

Leaving Government service was in many ways a relief. No longer a 'Heaven-born' game warden, I was able to take a broader view of conservation and, of necessity, be far more aware of political trends. Fear about a declining quality of governance was at the forefront of many minds. The first two big ivory scams in Uganda and the 'collectors permits' in Kenya were pointers to the future and ground for pessimism. In both Uganda and Kenya the corrupt initiatives had come from very high in the political structures. Let us now return to the fortunes of David Sheldrick and Bill Woodley who stayed at their posts after political independence.

* * * *

If the 1950s had been a decade of success in Tsavo East, the 1960s were years of crisis. At the outset things were on a high note. Internationally acclaimed for the successful anti-poaching campaign, happily married to Daphne and internationally very much a rising conservation star, the world was at David Sheldrick's feet. Had he wished, he could have joined any of several conservation departments in Africa or organisations in the same field elsewhere. Offers for his services were made, but he turned all down to stay in Tsavo East.

The first crisis came in 1960 and ran through to September 1961 in the form of a severe drought. Tsavo East started to look like a battlefield. Elephants tore the trees apart, the shrub layer disappeared and animals – particularly black rhino – died in their hundreds. The received wisdom at the time was that more watering points were needed to spread animals, especially elephants, more widely across the land. A *Water for Wild Animals* appeal was launched with the catchy slogan of 'Buy an Elephant a Drink'. I was in a minority in suggesting that 'Buy an Elephant a Bullet' would have been more appropriate, as I saw the issue as one of too many elephants. David's flair for publicity was at the core of this appeal and again his *modus operandi* was through other people and the media, but particularly through the national park headquarters and the Director Mervyn Cowie. Funds poured in for an ambitious scheme to pump water from the Athi River to the top of the Yatta Plateau and then gravitate it some 50 miles or so into the driest parts of northern Tsavo East. A swathe of 1,500 square miles (3,840 square kilometres) of park would become available to water-dependent animals.

The pumps ran for twenty-four hours when an unexpected problem arose: the volume of silt in the Athi was more than they could handle. Before the problem was resolved, four days of continuous rain in September 1961 ended the drought. This was followed in October and November with the heaviest rains recorded in Kenya up to that time. Rivers flooded higher than ever before, the huge water-for-wild-animals pumping station was irreparably damaged and never pumped another gallon. The money spent, the work done, the campaign and all that went with it was ironically washed out.

That the project had been ill-conceived took time to sink in. The evidence was clear. Veterinarians proved that rhinos had been dying of starvation, not thirst, and had needed

Ironically, floods destroyed the ambitious water-for-wild-animals project launched in Tsavo East in 1961.

food, not drink. Many died standing in flowing water – in the Athi, Tsavo and Sabaki rivers. Making new water points far away from existing water supplies was a short term solution. The real issue lay in Tsavo's aridity. The meagre and erratic rainfall guaranteed periods when plants could not produce the food on which animals depended. The drought of 1960-1 was well within expectations for that climatic regime. In a land where feast and famine was the natural order, animal population crashes should also be expected. Where nature clashed with the park managers' wishes was that these natural processes could produce local extinctions. If this natural risk was to be reduced or removed, then water *for plants* was the critical long term need. And *that* was impractical. All this has a clarity now that was not so obvious then. In 1961 the park was a bare thirteen years old and knowledge of its past was very thin.

The rhino die-off in 1960-61 made a point that is still not generally grasped. Stability in animal and plant populations is tied to climatic stability. If you make a national park where the rainfall is erratic and unpredictable, then plant and animal numbers will fluctuate in response. With species such as elephants and rhinos, which can live in a wide variety of habitats, it makes best sense to conserve them where climates are most stable. It makes least sense in an area like Tsavo, where rainfall is erratic and unreliable. Recall that the reason why Tsavo became a park is because no-one wanted it: and no one wanted it for that very reason. Frequent famines make all plant and animal management, including conservation, difficult.

When the rains broke in September 1961, the philosophical dialogue which the rhino die-off was stimulating died away. With resurgent plant growth following the rains it briefly seemed that the crisis was over. Yet despite the good rains, Tsavo East's woodlands continued to decline. Elephants were eating all the plants that rhinos ate and it was elephant

Tsavo's Aruba dam made in the early 1950s had major ecological consequences. Prior to this elephant could only use the area in wet seasons. Thereafter they could use it in dry weather too. Coincidentally, at the same time, they were denied access to water points to the east of the Tsavo Park, further intensifying use of the park around Aruba. (photo Joe Cheffings).

consuming rhino browse that actually caused the rhino starvation so dramatically accentuated by drought. Elephant numbers in the Tsavo National Parks had been increasing. The principal cause seemed to be rising human numbers outside the Parks displacing elephants and forcing them into the sanctuaries. I had personally observed the process.

In the early 1950s Aruba dam was thrown across the seasonal Voi River in Tsavo East, providing a major permanent water supply where none had existed before. Until then the nearest permanent water available to elephants was in the Sabaki River and in Rei, Mido and Garbiti waterholes not far to the east of the Park boundary. Deep clefts in rocky outcrops, these three formed natural tanks with capacities large enough to sustain around two thousand elephants through the dry months. Coincidentally, soon after Aruba dam appeared, humans settled around Rei, Mido and Garbiti, denying them to the elephants which switched to the new Aruba dam. In this instance an elephant community that had previously used Tsavo as wet weather range and the thicker vegetation to the east as their dry weather resource, reversed the situation. Aruba became their dry weather focus and the area to the east their wet season foraging ground. They went outside the Park when they did not need to use the dry season water points now monopolised by people. To the west of the park the very rapid Kamba agricultural expansion displaced an even greater number of elephants into the sanctuary. When I was posted to Makueni and Makindu in 1956, both areas were 'good' elephant areas. By 1963 elephants were occasional visitors. Most of those which had previously lived there had moved east into the Parks.

David Sheldrick first reported elephants destroying patches of woodland in parts of Tsavo East in 1957. At the time it seemed a local phenomenon and while he sounded a tocsin in his reports and stressed the need to keep an eye on the process, it worried no-one, including himself. In the 1960-1 drought, even as he pressed for water for wild animals, he knew that elephants were becoming a major problem. At the time the degree to which it was part and parcel of the rhino die-off was suspected but not as obvious as it became with hindsight. Hopefully elephant killed in the neighbouring game management scheme would lower overall numbers and reduce this pressure.

Clearly Tsavo's woodlands could not sustain the use they were receiving from elephants. What was first reported in 1957 had become a general trend. If it continued the woodlands would disappear in favour of open grassland. All observers agreed on this. The other side of this coin was that *if* woodland was to be conserved then Tsavo had too many elephants. The irony was not lost on David who, less than a decade earlier had sold all-out anti-poaching with the argument that the elephants faced local extinction. Naturally we all wondered how many elephants there were in Tsavo and its surrounding lands. Aerial observation might provide the answer.

Aeroplanes had certainly flown over parts of southern Tsavo during the First World War. Bror von Blixen (the Swedish hunter) used aircraft to look for big elephants in the 1930s and he left strips scattered about what became Tsavo Park with the most remote of all being at Dakadima Hill to the east of the park. The first serious attempt to count elephants in Tsavo from the air was made by an American Fulbright Scholar – Hal Buechner – in 1953. He was working in Uganda, but made a short visit to Tsavo. In 1,025 square kilometres (400 square miles) between the Voi River and the Mombasa-Nairobi railway line, he counted 3,000 elephants. Everyone including David Sheldrick had been surprised by so many and felt that it must represent the whole population of Tsavo both East and West temporarily concentrated.

The next count was made by Peter Jenkins and myself during an extended aerial reconnaissance in October 1961 after the first heavy rain. This covered both Tsavo East and West and some contiguous areas, revealing over 6,000 elephants. We concluded that if we could count 6,000 on a reconnaissance flight, then the total number of elephants must be far greater. David reacted typically by arranging a more thorough count three months later with the help of the British Army Air Corps. Total elephants rose to over 10,000. As each successive count raised the total and Tsavo's woodlands were melting away ever faster, something akin to panic seized the conservation community. A national park elephant committee was established upon which I was the only non-park member. Scientific opinion was canvassed from all quarters – the weight of it initially favouring intervention and a huge reduction of elephant numbers. The national park elephant committee's rôle was briefly subsumed by the Scientific & Technical Committee of the East African Wildlife Society which, in turn, was replaced by a government elephant committee under the Ministry of Tourism & Wildlife.

Coincidentally at the same time the research being carried out by Dick Laws and large scale elephant reduction was taking place in neighbouring Uganda. The elephant committee in Kenya decided that Laws was the best man to unravel the Tsavo problem and he was headhunted to take up the formation and direction of a new Tsavo Research Unit. His assistant was the erratically brilliant but undisciplined Dr Murray Watson. Among their first undertakings was to get a better idea of just how many elephants there were in the Tsavo ecosystem (which included land outside the park). By late 1968 the estimate was over 40,000. They were not in a single homogeneous population but in perhaps a dozen discrete groups, each with different characteristics.

The sequential increases in numbers estimated did not so much indicate rising elephant numbers as improving skills and counting techniques. As astonishing count result followed astonishing count result, it was easy to accept that the spectacular loss of Tsavo's woodland happened because there were far too many elephants. Out of this grew conviction that elephant numbers must be reduced, a view David Sheldrick firmly espoused. However halving 10,000 elephants was a very different proposition to halving 40,000. It was a sensational lode for the news media.

Hardly surprisingly, they had followed the rising elephant count results closely and reports that large scale elephant reduction might be necessary stirred public interest. David sensed that the media, which he had hitherto controlled rather well, might become a tiger by the tail. I have a letter in which he asks me to keep details of the reduction work from the press. His apprehension was justified. The question of reducing Tsavo's elephants blew up into a media extravaganza that ran out of control.

For reasons he never discussed with me Sheldrick reversed his position on reduction and argued against it. He fell out with Dick Laws, the scientist charged with carrying out elephant research in Tsavo and, while their personalities did clash, this was quite literally managed and over-simplified by the media into a case of one being pro- and the other anti-elephant reduction. I was distressingly caught between the two, both good friends. I supported Dick's position and felt that David had lost his objectivity. He used the media to attack Laws. Indeed, in my view he did not behave like a gentleman and impugned both Dick's reputation and mine because I sided with Dick. It was an altogether sad affair.

Looking back, if David had ground to argue against culling, it would have been fear that large scale elephant reduction would invite corruption. In the 1960s this was not something any white official cared to state openly. I think if he had made his case to us directly neither I nor Dick would have accepted this view as we were not as politically astute as David. Yet even if we had not, I do not think it warranted casting slurs on our integrity, and claiming that advocating reduction was purely to make money.

These events happened in the decade 1960-9 when power was being transferred between colonial and independent governments, and during the new rulers' first decade in office. Mervyn Cowie's experience vanished just when it was most needed to bring sanity into the events swirling about Tsavo. Instead, the two men who dealt with it were the new Director of National Parks, Perez Olindo, and the new Permanent Secretary of the Ministry of Tourism & Wildlife under which the National Parks functioned, Alois Achieng. Neither had the experience to handle the issue. Where once David had called the media shots, by the end of the 1960s they were beyond his control and out of hand.

If the 1960s had become progressively less comfortable for David, the 1970s were worse. In 1971 a drought encompassing the entire Horn of Africa struck. In Tsavo both rhino and elephant populations crashed from starvation, with thousands of dead animals scattered across the park. Tsavo was now quite literally desert as no attempt had been made to preserve its woodlands by reducing elephants. Whether this should or should not have been done was now academic. Nature had resolved the controversy and matters should have settled back into a more peaceful routine. As in all ecosystems, events in one aspect influence those in all other sectors. The drought causing park animals to die also cast its cloak across the park's Kamba neighbours. Their crops failed, their stock died and they were on the bones of their backsides as the saying goes, living off famine relief handouts, and under intense pressure to eat game. The Kamba along the Park's western border still hunted. They may have steered clear of Tsavo East since the 1956-7 anti-poaching campaign, but recalled when they once hunted there with impunity. Cautiously, they started to trespass once again. They found a veritable El Dorado. Tusks from elephants that had died of starvation were there not in tens, not in hundreds, but by the thousand. They were there to be picked up without the danger of hunting living elephants. Further, where once there had been thick bush that hid carcasses, the open vegetation now made them easy to find.

Kenyatta's order in the early 1960s that certain people be allowed to 'collect' ivory had, by the early 1970s, evolved into a larger number roaming the country with collectors' permits. The given rationale was that elephant died from many causes other than the hand of man, and left ivory. Hitherto the Government had offered a small reward for any such tusks handed in. It was so small that most found ivory entered illicit channels. By licensing collectors to find it and sell it on the open market, ivory would be diverted away from illegal outlets and greater efforts would be made to find all that was available. The reasoning was logical. The argument against such a policy was that it would give cover to tusks from elephant killed illegally. And that, too, was logical. The subject never emerged into public debate. Collectors' permits were handled furtively and they were kept away from media eyes so, correctly or incorrectly, they seemed shady devices to get around the laws.

Collectors only had to read the papers to learn about the Tsavo die-off. Their agents gathered in the settlements along the Park borders, adding to the drought's incentive for Kamba to scuttle across the boundary, find ivory, and scuttle out again. Once out of the Park, and with their booty in collectors' hands, they were safe. Indeed all they needed was a document from a collector saying that they were his employees and they were beyond arrest. Word spread like wildfire and before David Sheldrick and his Field Force could react the battle was lost.

Kamba collectors came in such numbers that no matter how hard the Field Force worked it could only catch a small minority of the illegal entrants. The chances of an individual not being caught were far better than fifty-fifty. Once this was appreciated the influence of the previous anti-poaching campaign that had been such a deterrent until then, evaporated like morning mist and the work of the 1950s was undone.

Perhaps the severest blow to David's morale came when, in 1973, his men arrested the first Somalis collecting ivory and rhino horn in the park. Although in north Kenya out of the media eye Rodney Elliott had been fighting the Somalis for nearly a decade, they had taken a long time to cross the Tana River and come south to Tsavo. Knowing Somalis and their history, David felt that this presaged the end of the Park and a great deal more besides.

Elephants stopped dying after rain came in 1972. The bonanza of ivory from starvation deaths started to dwindle in 1973. By the end of that year tusks from natural mortality were harder to find. Having discovered how easy it was to enter Tsavo, many Kamba naturally felt that if tusks from dead elephants were unavailable, why not go after the living animals? Ivory scavenging was simply transformed into active hunting. Initially the Kamba hunted with their traditional bows and poisoned arrows, but with a major difference. In the 1950s

Elephants converted large tracts of Tsavo East into an erg off which great dust clouds daily rose several thousand feet. The vegetation recovered after the elephant population crashed, but that such an erg formed in a national park was bizarre. (photo Janet Clark)

the hunters were skilled in making bows, arrows and, above all, in knowing where and how to get good poison. Now, many of the younger men had not been trained to hunt elephants, knew little about making weapons and were suckers to whom the most indifferent poison could be sold. Before 1957 a good hunter was almost certain of killing every elephant into which he put an arrow. After 1973 there were very few good hunters around. Many who were trying their hand had little or no experience. If elephants died from their arrows, it was as often as not weeks later from septicaemia.

By 1975 David Sheldrick was in despair. With only fifty rangers in his Field Force together with himself and his deputy, he could not halt what was happening. The Field Force operated in sections of ten men and at maximum extension could put five sections in the field. Usually, however, with some men on leave and others resting after patrolling, it could muster three or at most four sections at any one time. Against them Sheldrick estimated that between 450 and 500 poachers were entering the park every month (at this stage almost all were Kamba). Often in gangs of threes and fours, this meant that there could have been up to 100 poaching gangs in the park simultaneously and David's chances of catching even ten per cent were small. Yet day after day he tried.

In the 1960s, along with other wardens, Sheldrick and Woodley had learned to fly. By the 1970s each had an aircraft. David's tactics were to have a Field Force section search for poachers' tracks then follow them. When a section felt it was getting close to a gang, he would be called up by radio. Other sections would then be brought in quickly by road and placed at strategic points ahead of the apparent line of pursuit to await further instructions. David then flew ahead of the section following the tracks. The aircraft's presence made the poachers hide or stick to thick cover and move slowly, impeding their ability to keep ahead of their pursuers. If they were spotted from the air David could vector his men directly to them and the reserve sections be deployed to intercept the fleeing poachers. As with all of Sheldrick's work, his operations were models of military efficiency. Flying almost daily in 1975, he and his team arrested 212 men, recovered 1,055 tusks and 147 rhino horns. In the first six months of 1976 a further 201 arrests were made: virtually the same number in half the time indicating a steep rise in illegal entries. Yet this formidable performance made little impression.

In 1976 the National Parks were merged with the Game Department into a single Wildlife Conservation and Management Department (WCMD). This was a disaster for, *de facto*, the larger, more corrupt and inefficient Game Department swallowed the smaller National Parks organisation. The Board of Trustees that had given the parks some independence and protection from the civil service was abolished and park employees became civil servants. On 24th June 1976 Sheldrick wrote a report[75] on the situation in Tsavo to –

> *". . spell out in detail the poaching problem with which we are faced, in order to remove any possibility of misunderstanding between the field staff, Headquarters and the Ministry regarding the gravity of the situation."*

In it he gave a resumé of past anti-poaching activities in Tsavo, facts on the situation in 1976 and his recommendations to redress the problems. While a valuable historical document it was strangely deficient. As David and Bill proved in the 1950s, field patrolling is inefficient. On the basis of his results in 1975 and 1976, and if he was capturing less than ten per cent of those entering the park illegally, he would have needed a Field Force of 500 men, that is ten times greater than the fifty men he had, to have made any impact. Yet he only recommended increasing his Force by a further two sections of ten men. Knowing the man and the circumstances, he clearly tempered what he wanted with what he thought was attainable. It was the fallacious approach of any increase being better than no increase. David abandoned the military dictum: do not take on a task without the strength to achieve it.

He recommended, as an act of desperation, that rangers be authorised to shoot any armed person in the Park who failed to halt when ordered. Anyone carrying a weapon is armed and a piece of wood is a rude weapon. It was asking for the power to inflict capital punishment for trespass: of being allowed to shoot someone as he was running away. The majority of illegal entrants at the time – 1976 – were Kamba who ran away when spotted. Up until this point, even the Somalis who had come into the Park came unarmed. As was the case with all law enforcers, park rangers already had the right to defend themselves when threatened by those whom they were lawfully apprehending. This included shooting someone dead as in fact had happened on two occasions between 1948 and 1976. When armed Somalis started operating, circumstances changed dramatically: they were guerrillas bent of mayhem first and poaching second. As such they were a very different kettle of fish: they shot first so automatically that one could not take the risk of not doing likewise.

David knew that increasing the Field Force and being allowed to kill poachers running away would not solve his problems. What he needed was what he had previously proved

---

[75] Sheldrick D.L.W., 1976. *Report on Poaching in the Tsavo National Park.* Typescript.

so successfully: licence to go after poachers in their homes outside the Parks and wholehearted backup from the police, judiciary and administration. However, he was not allowed to operate outside the Park, police and administration support was minimal (he was frequently obstructed by them) and magistrates passed laughably light sentences on the men arrested in parks. David knew better than anyone that the collector's permits were emasculating game law enforcement. Yet he referred to them only obliquely instead of directly. He drew a line between what he thought his superiors would wear and what he knew was necessary and became a casualty of independence. His omissions were acceptance that what was needed was out of reach.

David had been formidable when the colonial government ruled, but did not know how to operate within the African system which replaced it. He, like so many of us, felt that if he spoke his mind and voiced his criticisms, he would simply lose his job. And in that he was probably right. His only tool was to get the international press working for him again.

Corrupt men in the newly created WCMD knew David's international standing made him a threat to be neutralised. His presence hindered the people who wanted to get at ivory and rhino horn resources and they wanted him out of Tsavo. He could have been retired on the grounds that he was not a Kenya citizen, but his departure might have caused that hateful phenomenon: publicity. The solution, 'promoting him sideways' into a harmless position (advising and planning parks elsewhere in Kenya), happened soon after the Game Department and National Parks merged.

Characteristically, game and park wardens regard the areas and parks in which they serve very possessively as *theirs*. David Sheldrick epitomised the characteristic to the level of caricature: if anything happened to Tsavo, it happened to him personally. No doubt this dedication made him the sort of warden he was. He put his heart and soul into Tsavo East. At its height it was a showpiece park with everything meticulously kept – roads, buildings, records. In a special way it *was* his park not just in his eyes, but in the public's too and to many the names Sheldrick and Tsavo were one.

It can be imagined what being transferred away from Tsavo must have done to David's psyche. By then he was a sick man with a congenital heart problem (his father had died in his fifties from the same complaint, his daughter Valerie died at forty-five of a heart attack and his son Kenneth only averted the problem through modern surgery). Heart problems are exacerbated by stress and David had been stressed severely, mentally and physically, for a long time. He died of a heart attack aged fifty-eight in 1977. Congenitally predisposed to this as he may have been, few who knew him doubted that his death was accelerated by his state of mind and that, literally, he had given his all to Tsavo East.

* * * *

Bill Woodley's ingenuous approach to life and his warm personality was reflected in the way he ran his three mountain parks: the Aberdares, Mount Kenya and Marsabit. Of all the game and park wardens I have met in over forty years in African conservation, none showed a greater interest in his neighbours or in people generally. No-one was quicker to redress the problems that wild animals from the parks posed to nearby farmers. Crop raiders were shot without false sentiment. Today, with electric fencing, people forget that this technology was not available until the late 1970s.

As with old soldiers, yesterday's wardens are soon forgotten. David Sheldrick's last resting place in Nairobi's Langata Cemetery: beneath an acacia his headstone lies hidden in a tangle of herbage.

Appreciating that shooting was more a political palliative than a solution to the general problem of crop raiding, Bill advocated barriers to prevent animals from leaving the parks. A deep ditch that they could not jump across was an obvious option. All that was needed was the manpower to dig it. His predecessor had experimented on a minor scale with some success. Bill expanded the programme and in the most critical area of human/wildlife conflict (round the Aberdare Park's Treetop salient) soon had twenty miles of ditch separating animals and people. The reduction of crop-raiding dropped from a source of more or less constant complaint to nothing.

The Aberdare National Park lies as a long, narrow, forested wedge with farmlands to both east and west. The traditional and quickest route between the two was by foot over the mountain. Once the national park existed this route was technically out of bounds and travellers between the farmlands had to go round the Aberdares – a long and tiresome journey. In keeping with his sense of good neighbourliness Bill allowed the Kikuyu who farmed on both sides of the mountain to walk through the park, bringing cattle and sheep across, provided that they kept to one specified road and paid a nominal fee. Saving several days travelling, this earned considerable goodwill.

In the same vein and knowing that many Kikuyu had kept honey barrels in the forests for ages before the park was formed, Bill allowed them to visit their old hives once a year. The practice did no harm to the park but again earned goodwill. In dry times when there was no grazing in the surrounding farmland, Bill even advocated allowing small numbers of cattle into the parks. This was a step too far for Mervyn Cowie who, appreciating Bill's intent, felt it could become interpreted as a right that conflicted with the national park principles.

Another striking decision made early in Woodley's days as warden of the Mountain

Particularly associated with Woodley, the Aberdare ditch that separated high densities of people from high densities of wild animals was successful but difficult to maintain in an area of high rainfall and steep slopes. It was replaced by an electric fence. (Photo Peter Davey)

Parks was recruiting ex-Mau Mau men as park rangers. His logic was simple. Having lived in the mountain forests, these men knew them better than anyone else, could look after themselves in that environment better than others and, of all people, were best placed to look after it. Further, with the respect that often evolves between soldiers who have fought one another, Bill held that many excellent men had been on the Mau Mau side. They may have been misguided, but then how many soldiers in any war really understand why they are fighting? Depending on where and who they are, they get swept along in the name of the cause – whatever it may be.

Nevertheless the decision to recruit ex-Mau Mau was robustly criticised by many white people: recall that it was 1959, the rebellion a mere three years away and independence still out of sight for many of us. Bill refused to back down. Not only was his surmise that the former terrorists would make good rangers correct, but it created a strong bond between him and his subordinates and was precisely the sort of step needed to heal the divisions that the civil war had created. The strongest proof of this bond was that while poaching rose dramatically across Kenya over the next twenty years, less took place in the Aberdare and Mount Kenya Parks while Bill was in command.

Woodley's wardenship of the Mountain Parks lasted just under twenty years from 1959 until 1978. It was not an easy time for white officials who had commenced their service in colonial days. Africanisation of the public services meant that they were passed over for promotion and had to serve under younger, less experienced and sometimes incompetent men. As expatriates they had no chance of heading their departments or even attaining its senior levels. Perhaps worst for those holding their jobs through senses of vocation was being unable to stop the all-round decline of standards.

Woodley could see what was happening in the Game Department. Yet for as long as the national parks remained a separate entity there was a chance that he could control poaching in the areas under his command. More than that seemed too much to hope for. Taking this philosophical approach and keeping his head down, Bill both survived and succeeded within his limited objectives. Even when the Kenya National Parks were merged with the Game Department and the Parks' higher standards were swept away with poaching spilling over into where, hitherto, the Game Department staff had been unable to operate openly, Bill did not make waves. Working quietly, his area was less affected than the surrounding forests and he might have been able to continue in this vein had it not been for events in Tsavo.

After David Sheldrick left Tsavo East in 1976, poaching ran yet further out of control. Armed Somali gangs started coming into both Tsavo East and West. Where they originally had scavenged ivory and poached leopards, they now shot elephants and rhinos wholesale. Seeing the Kamba as competitors they achieved what the authorities had failed to do and frightened the Kamba out of the parks. The Kenyan wardens seemed, at best, powerless to do anything about it and, at worst, in cahoots with the poachers. The government was not so much concerned by the poaching as the long term political ramifications of an aggressive Somali presence. This, with public outcry in the western media about poaching, stirred the authorities into action. What better than to post the surviving architect of the 1950s anti-poaching campaign back into the area? In 1978 Bill Woodley was transferred from Nyeri and the Mountain Parks to Tsavo West where he had started his service with the Parks in 1948.

* * * *

When the Game Department and Kenya National Parks amalgamated, a move was made to revive a Field Force along the 1950s lines and give it a countrywide responsibility. This new Anti-Poaching Unit (APU) was commanded by a man with neither law-enforcement nor military credentials who was not one of the Park's Old Guard. In the circumstances it is quite possible that inexperience was the very quality for which he had been selected to lead the APU. In its response to public pressure over poaching and the desire to counter Somali activities in Tsavo, the Government felt that the APU's presence was inadequate. While sent initially as the warden of Tsavo West, Woodley was also asked to take over anti-poaching responsibility in Tsavo West and the country eastward of the park to the coast and the Tana River. As a curious exception, he was told that he was *not* responsible for law enforcement in Tsavo East, which lay between Tsavo West and these coast hinterlands. He logically assumed something was going on in Tsavo East that his superiors did not want investigated.

Although it was not his responsibility, Bill inevitably saw a great deal of Tsavo East from the air as he had to fly over it whenever he travelled between Tsavo West and areas further east. Naturally he kept an eye open for signs of poaching and in the year after his appointment he reported the locations of more than 50 elephant and rhino carcasses in Tsavo East (just under one a week). However no follow-ups were made, there were no contacts with poachers and no arrests as a result of his reports.

In 1979 the warden of Tsavo East was transferred to be APU deputy commander which, given his lack of law enforcement success in Tsavo, reinforced suspicion that the posting was for reasons other than ability to arrest poachers. Not long after a new warden was

Mount Kenya is climbed annually by many mountaineers. Creating an effective mountain rescue team was another of Woodley's achievements. (Photo Peter Davey)

installed Bill happened to be flying to the coast when he received a radio call from Tsavo East headquarters asking him to investigate a fire on the Sabaki River. As it was the first such request ever made from Tsavo East, it was unusual. Bill located the fire: it was so small that he immediately suspected it had been lit to draw him away from other areas. Curiosity roused he abandoned his trip, flew to Voi, stated his suspicions forcefully and insisted that the warden come up with him on a reconnaissance. Flying low along the roads within a radius of thirty miles (forty-eight kilometres) of the park headquarters, they found forty freshly killed elephants so close to the roads that they had obviously been shot from a vehicle or vehicles. Such brazen poaching was now characteristic of WCMD staff as, confident in their monopoly of law enforcement within parks, they had no reason to fear being apprehended. Anyone else would have been more circumspect. This, obviously, was what the fire was supposed to have drawn Bill away from seeing! Having located the dead elephants, Bill Woodley did no more. It was not his job to follow up these finds. No arrests were made but eventually the assistant warden who asked him to investigate the fire was transferred away from Tsavo East.

The law enforcement paralysis was general. In the 1950s the anti-poaching programme had a police prosecutor attached to it to ensure the closest liaison between those who arrested poachers and those who prosecuted them. In 1978-9 prisoners were seldom properly interrogated after capture. They had to be handed over to the nearest police for prosecution and that was the last Bill usually heard of a case. Consequently it was difficult to build up an overall intelligence picture. Many of the subordinate staff were good men, experienced in the bush and not afraid to fight, yet they needed leadership which the wardens generally did not provide.

The epitome of their state of mind was illustrated by a warden based at Mackinnon Road. He and a section of his men were ambushed by a Somali gang. A Ranger Sergeant was shot dead. The drill in the face of a simple ambush was to rally and counter-attack: it is a basic military reaction. Counter-attack was the last thing on the warden's mind. He fled as fast as his legs would carry him, his men following pell-mell behind. Leaping into their vehicle they drove at high speed for their base at the roadside hamlet of Mackinnon Road, straight through it, then on to Voi police station forty-five miles away. In wild hysteria they announced their arrival in Voi by firing automatic weapons into the air, yelling and shrieking, to send an astonished citizenry scampering for cover. The hysteria was contagious and the police, assuming that at least Martians had come and the town was under attack, set a siren wailing, spreading the panic and consternation yet further. Worst of all, the warden's arrant cowardice earned him no censure, making yet another point.

The authorities refused to enforce discipline. When Woodley proved that eight rangers on the Tsavo River gate of Tsavo West had killed an elephant, none were prosecuted or even dismissed from the service. Instead, they were transferred elsewhere, three of them only temporarily. Such lack of action and support was demoralising, regardless of the competence of the men in the field. And the defects were not monopolised by Africans.

The APU commander did not like danger. Woodley was made aware of this one day when flying back to Tsavo with him as a passenger. Knowing that the Tsavo East warden had been searching from the air for a Somali gang, Bill enquired over the radio to learn how things were going. To his delight there was action directly ahead. The warden had located the gang and was vectoring his men on the ground towards it. Being low on fuel, he asked if Bill would take over observation while he replenished tanks back at base.

Bill's delighted acceptance dismayed the APU man. Coming up to the gang Bill radioed being in sight and took over shadowing it. The gang, aware that they were under observation, were moving fast and not keeping to close cover. To slow them down and give the pursuing patrol time to catch up, Bill buzzed them at scrub height. Even though he had no weapons with which to attack the Somalis, an aircraft that low made them split and duck. As it flashed over them, the gangsters opened up, but their sense of the deflection needed to hit a fast flying machine was poor and the chances of being hit were small. The sheer thrill of action was not in the anti-poaching leader's make-up and rendered him pathetic, his hands all over the controls, whimpering and begging Bill to let him take over and fly them to safety. Descending back towards the bandits he became cataleptic. As they shot over the top of them and climbed away this was replaced by gibbered entreaties to keep away, far, far away from the Somalis. When they eventually landed to refuel, he refused to get back into Bill's aircraft. Woodley recorded this tale humorously and without malice, for he liked the man. I recount it here because it is significant that the WCMD's key 'military' position was held by a veritable Duke of Plazatoro. Further, to know of the incident and keep silent would leave a record in which white wardens were without fault, which was untrue.

The record of these anti-poaching years was one of occasional success in a morass of missed opportunities and bad organisation. It was not helped by Bill Woodley's ill health and a fight with cancer in which he lost half his stomach. Most men would have retired honourably on medical grounds, but Bill kept going. The chance for direct personal action was rare but sweet. One evening flying ten feet above the ground, Bill surprised two

In the 1950s one warden – Dennis Zaphiro – owned and flew his own aircraft. In the 1960s many game and park wardens learned to fly and both Game Department and National Parks acquired aircraft. Bill Woodley with a Cessna 180. (Photo Phil Snyder)

Somali bandits at a waterhole fifty paces ahead. Instinctively he flew straight at them hoping to hit at least one with propeller or undercarriage. He missed because they threw themselves flat on the ground. Though he failed, the brief one-on-one encounter was exhilarating in its simplicity.

Bill laughingly referred to his most successful action as the 'Battle of Wounded Foot'. On 27th December 1981 thirteen Somalis armed with five rifles raided a railway maintenance gang's quarters on the line close to the Tsavo River bridge. When word came in, Woodley mobilised his rangers, put them on the gang's tracks, and flew ahead of them watching for movement. For the next week they followed this routine, repeatedly finding and losing the tracks. The only excitement arose when an assistant warden accidentally shot himself through the foot.

On the night of 3rd/4th January 1982 the Somalis raided a settlement – Ngeluni – near Mtito Andei. This time their tracks were easy to follow and led towards the Athi River and more open country. The Tsavo East warden, Joe Kioko, brought his aircraft up from Voi so that they had two machines observing ahead of the trackers. The direction of the tracks was towards a belt of thicket that lined both banks of the river and made an obviously suitable place for the gang to lie up. A group of game viewing tourists were nearby and while Bill was dropping a message on them to get clear of the area quickly, Joe called that he had spotted the gang. Seeing the rangers the Somalis had decided to run. They were heading for the Yatta Plateau and away from the river when Bill flew low over them and dropped a message commanding them to surrender. A volley of rifle shots was the expected Somali answer. The men were trotting in single file and Bill approached from behind about 100 feet above the ground. As the rear man disappeared beneath the aircraft's nose he dropped a hand grenade which exploded harmlessly behind the Somalis. When Bill looked back, however, they were no longer trotting, but running hard and shedding equipment.

Bill's next approach was again from behind, slightly lower and he did not let his grenade go until the front man of the file disappeared beneath the nose. Banking hard and looking down and back, Bill saw the grenade detonate just in front of the leading man who staggered, then peeled off on his own to disappear in the thin scrub. When the following rangers came up to the spot they deduced from the heavy trail of frothy blood that he had taken shrapnel through at least one lung. Assuming that he would not live, the patrol kept on the main gang's tracks.

Having no more grenades, but a ·264 sporting rifle and plenty of ammunition, Bill settled down to shoot at the gang from the air. While this is not easy from a moving aircraft, he reasoned that he might slow the Somalis down and improve the ground patrol's chances of catching up with them. Under fire from the air and firing back at Woodley's circling aircraft above, the tactic worked. Two Somali bullets passed harmlessly through the aircraft before Bill scored his only hit. Just as the gang reached the top of the Yatta Plateau one member took a bullet in the foot (as Bill pointed out when recounting it, the battle was living up to its name). The softnose bullet would have made a nasty wound. The injured man assisted by another who carried a rifle, broke off and hobbled away and both were soon lost to sight as the chase continued after the ten who remained together. Rain and failing light brought the aircraft involvement to an end, leaving the rangers on the ground to keep up the chase. They got close enough to exchange fire and a ranger was shot through the arm before darkness ended the action.

All through the day radio messages had been streaming back to Tsavo East headquarters in Voi to deploy rangers across the bandits' line of retreat. It was to no avail. No patrols were sent out. By the end of the day the gang had been in dire straits. Running since morning, without food or water, they were heading across a waterless wasteland towards the Tana River 100 miles away.

The chase was taken up at first light with the aircraft again scouting ahead and around midday the rangers came upon a lone, exhausted, seventeen-year-old Somali who had been unable to keep up with his comrades. All twenty rangers in the pursuing posse emptied the magazines of their automatic rifles and the Somali was lost in the pall of dust kicked up by the 500 rounds fired at him in the space of twenty seconds. It was that Somali's lucky day as not one bullet hit him. He was captured, but apart from two rounds of ammunition and a Somali ten-shilling note, they got nothing from him. Leaving guards with the prisoner, the rangers were off again on the gang's trail.

Predicting exactly where they would cross one of the internal Park roads, Bill requested an ambush be laid. Sure enough, the nine Somalis crossed the road just where Bill had predicted, but there was no ambush. The assistant warden detailed for the task disobeyed orders and deliberately avoided action. His defence – so reasonable to him that he correctly expected its unquestioned acceptance – was simply that Somalis were too dangerous to tangle with. Had there been an ambush, the exhausted gang could not have put up much of a fight. With no ambush they escaped from the park. Fifty miles on they came across some Orma herdsmen and asked for shelter. With no love lost between Somalis and Orma and after having lulled the gang by appearing friendly and getting the bandits off guard, the Orma speared five, and captured the rest together with their four rifles. So ended the Battle of Wounded Foot.

There was an appendix, however. For some weeks after the engagement a lone Somali with a rifle had been holding up Kamba households and stealing food around Mtito Andei. One day he appeared at a homestead and holding up a man with his nine year old son, demanded food. Seeming to acquiesce, the father turned to his son and speaking in Kikamba told him to go and get his brother who was nearby. The Somali thinking that he was being told to get food, allowed him to run off. The boy did as told and returned with his uncle carrying a bag which the Somali thought contained the food – as intended. Once close, both uncle and father set on the Somali and overpowered him, but not before he managed to get off two shots. One hit the boy, where else but in the foot? Fortunately it was not a serious wound.

The prisoner was the Somali leader from the Battle of Wounded Foot whose best friend happened to be the man Bill had shot from the air. Seeing his comrade's plight, the leader had ordered the gang to run on while he took his wounded companion to one side and lay low, counting on the patrol to keep after the rest of his men, as had indeed happened. Once the rangers had passed, he had helped his friend down to the Athi River where the two of them holed up, waiting for the wound to heal. During this period the gang leader had periodically raided the Kamba along the park border for food. As was not surprising with the sort of wound a soft-nosed bullet makes, gangrene had set in and the wounded man died three weeks later. Months afterwards rangers patrolling along the banks of the Athi found his skeleton, confirming the gang leader's story and closing the episode. The man who suffered a lung wound from grenade shrapnel was never found.

The Battle of Wounded Foot (or should it have been Feet?) was unusual in that an entire gang was accounted for. Seven died and five were captured at the cost of a warden who accidentally shot himself through the foot, a ranger hit in the arm and a boy wounded in the foot. The youth captured by the rangers and the men caught by the Orma were all handed over to the police. The event illustrated both the strengths and weaknesses of the system under which Bill was now working. In their dogged pursuit and the hardship they endured, the ordinary rangers on the ground had been magnificent. With the exception of Joe Kioko of Tsavo East flying the second aircraft, the educated wardens had not performed well.

David Sheldrick had summed up his impression[76] of the new breed of wardens thus –

> *"There is now singular dearth of young Officers within National Parks who are sufficiently interested or dedicated to take an active part in anti-poaching work. Most of them are not prepared to 'rough it' in the bush, and share the same dangers and hardships as the rangers. In the last few years, there appears to have been a complete reversal of attitude towards field work among young officers in the service. Whereas, in the past, Junior Officers were only too pleased to escape from office work and get out into the field, today it is just the opposite, and every device is employed to ensure that they do not have to go out on foot, or are deprived of creature comforts in any way. There are, of course, some exceptions, but this undoubtedly applies to the majority of young Officers with whom I have come in contact . As I have pointed out. . . the Officers must be given a military as opposed to a university type training."*

This observation was endorsed wholeheartedly by Woodley, Jenkins and the old-time field wardens. The standard African response to it was that it was racially motivated and an oblique way of saying that Africans are not as good as whites. There is nothing to be

---

[76] Sheldrick D.L.W., 1976. *Ibid.*

gained by beating about this bush. Sheldrick's references to Junior Officers was only incidentally a racial observation as in the past all the young officers had been white, whereas those in the present of which he wrote (1976) were all black. The difference in attitudes of which he wrote was real. The young colonial white was by inclination aggressive and self-confident. Yet this did not arise because he was white *per se*. In large part the self-confidence stemmed from a stable culture in which the rewards of being a game or park warden lay in society's admiration outweighing low income, coupled to genuinely enjoying outdoor physical activity.

Such an outlook was not unique to whites, however. It was also typical among the young men of the pastoral tribes who cared less about formal education than in the rewards and esteem that bravery and ability to look after the family herds brought. They, too were men of the outdoors, physical action and success. They, too, were brim-full of self-confidence, liked a scrap and they, too, had stable cultures. For this reason they were the last to adopt western education. Yet it was not from among them that Officers for the National Parks were recruited – as they should have been. Instead the Government demanded that members of its civil and associated services be educated in the academic sense of exams passed, of abilities to read, write and add.

There was a hollow perception in which civilisation was seen as wearing a well-pressed suit, competition was resolved by litigation, and status determined by money. The cultures where these attributes have taken hold are in transition toward western models and their measures of esteem are changing. Self-confidence is elusive. Achievement seems to stem from paper and working behind a desk and outdoor, physical work is held demeaning. It is in this milieu that the 'educated' people who have passed exams are found. And this is where the Government sought conservation officers. They tended to hate the wild and its physical demands. It was a frightening place in which Somalis merely added to an already long list of dangers. It is hardly surprising that they did poorly in the field.[77]

In 1982 ill-health stopped Bill Woodley from further flying and his final anti-poaching episode was over. In the four years that he had been back in Tsavo, six Somalis were shot in the field, five speared at the end of the Battle of Wounded Foot, 106 Somalis had been arrested in townships along the Mombasa-Nairobi road or in Mombasa itself, 150 tusks, fifteen rifles and 700 rounds of ammunition recovered from them. It was not an impressive tally given the scale on which Somalis had been operating. Only forty (38%) of 106 men arrested were convicted against over 95% convicted in the 1956-57 campaign, which speaks volumes.

Bill stayed on with the WCMD in Tsavo until he retired in 1991 after forty-three years' service in Kenya's National Parks. The last seventeen years of his life had been plagued by illness, yet it never dampened his humour. Not long before he died in 1995 I asked Bill if he believed that his service since independence had achieved much? Reflecting a moment he said that the parks seemed to have gone through three phases. The first was from 1948

---

[77] The foregoing comments applied particularly during the 1960s and 1970s – an era in which the vacuum created by the precipitous departure of the old regime had to be filled, willy nilly, by inexperienced and untrained men ill prepared in governance. Now, nearly three decades on, that rawness and lack of confidence is not so apparent and both selection and training for the job is more appropriate.

until Kenya's Independence in 1963 when he believed that the National Parks achieved everything that they set out to and working for them was a pleasure. The second phase was from 1963 until 1976 when the National Parks merged with the Game Department. In this period he found that he was unable to get on with his work as had previously been the case. The need for diplomacy, Government's rising corruption and declining efficiency removed many of the job's pleasures, but were not insurmountable. The change from the colonial to independent regime had gone remarkably smoothly, all things considered. More important, he felt that this period still gave grounds for confidence in the future of National Parks.

The third phase, 1976 to 1991, his final years, removed most of this optimism. Once the Game Department and Parks merged, he felt the national parks were irreparably wrecked. In 1989, when WCMD had become the quasi-Government Kenya Wildlife Service under Richard Leakey, Bill hoped that Leakey would succeed in creating an effective service, but was nonetheless pessimistic about his chances. His feelings led inevitably to my, "Well, why did you stay on?"

There had been no hesitation. "If everyone abandoned the parks, then their failure would have been certain wouldn't it? If we didn't stay on" (and the 'we' was reference to the parks' Old Guard of David Sheldrick, Peter Jenkins and himself), "who would have set an example? After all," he went on, "in 1980 none of us felt Parks would survive another decade, yet now in 1995, fifteen years later, they may not be wonderful, but they exist."

This was the imperial tradition: one did not give up simply because there was no light at the end of the tunnel. The ethic whereby Billy Woodley and David Sheldrick (and Peter Jenkins – though his tale has yet to be told) lived is captured in a passage by the historian Arthur Bryant[78] quoting Kipling –

> *"As to my notions of imperialism, I learned them from men who mostly cursed their work, but always carried it through to the end, under difficult surroundings, without help or acknowledgement."*

---

[78] Bryant A., 1942. *English Saga (1840-1940)*. Wm Collins Sons & Co. Ltd., London.

Somalis, a race apart, widely admired and equally widely feared, have expanded their territory steadily throughout the millennium that they have been known as Somalis.

(Photo Camerapix)

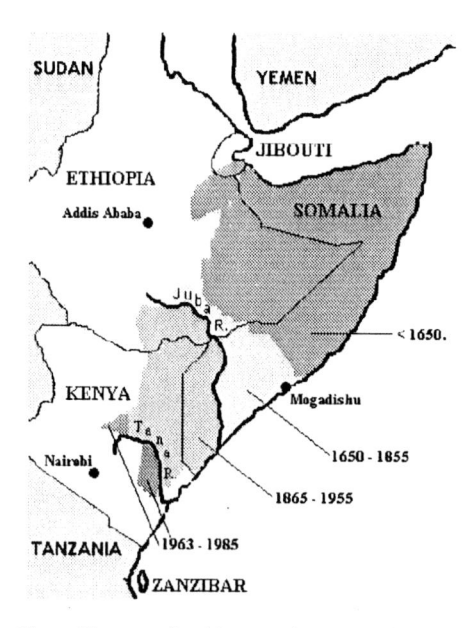

The Horn of Africa showing Somali distribution prior to 1650 and their advances between 1650 and 1855, 1865 and 1955 and between 1963 and 1985.

(Derived from Turnbull)

CHAPTER FOURTEEN

SOMALIS

David Sheldrick said it was the beginning of the end when the first Somalis were caught collecting ivory in the Tsavo Park. A trifle melodramatic, perhaps, but then David was very conscious of Somali history and territorial ambition. The soldier in him admired them and their fighting spirit and he had deliberately recruited them for his Field Forces over the years. Yet in the very qualities that he so admired, he foresaw nothing but trouble for Kenya. Harnessed to fundamentalist Islamic outlooks, and in their overweening senses of superiority, David felt the Somalis presented an enormous challenge to Kenya's political integrity in the coming century. I concurred with his predictions, which events since David's death a quarter of a century ago have borne out. If one is to understand their poaching, it must be placed in the context of Somali history.

* * * *

Somalis trafficked ivory for centuries before white men colonised Africa. They did so throughout the colonial era and, one way or another, gained a reputation for being poachers. Since Kenya's independence they have enhanced this reputation enormously. Most recently their activities in Kenya have been eclipsed by what has gone on in Somalia itself when a ragtag militia of ill-trained, ill-disciplined, arrogant anarchists under a local warlord humiliatingly chased a modern American military force out of Africa. Such a people deserve more than passing comment as the world will be hearing a great deal more of them in the near future, and no country more so than Kenya.

Today there are between eleven and twelve million people in the Horn of Africa who refer to themselves as 'Somalis'. They are fervently Islamic and have a common language. Their culture reflects a harsh, arid environment in which the people are mainly pastoralists depending on camels, cattle, sheep and goats. The land is so dry that the herds and people have to be spread very thinly: it could not be otherwise. And with such a distribution, large, close-living, highly structured and interdependent communities cannot develop. Consequently Somalis are individually resourceful and very independent. If they have an ethos outside Islam, it is ecologically driven and tied to the well-being of their animals. Such is their feeling for camels that it is acceptable for a swain to liken his beloved to a one. Somalis are indeed a breed apart!

Somalis consider themselves either Sab or Samaale. The Sab are the smaller of these two and as settled agriculturists are concentrated in the few places where farming is possible. They are socially cohesive and organised, but looked down on by the more numerous pastoral Samaale. Within the two basic divisions there are super-clans, clans, sub-clans (*rer*, groups of families) and so on down to the smallest unit which is the individual family (*jelib* meaning literally 'finger') in a system not unlike that which once prevailed in Scotland.

Some clans and super-clans have hereditary chiefs but many do not. The most striking political feature of the Samaale, however, is their lack of organisation. Everyone belongs to a group that can claim (or enforce) compensation if one of its members is injured or killed, or which must pay such compensation when one of its members injures or kills someone of another group. In its rudest sense this debt (*dia*) collecting or paying network is a form of communal insurance. Yet it is also the base of an immensely complicated and self-perpetuating tangle of feuds arising from unpaid debts that may extend back centuries. Changing circumstances can bring long ignored debts back into currency and, *vice versa*, a seething dispute can be held in abeyance if the parties' interests are served by uniting to attend to some other issue. Yet it will only be in abeyance and never eliminated until the debt is properly paid off. Consequently, and to make the point *ad absurdum*, most Somalis are technically in debt to other Somalis. The Arab saying, 'Me against my brother, my brother and I against our cousins, and our cousins and us against the world', sums the Somali outlook.

Ever since they became an entity Somalis have slowly expanded their territory which started in the north of Africa's Horn opposite Arabia. Stability and peaceful coexistence are alien to their history. Somali advance was through force of arms where they were strong enough, or diplomacy when they were not. The traditional approach to those too powerful to take on directly but whose land they coveted, was to go in friendship and offer them services in return for the right to pasture stock. They offered to be vassals or, in their term, *shegats* – explained to me thus –

> "Let us graze our herds on your land and our spears will be your spears, your enemies will be our enemies and your causes will be our causes."

Once well established they would turn on their hosts and slaughter them. If that opportunity did not arise then their ends were achieved through intermarriage and conversion to the Faith. Gradually the hosts became part of their vassal's clan. Today the process is apparent in two Kenya groups: the Gurreh and the Adjuran. A century ago the Gurreh of Mandera were a pagan Oromo-speaking group and thereby a Galla people. Today, through intermarriage with Somalis, they are of Islam, Somali speakers and are now a Somali clan. Their southern neighbours – the Adjuran – are still in an intermediate stage. Two of their sub-clans still speak Oromo and regard themselves as Boran. Two speak Somali and consider themselves to be Somalis. Less than a century hence, all will be Somali and, as true Somalis, claim lineal descent from the Prophet.

Somali expansion has been a piecemeal process over the past thousand years with clans fighting and feuding amongst themselves the while. Only once did they cohere briefly under the leadership of a remarkable man: Imam Ahmed ibn Ibrahim al Ghazi, known widely as 'Ahmed the Left-handed' (aka Ahmed Goran or Ahmed Gran or Ahmed Ghurre). He led the clans against Christian Ethiopia to become a national hero.

In the seven years between 1528 and 1535 he overran most of north-east Ethiopia, defeating the Christian Amhara in a series of battles that eventually put Kassala in Muslim hands. At that point the Amhara were helped by the Portuguese (who had come via Africa's east coast and the Red Sea). Between 1541 and 1543 they held the Somalis – losing some fights and winning others. In 1543 near Lake Tana there was a decisive battle in which Ahmed Ghurre was killed. Thereafter the Somali forces disintegrated and were driven from the highlands back into the hot lowlands from whence they had come and left to go on expanding southwards and south-west.

The next 'national' hero, Sayid Mohammed ibn Abdullah Hassan, was born in the 1870s. Deeply religious, he was unusual in his perception of what was going on in the outside world and disliked Somalia being split between Ethiopia (the Ogaden), France (Djibouti), Britain (British Somaliland) and Italy (Italian Somaliland) in the scramble for Africa after the Berlin Conference of 1884/5. Initially he held his peace and late in the nineteenth century Sayid Mohammed settled in the northern British Somaliland port of Berbera to convert people to his Salihiyah sect. Unsuccessful, he moved inland to try among the Darod Dolbahanta. They listened to and followed him. As he gained adherents and prestige he became progressively more influential in inter-clan affairs. Initially the British administrator approved as Sayid Mohamed seemed a moderating influence. Yet power went to his head. His followers organised themselves militarily and in 1899 occupied Burao.

Sayid Mohammed declared himself the 'Mahdi', intending to unite all Somalis against outsiders. The British with their recent memories of another Mahdi in the Sudan and the loss of General Gordon saw a new Islamic rebellion on their hands. Sayid Mohammed was his own worst enemy, however. As a true fanatic, he brooked no differences of view and his first targets were not outsiders, but Somalis who rejected his Salihiyah ideas. Any refusing to co-operate with him or accept his ideas were pillaged. Trade caravans were hijacked and looted and Sayid's extreme cruelty led to the name by which he became widely known – the Mad Mullah. Simultaneously he prosecuted war in Ethiopia, British Somaliland and Italian Somaliland and threatened Djibouti to the north and in the British East Africa Protectorate (Kenya) to the south.

The British were under no illusions: the Mad Mullah had to be defeated decisively. Yet 1899 saw the outbreak of the Boer War in South Africa and the British were loathe to have fights at both ends of the continent simultaneously. Consequently Sayid Mohammed was never tackled with the force and determination called for. Eventually four expeditions against him were launched between 1901 and 1904 with regional troops. He suffered setbacks, but never a decisive defeat.

For four years the Mad Mullah lay low. Then, in 1908, he went on the rampage again initially focusing on those who rejected his authority. Believing the arid interior of Somaliland was not worth fighting over, the British withdrew to the coast and chose the less costly policy of arming the Ishaq tribesmen who opposed Sayid Mohammed. In historian Saadia Touval's[79] words –

> *"Soon after the withdrawal, however, the interior was seized by complete anarchy. The tribes took advantage of the opportunity to settle old scores and to raid one another's stocks, instead of uniting to face the Mullah. The Mullah's followers were active too, trying to coerce the tribes to accept his authority. It is estimated that one third of the Protectorate's male population perished in these disorders."*

Late twentieth century troubles in Somalia seem more a recurrent Somali condition than anything unusual. The British regretted their policy of *laissez-faire* and raised a special

---

[79] Touval S., 1963. *Somali Nationalism*. Harvard University Press. Cambridge, Massachussets.

Somali Camel Constabulary to fight the Mad Mullah under the command of one Richard Corfield. In August 1913 while pursuing one of the Mullah's raiding parties, he and his entire constabulary were ambushed and annihilated. Arrogant in the wake of his victory, the Mullah wrote tauntingly to the British Commissioner at Berbera –

> "You. . . have joined with all the peoples of the world, with wastrels, and with slaves, because you are so weak. But if you were strong you would have stood by yourself as we do, independent and free. It is a sign of your weakness, this alliance of yours with Somali menials, and Arabs, and Sudanese, and Kaffirs, and Perverts, and Yemenis, and Nubians, and Indians, and Russians, and Americans, and Italians, and Serbians, and Portuguese, and Japanese, and Greeks, and cannibals, and Sikhs, and Banyans, and Moors, and Afghans, and Egyptians ... it is because of your weakness that you have to solicit as does a prostitute."

Apropos of this quote P. J. O'Rourke[80] wryly wrote that the Mad Mullah might have been describing the composition, effectiveness and moral stature of the United Nation's force that tried to settle the inter-Somali shambles of the early 1990s. The Mullah's victory over Corfield's force confirmed what the British had known all along. The only solution was a decisive defeat. They were learning the Horn of Africa's premier lesson: never be conciliatory with a Somali, meet force instantly with greater force and, if possible, kill the leaders. Any other course will be taken as weakness. Failure to appreciate it resulted in Britain's humiliation at the beginning of the 20th century and the United States' at its end. At the time of their humiliations both nations were the world's greatest military powers and both were humbled by petty Somali warlords – Mohammed Abdulla Hassan and Mohammed Farrah Aideed.

The British Government decided to break the Mullah, but had to wait for the First World War to end before it could do so. The final offensive against him commenced in 1919. Doing what they should have done at the start, massive military force was deployed. The Somaliland Camel Corps, irregular Somali levies under British officers, Indian troops and the King's African Rifles supported by six aircraft of the RAF took to the field as they should have done at the outset. Against a sustained campaign by organised forces, the Mad Mullah's men were beaten wherever they chose to fight or could be cornered. They deserted him and he fled into the depths of Ethiopia's Ogaden where, in November 1920, 'flu ended the twenty year rebellion of a man who, in many ways, characterised the Somali psyche.

In 1960 British and Italian Somaliland merged to become the Republic of Somalia. No sooner had they merged than Somalia laid claim to the lands occupied by their kin in Djibouti, Ethiopia's Ogaden and in the Northern Frontier District (NFD) of Kenya.

By the end of the 1960s Somalis occupied 36% of Djibouti, comprised 43% of its people and have lived there for centuries. They occupied 20% of Ethiopia, comprised only 4% of its population and have lived at least in northern Ogaden for several

---

[80] O'Rourke P. J., 1994. *All the Trouble in the World.* Picador, London.

centuries. In Kenya they ranged over 20% of the land, formed 1% of the people and are newcomers arriving well after 1865. In both Djibouti and Ethiopia the Somali claim is backed by long tenure, but this is not so in Kenya. Regardless, they pressed their claim, threatening that if Britain did not hand them 'their' bit of Kenya they would take it by force from the independent government.

And this is what they tried to do. First they broke off diplomatic relations with Britain to stress their seriousness (no-one in the Somali Foreign Office knew the protocol for doing this and they had to ask the British Embassy how to go about it[81]).

* * * *

In 1900 the British East Africa Protectorate was larger than Kenya is today. Its Jubaland Province stretched eastward from the present Kenya border to the Juba River in what is now southern Somalia (see p. 148). The port of Kismayu was thus in the East Africa Protectorate though it was shared with the Italians who used it as a base to administer land to the north of the Juba. To distinguish between these old and new borders I shall refer to land within present Kenya boundaries as Kenya and that between them and the Juba River as Jubaland.

The following account of the Somali arrival in Kenya is drawn heavily from a paper[82] written by Sir Richard Turnbull, the last British Governor of Tanganyika who, before that, had served many years in northern Kenya. He liked and respected Somalis and with their help had accumulated a verbal history of each group's movements.

Twelve hundred years ago somewhere in the north of the Horn of Africa a group of pastoralists emerged as the Somalis. Perhaps the arrival of Islam was the critical stimulus. At the time most of the Horn between the Ethiopian highland massif and the Indian Ocean was peopled by Oromo-speaking pastoralists known collectively as the Borana. From the outset the Somalis were at war with these Borana (whom they referred to as Wardeh or Warra Daiyah), and slowly took over their lands driving the survivors further south and west. These Oromo speakers moved into what is now Kenya east of Lake Turkana, and yet further south into the coastal hinterlands between the Kenya highlands and the coast. Today their descendants are the Sankuye, Gabbra and Boran (as distinct from Borana which covers them all) and the Orma of Kenya's Coast Province – who still share a language and pastoral way of life.

By 1650 the Jidu (Jidwak) Somalis had driven the Orma off the Baidoa Plateau north of the Juba and over the next one hundred and fifty years went on pushing them steadily southward. If it was not one group of Somalis, then it was another keeping up the pressure until by 1800 the Sab Rahanwein had forced them south of the Juba. Across the river the Orma rallied and amassed such strength that the Ogaden clans pressing to cross the river opted for diplomacy. According to Orma oral history, their elders debated long and hard before, in 1865, allowing the Ogaden Somalis across the Juba and access to the wells at Afmadu and the rich grazing of the Deshek Wama deep inside Jubaland. That decision has been regretted ever since.

---

[81] Anon, 1963. Blackwoods Magazine # 1771, Vol 293.
[82] Turnbull R., c. 1960. *The Darod Invasion* – a paper printed privately by the Government Printer, Dar-es-Salaam.

No sooner had the Ogaden Somalis moved over the river to Afmadu than Marehan Somalis thrust across the Juba higher up. At the same time the Orma were also under siege from the Masai to the south. And as if this was not enough, a smallpox epidemic broke amongst them in 1865 (tradition saying that it was deliberately introduced by the Somalis).

Between 1865 and 1875 the Orma lost Jubaland. Many of their men were killed or enslaved, their women were taken and the survivors fell back on the Tana River in Kenya. Briefly there was a pause in the Somali advance. True to tradition, the first clans across the Juba turned to settling old inter-clan scores set aside during confrontations with the Orma. More northerly clans hearing of the successes migrated across the Juba to share in this new territory of (relative) milk and honey. Soon Jubaland was not big enough for all of them. Once more the pressure built up to move yet further south and west into Kenya. Three distinct thrusts out of Jubaland developed: west to acquire the Wajir wells; south-west to take the Lorian swamp and the northern Uaso Nyiro River towards Mount Kenya; and south to the Tana River.

Sir Richard Turnbull, Britain's last Governor of Tanganyika, Somaliphile extraordinary, wrote a history of the Somali expansion into what is today Kenya.
(Photo Camerapix)

Access to the Wajir wells was obtained through coming initially as shegats. The moves up the Uaso Nyiro River towards Mount Kenya were military. By 1880 forays of combined Ogaden clans commanded by Abdi Ibrahim of the Abd Wak had arrived at the swamp and were feeling their way up the river. They were opposed initially by Samburu and Rendille and between 1880 and 1900 the Somali raiders were beaten more often than they won. Turnbull[83] gives two encounters in some detail.

In 1886 Abdi Ibrahim led a mixed group of Abd Wak, Aulihan, Mohammed Zubeir, Herti and Marehan to the base of the Nyambeni Hills – Mount Kenya's north-eastern outliers. Near a small wedge-shaped hill just to the north of the present Isiolo-Garbatula road they were ambushed by a party of Samburu and Wanderobo bowmen who inflicted such severe casualties that the Somalis retreated in confusion (the arrows were almost certainly poisoned). To this day the Somalis call this hill Bur Bilaya – the Hill of Shame. In 1896 Abdi Ibrahim again led a foray to Mount Kenya, penetrating deep into agricultural country near to where Meru town stands today. They were confronted by a force of determined Samburu and Meru bowmen who massacred the Somalis. Among the dead was Abdi himself, finally despatched by a Samburu spear as he knelt on his prayer mat, with one of his retainers holding his pistol arm.

Abdi Ibrahim was one of the great Somali fighters of the last century. Starting as a young man fighting fellow Somalis – the Elai – on the Baidoa Plateau, he was with those who

---

[83] Turnbull R. *Ibid.*

crossed the Juba, took part in breaking the Orma, and initiated the series of skirmishes that, while inconclusive, caused Samburu and Rendille to withdraw from the Lorian Swamp.

The most southerly thrust towards the Tana river in Kenya was in parallel to those up the Uaso Nyiro. In 1884 a raiding party over a thousand strong had burned the German mission station at Ngao. In 1889 there were parties raiding for cattle up and down the river and selling their booty in Lamu. Yet the first Somalis to settle on the Tana on any scale were 15,000 of the Telemuggeh (a collective name for the Abdullah, Abd Wak and Rer Mohammed) with 50,000 head of stock who came from Afmadu in Jubaland in 1909. They brought with them many Orma who had been captured when their people were dispossessed of Jubaland, but on arrival many of these slaves escaped and rejoined the free Orma living on the Tana. This sudden increase in manpower raised Orma hopes of getting their own back on the Somalis and capturing their stock.

All these developments were taking place as the British were installing their colonial regime. Committed in principle to keeping the peace, by inclination the British wanted to send the Somalis back from whence they had come. They did not have the military muscle to do this however. Consequently the administrators vacillated and in so doing, favoured the Somalis and made the Orma move south of the Tana. Before the Mad Mullah died every step taken was considered in the light of whether or not it might bring him and his followers down from the north. The outcome was the British decreeing that the Galla people who had first let the Somalis into the Wajir wells as shegats must themselves vacate the region. By the end of the colonial years the town of Isiolo on the northern foot of Mount Kenya was a Somali town, and the Somalis were well established on the Tana River. More ominously, they claimed all of Kenya's NFD as theirs by right of tenure that stretched back centuries! In reality they were strangers who entered the area when the whites started settling the Kenya highlands.

The British respect for Somalis was learned at some cost. In April 1898 the Ogaden Somalis massacred the British garrison at Yonte. The 'Ogaden Expeditionary Force' was formed and sent to teach them a lesson. They ambushed it, killed twenty-eight members, stole its weapons and sent the rest fleeing back to Kismayu. The Provincial Commissioner of Jubaland, A. C. W. Jenner, then tried to bring the Ogaden to heel. After they had murdered some Gosha tribesmen he ordered them to pay compensation. At the same time he had two Ogaden flogged. Shortly afterwards Jenner went on to Liboi in Kenya with thirty-five armed policemen. At four in the morning 300 men of the Aulihan and Mohammed Zubeir sections of the Ogaden stormed the camp, castrated and beheaded Jenner, and treated most of his policemen in the same way.

A punitive column of battalion strength marched on Afmadu to avenge Jenner and seized Ahmed Maghan, the head of the Ogaden, and set off with him to Kismayu to stand trial for murder. On 16th February 1901 this column was attacked and routed. A lieutenant-colonel and thirty-eight men were killed and Ahmed Maghan escaped. A fine of 30,000 head of cattle was imposed on the Ogaden, but they refused to pay, so it was reduced to 5,000 head – which made them laugh and still refuse.

Some of the British field officers favoured force against the Somalis (provided that they were given sufficient troops), but they were reined-in by their superiors. Sir Charles Eliot – the Commissioner (i.e. Governor) of the British East Africa Protectorate – advised leaving the Somalis to themselves, for as long as they left others alone. At his level and among his

It is a compliment for a swain to liken his beloved to a camel, so deeply is this animal embedded at the heart of Somali pastoral culture.

superiors in London the prospect of a full-blown Somali war was to be avoided at all costs. No incident brought out this attitude more clearly than when K. MacDougall, Jenner's successor in Jubaland, took action. With 200 men he marched on Afmadu, surprised the Ogaden, seized 4,000 cattle and brought them back to Kismayu. For this success he was severely reprimanded for risking war.

The Aulihan caused trouble in 1915 and ended the year with a raid into Meru District where they killed Samburu men, women and children, captured many as slaves, took thousands of cattle and chased away a District Officer sent to stop them. In 1916 330 northern Aulihan raided the Marehan, killed nine and stole 800 camels. Elliott (no relation of Rodney Elliott mentioned elsewhere in this book), the DC at Serenli in Jubaland, gave the Aulihan Chief Abdurrahman Mursaal – reputedly the most trusted and least trustworthy Somali in East Africa – three days to pay blood money for the men killed and return the stock, or face arrest and seizure of Aulihan stock. The Aulihan responded by attacking Serenli, killing Elliott and all his troops. Fearing that this heralded the dreaded Somali uprising, British Administrators temporarily evacuated Wajir. With Quixotic chivalry the Mohammed Zubeir Somalis looked after the station for the timid British Lion. When the Administrators returned, the post was handed back to them spick and span with nothing missing. The Somalis were not without humour!

The single step the British took to resolve the Somali problem was in 1923 (after the Mad Mullah was dead) to cede Jubaland and all Somalis in it to Italian Somaliland. Thus did they shut a stable door long after the horse had bolted.

* * * *

Where present Kenya is concerned, the first modern Somali political organ was the Somali Youth League (SYL). Ironically, it had been started during the war in occupied Italian Somaliland by British officers as a sort of Boy Scouts' Association – the Somali Youth Club. It was not benign for long. Within a short while it had little to do with youth, had a radical political agenda, was xenophobic and aimed to unite all Somalis in a common cause for a single country that would take in not only British and Italian Somaliland, but Djibouti, Ethiopia's Ogaden region and most of Kenya north of the Tana. The SYL stimulated the never well hidden Somali sense of racialism, and advocated Somali rights to migrate where and how they wanted.

Within two years of the Second World War's end, the SYL was spreading sedition and preaching anarchy in northern Kenya. Towards the end of 1948 it was proscribed and seven of its 'young' leaders (all over 40 years old) arrested and exiled to Turkana District where there were few Somalis. Yet whether or not they were politically active, the Somalis never stopped consolidating their status in Kenya. By the end of the 1950s they were the dominant people in northern Kenya east of the main Nairobi-Addis Ababa highway. Technically they were not supposed to graze their stock further west than the boundaries of Wajir, El Wak and Garissa districts. Somalis hemmed in the Meru people who occupied Mount Kenya's north-eastern quadrant. They were looking across the Tana and in Charles Trench's[84] words –

"The Somali line here was the river [Tana], fordable in many places. Poised along it like runners at the start of a race were three Ogaden sections: Aulihan, Abd Wak and Abdullah."

Plans for Kenya's independence were being set in the 1950s and this was seen as the next big Somali opportunity. Again quoting Trench –

"To the Somalis of the NFD, however, independence meant independence from Kenya, be it ruled by whites or blacks, and union with Somalia. Outraged by the prospect of being handed over to the government of tribes which from the bottom of their haughty hearts they despised, they responded unanimously to SYL propaganda, carrying with them most of the semi-Somali tribes and many Muslim Boran. ... in no way could the British Government countenance this. Allowing north-east Kenya to secede would give the worst possible start to British relations with independent Kenya. It would be almost as unpopular with Ethiopia, and good relations with Ethiopia were considered more important than with Somalia. Finally it would greatly facilitate the southward and westward pressure of Somalis into Kenya."

The Somalis said that if the NFD became part of independent Kenya, they would fight for its secession and unification with Somalia. The weary historian's reaction might well be so what? When have Somalis not been fighting? When in their abrasive history have they not been trying to expand southward and westward? Yet they were true to their word and did not wait until Kenya was independent before commencing hostilities. In 1963 just before Britain pulled out of Kenya, SYL agitation flared into full rebellion. Two Somalis – Mahamud Mohammed Farah and Mohammed Gelle – ambushed and killed the DC Wajir, the DC Isiolo, the DO Mandera and the Boran Chief Galma Dida near Garba Tula[85]. Both fled Kenya and became policemen in Somalia: their acceptance there being an open statement of Somalia's support for their action. Thus Kenya became independent with a guerrilla war on its hands.

Kenya's new rulers had not paid much attention to 'the Somali problem'. Their minds had focused on replacing British rule and as the Somalis comprised a mere one per cent of the population, they did not constitute a block of votes worth courting. In the regional

---

84 Trench C.C., 1993. *The Men Who Ruled Kenya.* Radcliff Press, London.
85 Anon, 1967. *Kenya-Somalia Relations: narrative of four years of inspired aggression and direct subversion by the Somali Republic against the Government and people of the Republic of*

government system that Britain prescribed for independent Kenya, the NFD would have been largely autonomous which, many hoped, would cater for Somali wishes. What really stuck in Kenyan gullets was that although the Somalis were only one per cent of the people they wanted to take away twenty per cent of the land. Further it was not land that they had occupied from time immemorial, but which they had mostly occupied during the colonial era.

There were other reasons for holding on to the NFD. In the decade or so prior to independence the Shell Oil Company spent heavily prospecting for oil in north-eastern Kenya. While it had not found oil, it clearly felt the signs were encouraging. With this in mind Kenya did not want to surrender a potential oil field to the Somalis.

Between June 1963 and April 1967, bands of Somali guerrillas – known appropriately as *shifta* (an Ethiopian term for bandits) – ran up the following tally[86]: they killed sixteen officials (including two DCs), 374 civilians and abducted ninety-nine, blew up thirty-seven vehicles with landmines, destroyed six bridges, ambushed sixty-one convoys, fired on six towns, attacked thirty-five villages, looted seventy-five trading centres and stores, raided 208 *manyattas* (pastoralists' field camps), fired on fifty-three police stations and two army camps and rustled stock on at least eight occasions. Put in another light the Somalis killed someone every fourth day, raided a manyatta every seventh day, blew up a vehicle every thirteenth day, abducted someone every fourteenth day, looted a trading centre every nineteenth day, ambushed a vehicle and shot up a police post every twenty-third day, attacked a village every fortieth day and fired on a town or destroyed a bridge every eighty days.

There were 385 engagements between the government security forces and the Somali shifta, 437 Somalis were killed, at least four were wounded and captured and 406 surrendered. No comparative figures of casualties on the Government's side were ever published ('for security reasons'), but they were substantial.

The number of Somali casualties implies a military competence on the Kenyan side that is a trifle exaggerated. In Trench's words[87] –

"*. . the Kenya Army entered the district* (NFD) *in force and suppressed the rebellion with far more severity than the British had ever used.*"

The severity to which Trench refers included hitting 'soft targets' in the form of old people, women, children and livestock. While some Kenyan soldiers and policemen fought aggressively and gave Somali shifta as good as they got, many were disinclined to take them on in the bush, man for man. It was easier and less dangerous to shoot up those who could not hit back. There was method to the system. As the British had learned, the Somali weakness was dependence on livestock. Confiscate it, as the British did, or kill it as the Kenyans did and the guerrillas are crippled. Harassing their women and children made the point yet more strongly. Militarily it made good sense to hit this weak spot hard. While it may not have been sporting it was the most effective way of bringing guerrillas to heel and went a long way towards curbing the Somalis.

---

[86] Anon, 1967. *Ibid.*
[87] Trench C.C., *Ibid.*

The new Government in Somalia supported the guerrillas in Kenya with arms and ammunition as well as with constant propaganda. In October 1963, two months before Kenya's independence, the following was part of a broadcast[88] to northern Kenya –

"How wonderful it would be for us to take up arms and stir up a war . .
"The Colonialists are grossly torturing the NFD people and killing them en masse . . Let us retrieve our land from the enemy . . our land will never be united until we die for it."

Such words were broadcast daily. The Somali government set up training camps for the guerrillas, provided them with arms, ammunition, land-mines and explosives. Internationally the Somali Government was openly hostile towards Kenya. Ahmed Yusuf Dualeh, its Foreign Minister, said publicly that people should not take too much note of the Organisation of African Unity's resolution for preserving present boundaries in Africa. The Somali campaign to detach the NFD from Kenya and join it to Somalia caused considerable inconvenience and retarded development. Yet it was striking not so much for what it achieved as what it did not achieve. Had the Somalis a planned strategy, they could have, at the very least, made the Kenya Government withdraw from much of the NFD and rendered it a 'no go' area. They failed because they could not cohere. As always there were feuds to be settled that led Somalis to not only fight Somalis, but also inform on one another to the Kenya security forces. And it is difficult to believe that many of the young men joined gangs as much to have fun as through political conviction. The sons of a Gurreh Chief and a Degodia Chief both joined a gang and went 'active' despite the fact that their fathers were in Kenya Government service. After a year or so when they had had their fill, they 'surrendered' and went back to civilian life.

In 1967 a thousand shifta paraded at their training camp at Lugh in Somalia. They said that if Somalia would not openly support them by attacking Kenya, they would make peace. Whatever answer they got, it clearly did not please them for they rioted and killed two Somali policemen. One of the shifta was killed in turn and six arrested after this incident, stressing the difficulty that even Somalis have ruling Somalis!

Eventually in 1968, under the aegis of the OAU, Somalia was persuaded to stop its support for the guerrillas in Kenya and the 'war' officially ended. It did not eliminate the shifta, however, but merely left them on their own. Denied supplies they toned down their political agenda, which was nonetheless still close to their hearts, and turned to sheer banditry. In 1977 the Somalis in Ethiopia's Ogaden region rebelled against the Ethiopian government and the Somali army went to their assistance. Initially successful, they were soundly beaten by Ethiopians supported by Cuban and East German troops. Deserters and disenchanted soldiers after the Somali Army's defeat in Ethiopia drifted south into Kenya to join up with the shifta gangs already there. Bringing their arms with them, they sought their fortunes in Kenya through brigandry and poaching for rhino horn and ivory.

In the late 1970s Somalis at last crossed the Tana River openly with their stock, claiming that, as Kenya citizens they had a right to go anywhere in the country.

---

[88] Anon, 1967. *Ibid.*

Government, having stated *ad nauseam* that tribalism was bad, was hoist with its own petard. By sheer weight of numbers, the Somalis had gradually pushed both Orma and Kamba out of the grazing lands south of the river. And while they had done so peaceably, the shifta gangs (whom the other Somalis *naturally* deplored) went ahead of them as a band of skirmishers preceding an army, robbing and stealing from other citizens, predisposing them to move away from the advancing horde.

Ahead of them now lay the Tsavo National Park and the Galana Ranch which, rumour had it, they would have liked to buy. If they got it, they would be on the Sabaki River; after that, the Kenya-Tanganyika border.

This is not just speculation. In 1985 I was in a Mogadishu Government office looking at a map on the wall with some amazement because it showed Mombasa in 'Somalia'. Sensing that I was observed, I turned to find several officials watching me. Addressing them I had said lightly, "Your cartographer slipped a bit on that map, didn't he?"

"Not at all," came the reply.

"But I thought your claims stopped at the Tana?"

"Oh, they did," came the reply, "but that was in 1963. Times have changed."

"Come off it! You'll never get away with that," I said, pointing at the map.

"Oh, we will. We definitely will. If not in five years, then ten; if not in ten years, then fifty and if not fifty, then one hundred. But in the end we will."

"How?" I asked and was answered in a word.

"Islam."

I thought no more of it until 1992 when some infidel Kenya police went into Mombasa's mosques to arrest dissidents without the good sense to remove their boots. At that I conceded the Somalis might have a point. Such incidents could inflame Islam and become the cornerstone of a case for Kenya's Muslim coast seceding and joining their fellow members of Islam. They would be supported by Iran, Iraq, Libya and Saudi Arabia. All it needs is for a few more careless acts on the part of the Kenya police to 'internationalise' an Islamic issue: what the Somalis said was not so far-fetched.

In 1994 while I was on a consultancy in Tanzania's Loliondo District, a gang of Somali shifta operating out of a Somali settlement at Shomboli on the Kenya-Tanzania border made a series of raids across the district. In the drive south and west they were now looking beyond Kenya!

Turnbull wrote in 1960[89] –

> "...*There would almost certainly have been an invasion of the Tana by the Bartirreh, Maghabul and Herti (from Jubaland)*"

if the Italians had not stopped it. That migration in the form of many thousands of refugees has happened. They may have come as refugees, but they will never be going back to Somalia. Touval[90] wrote in 1963 apropos of the Somalis –

> "*The stage seems set for an inevitable conflict . . .*".

---

[89] Turnbull R. *Ibid.*
[90] Touval S., 1963. *Ibid.*

That conflict has taken place. Trench[91] wrote –

> *"The last has not been heard of the Kenya Somalis' determination to unite with Somalia, however chaotic Somalia might be."*

He, too, will be proved right. As I will show in subsequent chapters, Somalis have had an immense influence on conservation, ivory, rhino horn and all to do with them. Yet to treat the Somalis as simple miscreants or poachers, as has widely been done, is to avoid the real issues behind their activities. They may act as bandits, they may poach, but these are minor aspects of their overall agenda. They are driven by the ecology and politics of the Horn of Africa. It is these that need to be understood before solutions to any 'Somali problem' – be it poaching, brigandry or whatever – will become apparent. Negotiations have never delayed an inexorable thousand-year advance, except in the very short term. Conservation is beset by many intractable problems the length and breadth of Africa. That involving Somalis is but one. If they have solutions, a basic first step towards them has to be comprehension in depth. Like the contents of Pandora's box, this may produce more than conservation can handle.

---

[91] Trench C.C., 1993. *Ibid.*

Dark shading, Elliott's Maralal District; light shading, the Somali shifta districts of Marsabit, Mandera, Wajir, Isiolo and Garissa.

CHAPTER FIFTEEN

SOLDIERING AGAIN

When David Sheldrick said that the arrival of Somalis in Tsavo was the beginning of the end, he of course knew they been plaguing northern Kenya since the dawn of Kenya's independence a decade earlier. His erstwhile Sergeant Major Mohammed had taken up the shifta cause and occupied himself training guerrilla recruits in basic military tactics. As the Parks employed some Somalis as rangers, such news filtered back to the wardens under whom they served. No doubt David hoped that the Somali threat would confine itself to Kenya's north as, until their appearance in Tsavo, there was little suggestion of territorial ambition south of the Tana even though others had predicted it. Yet up north, largely overlooked by the media, those who had battled the shifta for ten years did not doubt the almost limitless extent of Somali political ambition. Rodney Elliott was one of them.

* * * *

Rodney Elliott stayed on with the Kenya Game Department until 1971. In some respects he had his cake and ate it. Paid his 'golden bowler', he left the colonial service but was then employed to remain in the Department on two year contracts. For most of that time he stayed at Maralal, the original post of his choice.

Maralal was in Kenya's north but not technically considered part of the NFD. Its principal tribe was the Samburu – northern cousins of the Masai – who, by and large, were law-abiding people. Yet all the districts to Maralal's east and north – Marsabit, Moyale, Wajir and Garissa Districts – were NFD proper and the last three were Somali country. When Rodney arrived in Maralal in 1955 there were a few Somali traders scattered about the district. As apparent from the records of Ritchie and MacArthur in the 1920s and 1930s, Reece and George Adamson in the 1940s and 1950s, they had always trafficked rhino horn. Yet by 1960 as Somalis realised that British rule would shortly end their activity seemed to escalate.

On the 19th July 1960 Rodney wrote to Gerry Dalton, the warden who looked after the Marsabit National Reserve –

> "...have been looking into the appalling slaughter of rhino ... around Kilitamon Lugga. Four killed and two wounded ... a further one killed in the Wamba Lugga".

In notes for inclusion in the District Commissioner's Annual Report dated 30th December 1960 Elliott wrote –

> "The worst year yet. Several cases of professional poaching and far more rhino killed than ever before. The average number of convictions in previous years amounted to 20. This year the total is 73."

Most of this was Somali inspired. Illustrating anticipated trouble the DC took steps to close his District. Thus he sought permission from the Road Authority in Nairobi[92] to erect a permanent barrier under Police control on the Isiolo/Marsabit road. His reasons were –

> *"(a) to control the increasing amount of illicit trophies,*
> *(b) to control export of hides and skins from this district,*
> *(c) to control the illegal entry of vehicles into and out of this district without passes from the Administration, Veterinary Department or Game Department,*
> *(d) to control the illicit export of livestock from this district."*

A month later the DC asked Rodney and three other Government Officers[93] –

> *"With a view to implementing aspects of the District Security Scheme, I would be grateful if you would volunteer as a Class 'B' Police Reservist."*

Overall, it is clear that from 1960 onwards the NFD was becoming more lawless. As of 30th March 1961, all visitors to Samburu District, other than officials whose duty station it was, were required to get an Outlying District Ordinance pass to enter it. The District had in effect become part of the dangerous NFD.

* * * *

As already recounted, on the eve of Kenya's independence the Somalis made good their threat to try and force the secession of the NFD from Kenya to become part of the new Republic of Somalia. The sudden departure of veteran British civil servants left affairs in the hands of young and inexperienced Africans, most of whom had no military training. Many feared Somalis and when posted to the NFD had one thought in mind: to get out of it as quickly as possible. Unlike the other Departments under 'new' inexperienced management, the Game Department under Elliott in Maralal was headed by an experienced soldier. His men were already trained along quasi-military lines and proficient with firearms. In the time-honoured rôle of hunters as frontiersmen fighters, they were expected to take on the shifta gangs and did. Throughout the shifta war Game Department morale in Elliott's area remained high in marked contrast to other sectors of government.

The Somalis mined the roads, blew up bridges, rustled livestock, ambushed travellers, murdered herdsmen and civilians out of hand, routinely raided trading centres, and shot up police stations, villages and towns. Outposts were at risk and government withdrew into the more defensible towns or administrative centres where a siege mentality prevailed. Government became a largely paper exercise, administrative standards dropped steeply and official responses to shifta attacks were reactive, steps being taken in response to Somali initiatives.

---

[92] District Commissioner, Maralal's letter to the Road Authority in Nairobi, *W. 5/81* of 24th January 1961.
[93] District Commissioner, Maralal's letter to Elliott, *Conf. 4/5* of 15th February 1961.

The prevailing situation is made clear in the orders issued by Bill Woodley who was responsible for the national park staff on Marsabit Mountain. He wrote –

> *"All vehicles will be formed up in 'laager', on high ground and as far away as possible from hillocks, luggas, gullies, groups of trees with the object of achieving maximum visibility and sound fields of fire to ensure adequate protection of the camp.*
>
> *"Stand-to drills will be observed for an hour at both dawn and dusk each day, and all personnel must be allocated positions for stand-to, and be thoroughly acquainted with what action to adopt in the event of any circumstances* [sic]*. Guards will be maintained at all times. Slit trenches or sangars will be sited between each vehicle on the perimeter."*

Although Somali incidents were sporadic and unco-ordinated, travel became awkward and vehicles had to move in convoys guarded by armed police or soldiers. For the most part the Somalis kept to those areas they claimed should be part of Somalia and did not include Samburu District. Even so they made occasional forays into it and beyond.

On such an instance they raided the small tourist lodge at Loyengalani on the south-east shore of Lake Turkana. There were no tourists there at the time, but they tied the Lodge Manager, Guy Poole, and the local Catholic missionary to chairs then shot them. An Italian friend of the Catholic father was taken prisoner and forced to drive the bandits away in the lodge Land-Rover until it ran out of petrol. He was never seen again, but months later a Somali turned up in Marsabit town and displayed a small piece of wrinkled raw leather that he said had been cut off the Italian before he died.

In 1968 the shifta war was supposed to have stopped after President Kaunda of Zambia mediated between the Governments of Somalia and Kenya. Somalia withdrew overt support for the guerrillas in Kenya and their supply of land-mines dried up. Yet things did not improve greatly in Maralal or the NFD generally. The banditry continued and civil servants still tended to stay in their urban redoubts, hoping to stay unscathed until their tours were over. The Samburu continued to suffer. Turkana took advantage of a weakened Administration and raided them from the west. Boran gangs joined in the banditry, but the highest in profile were still the Somalis, who started coming further into Maralal District.

One issue particularly exercised the authorities in Nairobi: the welfare of tourists visiting the Samburu Game Reserve. Tourism was a major foreign exchange earner and if any visitors were killed by Somali shifta it would adversely affect Kenya's revenue. Government was so concerned that it considered closing Samburu to tourists. In 1967 a meeting to discuss this possibility was attended by senior men in the civil service and armed forces, the most junior present being Rodney Elliott.

The mood was to avoid risk. Policemen, Administrators and soldiers all favoured caution and it was only towards the end of the meeting that the Permanent Secretary in the Ministry of Tourism and Wildlife under who the Game Department came, asked Elliott for his opinion which he gave[94] –

---

[94] Elliott R.T.E. Pers comm.

The Government was deeply worried by the possibility of tourists being attacked by shifta gangs.
(Photo Peter Davey)

> "The issue is where do you want to draw the line? If we give up here and don't allow visitors beyond Isiolo, the shifta have won a round. If they continue their activities, the next front line will be drawn in Nanyuki further south and the process will go on until eventually people will not be allowed to travel north of Nairobi.
>
> "I think that we should take them on where they are now. We should draw the front line before they get as far south as the Samburu Reserve. Go after them in the field, chase them and kill them and make them too frightened to extend their activities into Samburu."

The police and army objected on the grounds of inadequate manpower and logistics. After listening to their arguments Rodney was again asked to comment.

> "Give me a Ranger force of forty men and I think that the Game Department can do it – at least where the security of the Samburu Game Reserve is concerned."

The upshot was that he was given the extra men and equipment he asked for and a free hand to go after the gangs. So started a successful three year anti-bandit operation and the first of Africa's post-independence anti-poaching 'wars'. Theoretically there would have been every advantage in combining all the available forces – army, regular police, the police paramilitary General Service Unit (GSU), Administration police and Game Department staff – under a unified command. Co-operation was ordered from above, but such as took place in the field was largely through independent initiatives by men on the ground. One feature of Elliott's security work was that it gave him direct access to levels in government that he would otherwise have had to approach through the Chief Game Warden.

After independence the Game Department had expanded, outwardly acquiring the form of a more conventional department. Rodney became a Divisional Game Warden of the Northern Division with headquarters at Maralal. Under him were four wardens; one at Wamba, one running the Samburu Game Reserve, one at Isiolo where Adamson had once been stationed, and one in his Maralal base. All had clerical staff to assist them as well as field game scouts. All his wardens were in radio communication with both him and one another. Even with forty scouts Rodney had insufficient manpower to chase the shifta wherever they turned up. Thus in many cases the Department took to the field in mixed units combined with both regular and administration police and 'home guards' who were local people Government had armed along the lines first established during Mau Mau.

The shifta concentrated on getting money. The easiest sources were leopards for their skins and rhinos for their horns. Both produced valuable commodities that were easy to carry and conceal. Elephant tusks on the other hand were bulky and difficult to move without either wheeled transport or numerous porters. The colonial Outlying District Ordinance was still in force and blocked free use of the roads so, initially, ivory was not poached on any scale.

The shifta caught leopards in traditional Somali style. A circular bush enclosure with a narrow entrance was made and a meat bait hung in the centre. A gin or 'jaw' trap that seized an animal by the leg was then lightly buried in the entrance and chained to a tree or heavy log. Any predator scenting the meat would approach the enclosure, try to get at the meat through the entrance, put a foot into the trap and be seized by the leg. These trap enclosures were of course the shiftas' Achilles heels as they were difficult to conceal and could be ambushed.

The men pursued in and around Tsavo in the 1950s were poachers and only poachers, and as members of local communities where they were vulnerable to conventional police investigation. The shifta, on the other hand, were primarily guerrillas who, amongst their many activities, poached. The communities from which they sprang were often in Somalia or Ethiopia's Ogaden area and beyond Kenya's ability to police. Importantly, because they tried to kill all officials with whom they came in contact, the only counter-tactic was to go after them in the field and shoot first.

In 1966 there were at least ten shifta raids into Maralal District in which twenty-seven officials or civilians were killed and 6,800 cattle and 145 goats were stolen. There were no shifta casualties. In 1967 when Elliott's scouts started general security duties they had at least seven skirmishes in which they killed or wounded fourteen shifta without casualty to themselves. In 1968 as their tactics improved they had twenty-five skirmishes, killed or wounded fifty shifta for one Samburu civilian killed. In January 1969 a laconic note in Elliott's hand noted that there had been thirty-three engagements in which sixteen rifles, 266 rounds of ammunition and sixty-eight jaw traps had been recovered. In addition the Samburu had reoccupied grazing areas that they had earlier vacated due to shifta rustling. He failed to mention that thirteen shifta died and one government servant was wounded. In 1970 skirmishes fell to four in which there were two casualties on either side. In 1971 there were another four clashes, three shifta were killed as were three Samburu civilians.

Before 1967 the shifta called all the shots and inflicted more casualties than they suffered. Once Elliott was organising responses, the initiative passed to Government, shifta casualties increased dramatically exceeding those that they inflicted. Their initial response was to increase the sizes of their gangs with sometimes as many as 150 men operating together. It made them easier to find and, providing that those chasing them were resolute, increased their casualties. Cattle thefts stopped and between 1969 and 1971 the shifta became timid and more difficult to contact.

Giving all credit for the successes against the shifta to Rodney Elliott would be inaccurate and unfair. Wardens who worked under him and the subordinate staff all deserve recognition. Some of them such as Sergeant Lerumben Lenjatin put up outstanding individual performances. On 21st August 1964 he was leading six game scouts on patrol when a village nearby was raided. A shifta gang killed several villagers, burned the houses to the ground and made off with all the cattle. Lerumben followed, caught up with the gang and though outnumbered, attacked it, killed three shifta, recovered a rifle, put the gang to flight and recaptured all the cattle.

On 22nd October 1967 he led a large posse of five game scouts and twenty-four policemen into contact with a big shifta gang. Again, though greatly outnumbered, he attacked and his men killed nine shifta. On 20th April 1968 Lerumben was in Kichich (also spelt Kitich) outpost in which there were four other game scouts, two administration policemen and some unarmed labourers. At ten o'clock in the morning some 150 shifta attacked the post and poured a hail of bullets into it over the next two hours. They intended overrunning it, but so spirited was the defensive fire that they could not close in and the attack petered out. No-one inside the post was hurt because Elliott had ensured it was well served with dug-outs, trenches and fox-holes into which all the occupants retreated and from which they could shoot back from relative safety. The shifta did not seem to appreciate that the enemy were at or below ground level, and fired a great deal of ammunition uselessly into the empty buildings. A number of shifta were killed or wounded and carried away.

Lerumben sent a runner for help. This arrived that same night and at dawn next day he led a mixed force of twenty seven made up of ten game scouts, five regular policemen, five administration policemen, five armed home guards, a forest guard and himself. They caught up with the gang only ten miles away. Leading his group into the attack, and though outnumbered more than five to one, his men killed fifteen and put the rest to flight, leaving their belongings behind them. On the 25th April 1968, three days later, he ambushed and killed two more shifta. No commander could have asked for a more competent and aggressive subordinate than Lerumben Lenjatin.

Another aggressive warden was ex-policeman Peter McClinton, based at Isiolo. Extracts from a report dated 17th July 1968 follow –

> "On 9th July I arrived with my men on the Kipsing luggah . . To cover our approach a patrol moved upstream and another downstream. The upstream patrol was seen . . by three shifta . . who took to their heels with the patrol in pursuit . . the shifta had been carrying three rhino horns . . these they abandoned and we recovered . . Leading the patrol downstream along the tracks of shifta I came on a number of small luggahs running into the

As leopard skins were valuable and leopards abundant and easy to trap, it was natural for the shifta gangs to go after them. However, a trap line which had to be visited frequently and which was difficult to conceal, provided ideal opportunities for ambush and was the bandits' Achilles heel.

(Photo Peter Davey)

> *Kipsing. As these were often used as camp sites by the shifta they were approached most carefully. Unfortunately the patrol left flank man came on one of the shifta . . there was an exchange of shots and three men were flushed. In the pursuit one shifta was probably winged. The blood spoor, though slight, was definite. It was followed but became confused with that of a wounded rhino obviously injured that morning . . by the shifta. As the three men ran on widely separate courses, I called off the pursuit. The three had been guarding a campsite . . three ammunition clips were recovered, many cooking utensils, two gin traps, . . a scribbling book with names of members of the gang, . . 21 giraffe tails. The following morning I took up pursuit . . another shifta camp was discovered but had been vacated the previous day . . 14 shifta had moved along the Kipsing by moonlight . . a second group eight strong had crossed the Uaso Nyiro . . animals taken in the traps and discarded . . included mongooses, jackals, African wildcats, caracals, civets, hyena, genet cats and serval cats."*

Another McClinton report in much the same vein covered the following month –

> *"On Monday 28th August ... a patrol of six game scouts encountered a poaching gang of 12 shifta ... without being seen themselves. The patrol followed the shifta who were leaving the remains of a freshly killed giraffe.*

Gradually the patrol closed the distance ... and as the shifta disappeared into a narrow luggah our men ran quietly forward to get within striking distance ... unknown to the Scouts more shifta were away on rising ground to their right flank. About the same time the Scouts and the shifta on their flank saw each other. . . the enemy forces opened fire... as the firing commenced men ran out of the luggah in all directions. Several were armed and stopped now and then to fire upon the patrol. The patrol was outnumbered about five to one. The patrol wounded two shifta ... Being so heavily outnumbered and out-gunned our patrol did not take up the pursuit, but returned to the luggah ... and found large quantities of clothing and cooking utensils etc. These were ... burned. The patrol recovered four rhino horns, three gin traps and a giraffe tail. The following morning . . a stronger patrol set out to follow ... I was able to get a message ... requesting assistance from the GSU ... a platoon of which was dispatched to intercept any shifta crossing the Kipsing luggah ... Our Scouts came across the body of one of the shifta who had apparently died during the night [from the wounds they had inflicted the day before] ... The GSU platoon encountered another gang of shifta ... four were killed and a rifle and 41 rounds of ammunition recovered ..."

These reports reflect the pattern of scrappy, untidy, skirmishing that typified Game Department activity at this time. There were others in the police, GSU, Administration police, army and Home Guards who displayed admirable fighting spirit. What distinguished the game scouts, however, was their morale and aggressiveness and the influence this had on the other services with whom they mixed.

Elliott reported that anti-shifta operations brought out the general low quality of leadership in the Government's forces. It was the same point that Sheldrick was later to make in reference to the calibre of new wardens in the WCMD. In colonial days game scouts were chosen from the martial pastoral people who were hardy, bushwise and, given the traditional rôle of cattle rustling in their cultures, knew something about fighting and weapons. They may not have been either literate or numerate, but they were well educated in bushcraft and how to fight. It was as unreasonable to discount this form of learning in their work as it would be to discount knowledge of anatomy in the competence of a surgeon. Yet this is precisely what had happened. The same complaints were made about army and GSU reluctance to leave roads and move on foot and on June 11th 1968 Rodney Elliott formally complained through the Chief Game Warden to the Permanent Secretary of the Ministry of Tourism and Wildlife that the GSU simply wasn't performing properly. On 13th June he wrote again to the Permanent Secretary directly –

"*The actions of the GSU Platoon ... have been so unmilitary in the last two actions ... that I wonder whether they are not under orders not to engage the shifta?*"

Because of the unsatisfactory Army and GSU performances, Elliott proposed that the Game Department establish its own permanent para-military anti-poaching force along the lines of the original 1950s Field Forces set up by Sheldrick and Woodley, but with an enhanced military capacity. In 1969 his proposal was adopted. It was not very effective because it was inadequately equipped and led by men lacking military experience and, even more, really aggressive spirit ... but that is another tale.

Elliott's performance gave him respect within government circles, particularly those in the field in the NFD. Nevertheless there was a certain ambivalence towards him. While liking the anti-shifta results, officials in the first flush of independence were sensitive to their own weaknesses being shown up. More embarrassing were Elliott's attempts to prosecute soldiers and policemen who poached.

The whole situation was complicated by rising disregard for game law within the Game Department itself. The normal channel for complaint about other arms of Government was through the Chief Game Warden who was then supposed to take it up with his Permanent Secretary in the Ministry of Tourism & Wildlife, who would pursue the issue with the appropriate Ministry under whom the offending party fell. Seemingly Rodney's formal complaints never got beyond the Chief Game Warden. From his anti-shifta activities Elliott had established a direct line to his Chief's superior, the Permanent Secretary – Alois Achieng – in the Ministry of Tourism and Wildlife. Where security issues were concerned this did not offend the Chief Game Warden, but he resented Rodney going directly to Achieng over law enforcement.

Into this delicate situation came a morose sometime journalist called Anthony Cullen who was passionate about wildlife. Tony first made a name in the 1950s through his articles on wildlife in the *Kenya Weekly News*, particularly during the anti-poaching years 1956-57. Indeed in rousing public support for anti-poaching measures he had, arguably, been the media's most potent element. At independence Cullen left Kenya saying all was lost. A short spell in Britain convinced him that this might not be so and he returned to Kenya. For his journalistic skills he was employed as a speech-writer for President Kenyatta. In this position he could put information in the presidential ear which, in terms of a civil servant, made Tony influential and a minor power. Two ambitions were attributed to Cullen. The first was to amalgamate the Game Department and the National Parks. The second, was to head such an amalgamated service.

In colonial times Cullen had been an honorary game warden and, in this capacity, he had once stood in when Rodney Elliott was on leave from Maralal. Consequently they were acquainted. Thus when Cullen engineered a transfer from speech-writing in the President's Office to the Ministry of Tourism and Wildlife to pursue his Game Department and National Parks goal, Elliott saw him as an ally to resolve problems with army and police poaching.

Cullen took the bit in his teeth and in the Permanent Secretary's name wrote an instructional letter[95] directly to Elliott as Divisional Game Warden, Northern Division, that bypassed the Chief Game Warden. It is worth quoting in full –

---

95 Letter WCM/27/101 of 9th October 1970, Ministry of Tourism & Wildlife, Nairobi.

"POACHING BY SECURITY FORCES

During the meeting and discussion which took place at my request in this Ministry on October 7, you very properly reported to me some continuing instances of game – including giraffe, rhino and oryx – being unlawfully killed in your Division by members of the Police or G.S.U. or Army. Some particular recent occurrences have been in the Archer's Post and Laisamis areas, but incidents have been widespread.

I make no comment, since such matters have passed into history, on an instruction by a previous Permanent Secretary that no action was to be taken in respect of certain notorious cases in the past.

But I am sure that my present Permanent Secretary would wish to make urgent representations – or instruct me to do so – in respect of any current or future instances of the wanton destruction of wildlife. His Excellency the President has made it clear many times that the laws of Kenya apply without discrimination to all persons – in or out of uniform – from a Minister to the humblest of the unemployed.

If, therefore, further incidents involving (allegedly) the illicit killing of game by members of the Security Forces come to your notice, a full report – outlining all relevant details and supporting evidence – should be sent to this Ministry, quoting the reference of this letter as authority.

Signed, Anthony Cullen, for Permanent Secretary for Tourism & Wildlife."

In the Permanent Secretary's name, in poaching cases involving the security forces Cullen had ordered Elliott to bypass the Chief Game Warden. This letter may have been the authority that Elliott needed go over his immediate boss's head, but it was hardly an exercise in diplomacy or tact. It also recorded the fact that a previous Permanent Secretary had ordered no action be taken in certain cases involving members of the security forces.

The letter was also a threat to officials in the Ministry; its unstated import being that if there was any further stepping out of line, Cullen's links to the President would be brought into play. It reflected a misplaced self-confidence. Cullen did not stay long in the Ministry of Tourism & Wildlife and never headed the combined Game Department and National Parks when they eventually merged. And the politicking between Elliott and Cullen hardened attitudes within both Ministry and Game Department that, as was decided five years later with Sheldrick, Elliott was too loose a cannon to leave unsecured, no matter how good on anti-shifta work he might be.

The anti-shifta activities that preoccupied Elliott in his last years at Maralal had not blinded him to a general and growing malaise in the Game Department. It had acquired a progressively greater load of clerical 'baggage' that involved all wardens, but particularly the senior divisional wardens, in much red tape and pointless bureaucracy. Nothing more stressed the great gap between the old-time game wardens who saw work entirely in the field and their incoming African replacements who believed that work was done at a desk. In part this was not peculiar to the Game Department, but characteristic of Government overall. Overstaffing reached such levels that in the early 1990s the World Bank and

Oryx can go for months without free water and, adapted to dry land where man cannot go, have lost less territory to man than species that need water. (Photo Peter Davey)

International Monetary Fund insisted on massive cuts in the civil service as a condition for further assistance to Kenya.

Declining civil service efficiency facilitated corruption. No sector of Government was subject to greater criticism in this respect than the Game Department. The professional hunting fraternity, in particular, was openly reporting departmental collusion in both poaching and illicit trophy dealing. The case[96] of Major N. Nzembei of the 3rd Battalion, Kenya Rifles, illustrates the drift. In January 1968 he shot three gerenuk, five oryx, two rhino, seven Grevy's zebra, three giraffe, one lioness and two caracal. Rodney obtained eye witness statements from two privates in the Kenya Army – Ncholei Baali and Lolimiri Ekaran, from a Home Guard – Purkei Lesankurukuri, and from two civilians Nkasana Lekadaa and Kabukoki Lolokoria. These men had either seen the animals shot and/or their trophies in Major Nzembei's possession. Acting game warden John Nairi and game scouts Nderobo Loltianya and Gishilian Leserewa recovered evidence of the kills in the field. It was incontestable and the case as strong as any that ever went to court.

Rodney reported the evidence to both Chief Game Warden, and Permanent Secretary Alois Achieng and they were emphatic that he prosecute Nzembei. Initially the man's Commanding Officer tried to get the case dropped. When he realised Elliott was proceeding, he requested the case be delayed. Major Nzembei was on the verge of retirement and if convicted of the game offences before this happened, he would lose all his pension and retirement benefits earned from an otherwise exemplary army career. Appealing as soldier to soldier, he won Rodney's agreement not to commence proceedings until after the man's retirement date. Yet after Nzembei retired, Rodney found obstacle after obstacle blocked taking the case to court. The Permanent Secretary, and the Chief Game Warden, expressed their disappointment and continued to support Rodney. It was only after the Permanent Secretary Achieng himself had been convicted for embezzling funds destined for fishermen that Elliott gained access to the Attorney General's office.

---

[96] Files in the author's possession.

Almost incidentally he found evidence that it had been at Achieng's request that the Attorney General stopped the Nzembei case going to court[97]. Thus while in public he and the Chief Game Warden had staunchly upheld the game laws, behind the scenes they had them set aside. Nzembei was never prosecuted.

Yet another issue upset Elliott's superiors. In the past, as a long-standing practice, whenever the Game Department had wanted to capture some animals, it had used the country's licensed trappers to undertake the job. Instead of paying them money, it let them have half the animals caught on a one-for-one basis. Suddenly an edict from above ordered Elliott to let a trapper catch zebra in Maralal on a two-for-one basis – the trapper taking the two and letting the Game Department have the one. Some of these animals were to stock a private ranch in the Rift Valley belonging to an Assistant Minister in the Ministry of Tourism & Wildlife, J.M. Kariuki. Elliott objected.

Thus while he had done much to reduce shifta activities long before Somalis appeared in Tsavo, enforcing the law where Government officials were concerned made him an embarrassment. On the 17th February 1971, with only a year to serve before retiring in 1972, Rodney was transferred from Maralal and the north he knew so well, to be Divisional Warden of the Central Division based in Nyeri between the Aberdares and Mount Kenya.

The intention was to throw Elliott off balance. He would have to hand over the awkward enquiries he was making in Maralal to other parties and after barely finding his feet in a new station his service would end. By the time he was familiar with Nyeri it would be too late for him to do much damage: or so it had been hoped.

In another time and place Rodney Elliott and the men who had served under him when he was divisional warden at Maralal would have been publicly acclaimed. Yet his skirmish war was largely overlooked by press and conservationists.

Zebra hides were in such demand that many 'zebra skin' articles like handbags, were in fact made from white cow hides upon which black stripes had been dyed.
(Photo Peter Davey)

---

[97] .Elliott R.T.E. Pers comm

CHAPTER SIXTEEN

THE DOL DOL ANTI-POACHING SAFARI

By 1970 public outcry over developments within the Game Department were embarrassing Government. This, coupled with anonymous information from within the Department, caused the police Criminal Investigation Department (CID) to formally open investigations into its activities. These were well under way when Rodney became Divisional Warden, Central Division in 1971. Coincidentally moving Elliott to keep him quiet at this moment, was a bad error of judgement.

* * * *

Aspects of the administration he was taking over left Elliott uneasy and he was particularly curious about permits being passed out to ranchers for game trophies shot 'on quota'. By 1971 most ranches in the Central Division's jurisdiction had annual quotas of game whose trophies could then be sold. However, before selling them, landowners had to get sale permits from Game Department headquarters in Nairobi. Applications were routed through the local game warden, who forwarded them on to the Divisional Warden, who in turn forwarded them to Nairobi. Going through the files, Rodney's attention was caught by a request for twenty-one elephant ear skins (which made good leather) in the name of Ol Pejeta Ranch. As elephants were *never* on landowners' quotas, this seemed fishy.

The ranch manager, New Zealander Rod Herd, was asked to explain. Herd confirmed that no elephants had been shot on the ranch, no elephants were on the ranch quota and no application for sales permits had ever been made for elephant ear skins. The warden at Nanyuki had submitted a false application for sale permits in Ol Pejeta's name.

The matter was brought to the attention of American millionaire, Court Parfet, who at the time, held shares in Ol Pejeta. Parfet, keen conservationist, took the matter to his confidant the Attorney General, Charles Njonjo. In turn, Njonjo demanded the Chief Game Warden explain how sale permits had been issued for elephant ear skins from Ol Pejeta, when no elephants had been shot there. The Chief Game Warden, not knowing where the request originated, had to at least go through the formality of asking his Licensing Officer who, in turn, could only get this information from the Divisional Office that had approved the request for them. The Divisional Warden could only get it from the warden of the station who made out the original request which, in this case, was the Game Warden Nanyuki.

Such was the situation in the Game Department at that point that, had Elliott investigated his subordinate in Nanyuki (who was related to the Chief Game Warden) off his own bat, he would have been stopped. Arranging the enquiry through Attorney General, one of the most powerful men in Government, it had to be attended to and the Chief Game Warden was 'looking in the wrong direction' when trying to figure out what the enquiry was about. Machiavellian, maybe, but Elliott arranged matters so that he was instructed to look into something that, had the truth been known, he would have been forbidden to touch.

Rodney's first step was a routine station inspection with no intimation of his suspicions. Reading through the Nanyuki trophy register he noted a larger number of lion skins than fitted with any of the control work that had been done in Central Division. Of the 21 elephant ear skins ascribed to Ol Pejeta there was no evidence at all. Elliott left, seemingly satisfied with what he had seen.

He went directly to the Police CID team already looking into Game Department corruption. His thinking, if the policemen took over the investigation it would be difficult for anyone in the Game Department to close it down. The police agreed to help. A week later Elliott again descended on Nanyuki. This time taking his divisional clerk, whom he sat down with the Nanyuki station clerk, ordering him to check the Nanyuki Station registers against the Divisional registers, item by item. The Nanyuki warden realised that this was more than a routine check, but could not interfere as Rodney detained him to explain the discrepancies noted during his first inspection.

As planned, half an hour after Rodney started his enquiries Superintendents Giltrap and Gontier and Chief Inspector Mutha of the CID arrived, announced police interest in the Department's affairs and took the warden aside for questioning.

Shortly after this Rodney's divisional clerk noted a page missing from the Nanyuki trophy register. Specifically, it was the one covering all the lion skin entries that had earlier interested Elliott. The station clerk who kept the register was told that he was in deep trouble and without further ado explained that his warden had ordered him to tear out the page and burn it. Taking Rodney and the divisional clerk to the station rubbish dump, they found the burned page with enough of it intact to show the lion skin entries.

When Giltrap interrupted interrogating the Nanyuki warden to see what Rodney had found, the man made a break for the telephone and in panic called the Chief Game Warden for help. Minutes later the phone rang with the Chief on the line asking for Elliott. Angrily he wanted to know what was going on, why the CID were involved and why Elliott had not kept the investigation purely departmental? Giltrap, listening, took the 'phone from Rodney and told the Chief Game Warden that he was interfering with a Police investigation and if he did not desist, would be charged with attempting to pervert the course of justice. Alone, this guaranteed the termination of Rodney's service with the Game Department at the earliest possible moment.

The CID men arrested the Nanyuki warden and charged him with destroying evidence. This panicked the rest of the station staff who could not spill the beans fast enough. An extraordinary tale unfolded revealing that the entire Nanyuki staff had been undertaking extensive, organised poaching. One instance – the Dol Dol Anti-Poaching Safari – sums up what the situation and is best described in extracts from one of the written statements[98] –

> *James Omuse Stanley Amani, personal number 5262, Establishment Number 7057(7057) states I am a [sic] Assistant Game Warden (Trainee) of the Kenya Game Department and I was stationed at the Nanyuki Station of the Game Department. I remember on Sunday the second of May 1971 at about 4.30 a.m. in the very early morning we left Nanyuki Game Station by order of the Game Warden in charge ... to go on patrol to the Dol Dol area with the following Game Department staff [a list of 15 men given together with truck registration].*

---

[98] Statement by Assistant Game Warden James Omuse Stanley Amani in the Dol Dol Safari case file in the author's possession.

At about midday we arrived at the camp site at Mlango wa Mamba (Luwanyiro) just beside the Uaso Nyiro River and made camp there. On the 4th May we went out on patrol and the Assistant Game Warden shot one common zebra ... The Assistant Game Warden and myself and one game scout called Muthamia Kangane ... then shot one more common zebra near the same place as the first zebra was killed. ... Next we all went back to camp with the two zebra skins ... where they were salted ... and placed out to dry.

On the 7th May we again went on patrol, and game scout Akare Morijo shot one common zebra ... game scout Tagale shot one more common zebra that day ... we went on a further patrol and arrived at Tale Hills ... and saw two elephants ... the Assistant Game Warden assisted by an Acting Corporal and game scout Chepkorio shot one ... we removed the tusks on the spot. There were about eight persons present when the tusks were removed. game scout J. Ngugi Nganga took a photograph of this elephant and of us all and I have a copy of this photograph which I now give to you Mr R. T. Elliott Divisional Warden.

On the 13th May 1971 the Assistant Game Warden carrying the super grade rifle ·458 and I carrying a 30/06 and some game scouts with ·404s and ·303s went down to the Kipsing Hills for a patrol, and the Assistant Game Warden assisted by Corporal John Kirimanya shot one bull elephant ... On the morning of 16th May 1971 we again went out on patrol ... and ... shot ... two more common zebras. On 17th May 1971 (we) ... shot one bull elephant. On the 18th of May 1971 we again went to the Kipsing area ... we saw a group of elephant which were cows and calves so we left them alone. We then saw a group of eight to nine elephants and among them we selected only two to shoot to kill. ... We fired together ... and one dropped on the spot and we followed the second a few hundred yards and game scout Tangale finished it off. When the other game scouts came forward and told us that there was another elephant lying down nearby which must have been killed by our shots also. ... On the 22nd May we all moved back to the Nanyuki Station together with all the game trophies and the complete camp. I further state that I know of no reason that any of these animals were shot as they were harming no one.

Read back and agreed – signed by AGW(T) Amani. Statement recorded by R. T. Elliott, Divisional Warden.

Amani then added a postscript –

I further wish to add that although this safari was called an anti-poaching safari I realise that it was in fact a poaching safari ordered by my superiors and can only think that it was done for their benefit. I also realise that much of the shooting done on this safari was in Controlled Area Block One of the Northern Division and not in the area of the Game Warden Nanyuki or the Central Division and the place where the elephants were shot is miles from any cultivations [sic] or humans.

Impala, graceful, handsome, the quintessential antelope.
(Photo Peter Davey)

Rodney received similar signed statements from every member of the Dol Dol Anti-Poaching Safari. Amani had left out a detail. While out hunting for trophies, the Assistant Warden in charge of the safari, came across two civilians who were also hunting illegally with two impala and a zebra in their Land Rover. With supreme irony, the Assistant Warden arrested them, took their rifles, placed an armed escort in their vehicle and ordered them to drive to Nanyuki police station. Arriving at Nanyuki the two men refused to go to the police station but instead drove directly to the game warden's house. Those following behind were unaware of this and dropped the confiscated rifles at the police station before proceeding to the Game Department. Arriving there they were dumbfounded to be told to go back to the Police Station, collect the rifles and bring them to the Game Department's armoury and then go back into the field to continue the Dol Dol safari.

It transpired that one of the captured rifles was in fact a Game Department weapon issued to the Nanyuki game warden personally! The two men arrested were working for him, unbeknown to the rest of the station staff. The man responsible for the trophy register produced further damning evidence. He confirmed the warden had said, "No trophy should be entered in the book without instructions from me ... that is an order." He described how ivory, leopard skins, zebra and antelope hides came in from all parts of the Nanyuki area but were not entered in the register. He and other members of the staff told how such trophies were either taken to Nairobi by the warden or collected by trophy dealers from Nairobi. This part of the Game Department had clearly abandoned any pretence of undertaking its duties and, instead, was committed to full time poaching. It was now obvious how the elephant ears had been obtained!

This outpouring of evidence caused a furore. The Chief Game Warden was enraged, but the alarm spread on right up beyond the Permanent Secretary in the Ministry above him. Ostensibly to 'inspect' the Nanyuki Station, the Assistant Minister for Tourism and Wildlife, J.M. Kariuki, personally visited Nanyuki where the warden was out on bail. He had never visited any station before and it was the only time in Elliott's career that an inspection by an Assistant Minister was made in this way. Outwardly proper and formal towards Elliott, Kariuki none-theless spoke at length to the station staff when Elliott was not present.

Had all possible charges been pressed, the Dol Dol Safari alone could have earned the Nanyuki warden sixteen years in jail and a fine of shs 79,000. Yet the CID did not charge him on all counts that could have been brought against him. Instead he was charged only with the destruction of evidence. The policemen argued, "Get him safely locked up on this charge and we can take our time in preparing the other cases against him and really stitch him up." At no time did the warden seem worried by the prospect of going to court. On the night before the trial he openly threw a party for the prosecution witnesses. In court next day, they all repudiated their statements to Elliott and the Police, saying that they had been coerced into making them. The defence lawyer asked for the case to be dismissed for lack of evidence, and the warden walked free. In the year 2000 he was a very senior civil servant.

No further action was taken over the Dol Dol Anti-poaching Safari: mainly because it was felt, probably correctly, that the witnesses would also not stand by their statements. With hindsight, it was a grave mistake not to have thrown the book at the Nanyuki warden for the Dol Dol safari at the outset when there were too many witnesses to have subverted satisfactorily. The failure of this case did immeasurable harm to departmental morale and was a seminal point in the organisation's fortunes. The old time wardens, of whom Rodney Elliott was the doyen, were clearly powerless to prevent members of the Department from using game as they wanted.

The Chief Game Warden then went on the attack. Accusing Elliott of disloyalty, criminal actions and failing to obey orders[99] he sought his instant dismissal. As this could have attracted public attention to corruption, an issue Government preferred to keep out of sight, in a letter headed *Station Inspection – Nanyuki Station*[100] the Permanent Secretary in the Ministry of Tourism & Wildlife ordered –

> *"I now direct that this correspondence be brought to an end."*

Government had heard enough of the Dol Dol Anti-poaching Safari. To placate Rodney the Permanent Secretary wrote to him personally[101] –

> *"I am pleased to inform you that you are cleared of the accusations contained in the correspondence under reference . . ."*

* * * *

Corruption in the Game Department did not involve just one warden or one station, and it was reasonable to suspect that the Ministry of Tourism and Wildlife was either a party

---

[99] Letter A. 1/3 of 20th July 1971 from Chief Game Warden to Elliott and copied to the Permanent Secretary.
[100] Letter 207/A/106 of 15th December 1971 Permanent Secretary to the Chief Game Warden and Elliott.
[101] Letter EST 704 of 1st July 1972 Permanent Secretary to Rodney Elliott.

to what was going on, or grossly incompetent in its supervisory duties. Firearms purchases should have had Police Central Firearms Bureau authorisation, but did not, flouting some of Government's most stringently enforced rules. That the Game Department got away with this suggested either immunity from normal regulation – i.e. support from very high in the Government – or Government's collapsing ability to enforce its own rules. It certainly confirmed Ministry of Tourism & Wildlife collusion with the Game Department because payments had to be authorised by and made through the Ministry.

Knowing this, the CID passed little information to the Ministry much to the discomfort of the latter's officials. Eventually the Permanent Secretary wrote to the Director of Criminal Investigation asking for information on corruption in the Game Department[102].

The response came two months later as a secret report to the Permanent Secretary[103] by the Director of Criminal Investigation, summarising much of what was on police files. So many Game Department activities were queried by the CID that it is not possible to present them all here.

There were administrative issues: for instance the Chief Game Warden forbade any of his officers to speak to the police without his specific permission. This was beyond his authority. Officers who had done so were threatened with disciplinary action. This was *prima facie* interference with police investigations. He used the Department's Investigations Officer, not to check on illicit trophy dealing as was his duty, but to obtain confidential reports on other members of the Department whose loyalty to the Chief was suspect.

The police knew that ivory was no longer being sold solely through the Mombasa auctions, but privately to Nairobi dealers for amounts well below the auction prices. When ammunition expended by the Department was related to animals claimed shot on control, it was running at twenty-two bullets per animal killed. The police concluded that more animals were being shot than were being reported and the undocumented trophies were being taken by wardens on their own account. Dealer's Permits to trade in game trophies were being handed out without checks. An Acting Chief Game Warden had admitted to using the Game Department aircraft for private hunting safaris. He also confessed to using Government transport, firearms and ammunition in his personal hunting. The Chief Game Warden categorically refused to account to the police for the sale of tusks in his own name.

Centre of another scam were cheap licences for forest elephants. Toward the end of the colonial era the Game Department wanted to spread the elephant hunting load from the savannas and into the forests. However, many people did not like hunting elephants in thick forest: it was altogether too exciting. Also, large tuskers were less common in the forests than they were in savannas. To induce hunters to overcome their disinclinations, the Game Department reduced the cost of a licence to shoot an elephant in forest from £75 to £25 and in addition allowed a person to take several such licences in a year. While the intent – to get more forest elephants shot – was innocent enough, the manner in which it was done was irregular. Instead of prescribing a new licence through the normal legal processes, the Department took a short cut and simply issued Chief Game Warden's

---

[102] Letter MTW 87/80/01 CONF. 207/A/117 of 17th February 1972 Permanent Secretary to Head of the CID.

[103] Report INQ/33/71/3 of 7th April 1972 Director of Criminal Investigation to Permanent Secretary.

FORM 3   ORIGINAL

REPUBLIC OF KENYA
THE WILD ANIMALS PROTECTION ACT
(Cap. 376, Laws of Kenya)

CHIEF GAME WARDEN'S PERMIT

(Issued under section 14 of the Wild Animals Protection Act)   N⁰ 1293

Name (BLOCK CAPITALS) MR. ~~G.~~ RATH I. PARKER
Address (BLOCK CAPITALS) Box 40658, NAIROBI
Date of issue 4/7/72 Date of expiry 3/7/73
Fee (if any) paid Free
This permit entitles the holder to Collect Natural History Specimens for National Museum of Nairobi (Birds and Small mammals) from all over Kenya

This permit is issued subject to the following conditions:—
(1) It may be cancelled without notice and without any reason being assigned.
(2) It is issued subject to the provisions of the Wild Animals Protection Act, and all amendments thereto unless otherwise expressly stated.
(3) This permit is not valid for any area under the control of the Kenya National Parks Trustees unless endorsed by or on behalf of the Trustees.

CHIEF GAME WARDEN

Signature of Holder

GPK 712—20 Bks.—3/69

This small slip of paper, the Chief Game Warden's Permit, was the legal loophole through which officials of the Game Department pillaged the resource they were supposed to protect. That illustrated was held by me as authority to collect scientific specimens for Kenya's National Museum: the use for which these permits were originally intended..

Permits for forest elephants in lieu. The opportunities the system offered were immediately picked up by the post independence wardens and reflected in a huge jump in the number of forest elephant licences issued to wardens. Some wardens never left their desks but hired Asian hunters to get their elephants for them illegally. And most were not shot in forests, but in the savannas.

At the heart of what was happening was the enormous power vested in the Chief Game Warden by the Wild Animals Protection Ordinance (WAPO) about which I commented in Chapter 7. In addition to the various powers of arrest etc. that impressed me when I joined the Department, Section 14 – Chief Game Warden's Permits – allowed the Chief Game Warden to authorise any person to kill any animal by any means (except in national parks and on private land) and for whatever price, if any, that he might fix. Section 15 authorised the Chief Game Warden to issue a permit to capture any animal: again the period of its validity, the conditions and cost were his to fix at his absolute discretion. Section 39 (3) of the Ordinance allowed him to waive Government's right to any game animal, trophy or meat at his absolute discretion. Section 55 allowed the Chief Game Warden to delegate or assign to any officer of the Game Department or to any honorary game warden (who need not be a civil servant) *"any of the powers or duties conferred upon him by this Ordinance."* The latter authority was perhaps the most astonishing of all as it allowed the Chief Game Warden to pass on to anyone of his choice all or any of the amazing powers the law conferred upon himself – including the issuance of Chief Game Warden's Permits. And this is precisely what the Chief Game Warden had done.

As the CID reported, he challenged them to find his signature on any Chief Game Warden's Permit. He had delegated this authority and, clearly conscious of the ramifications, kept his own name off them. That he knew this would lead to public outcry is implied by sale permits for his own personal trophies being made out in names other than those by which the public knew him. The police pointed out that as he had not broken any law, there was little point in prosecuting him. A naïve public was wrong in thinking that the Game Department was acting illegally. As the law stood, quite literally, the Chief Game Warden could legally do whatever he wanted to.

With hindsight, the most striking aspect of what Rodney Elliott and the police exposed is not what the wardens did, but the extraordinary game laws written during the colonial era. Earlier I observed that the public saw game wardens as men with special vocations and high moral stature. Nothing makes this point better than the game laws and their implicit assumption that as 'good' men, wardens would not abuse them.

Yet even though the police would not bring a legal case against the Chief Game Warden, those in the higher echelons of Government were well aware that the lax laws were being blatantly abused. They could have taken action. On grounds of maladministration he could have been sacked. Nevertheless, despite the international scandal of keeping him in office, that is where he stayed. Alone it was proof of protection in high places.

In 1971 and '72 Rodney Elliott's hand was not easy to detect behind police interest in the Game Department. Yet it was there as a critical element in attempting to pull the Department back up the slope it was descending at such pace. Perhaps Elliott's influence was most forcefully stated in the negative; when he left the Department in late 1972 the CID investigations also sputtered to a halt.

CHAPTER SEVENTEEN

DOUBLE-O SIX AND SEVEN-EIGHTHS

After leaving the Game Department in 1972, Rodney Elliott did a two year stint as a park warden before being employed by the owner of Solio Ranch – Court Parfet – as its game manager. In this position he observed the merging of former Game Department and National Park organisation into the one Wildlife Conservation & Management Department (WCMD), without being personally involved. From his numerous contacts in the conservation field he was aware that it made things worse for wild life – not better. In August 1978 President Kenyatta died and was succeeded by President Daniel Toroitich arap Moi. The 'godfathers' who had arisen under the Kenyatta regime dispersed and some felt that the moment was ripe to bring conservation under control.

* * *

In October 1978 President Moi visited Solio as Court Parfet's guest. Given Court's conservation interests, it was inevitable that they would discuss poaching. When they did Parfet made the President an offer: he would loan Rodney Elliott back to the Government to clean up the new Widlife Conservation & Management Department (WCMD) created by the merger of former Game Department and National Parks and whose head was now the Director. The President accepted the offer there and then, ordering the Commissioner of Police, Ben Gethi, to take the matter in hand. This was the beginning of what Rodney referred to in his personal notes as the 'Wildlife Rescue Operation'. And it is clear both in the literature and in verbal recollection that being catapulted from Court Parfet's gamekeeper to Presidential Wildlife Plenipotentiary caught him unprepared.

Rodney met with the Commissioner of Police on 21st October, 6th November, 28th December in 1978 and again on 12th January 1979, submitting several memoranda on the reorganisation the WCMD needed and the conditions under which he was prepared to work. Knowing how the head of the Game Department had been protected in the early 1970s, Rodney was dubious about having any greater success the second time around. He thus wanted assurance that –

> his recommendations would be accepted,
> he could withdraw at any time if he felt he was not progressing,
> providing that all conditions he laid down were agreed, he would start on the 1st March 1979
> it was agreed that expatriates recommended by Rodney would be engaged (to work in the WCMD) with the seniority he specified,
> he was provided with a Range Rover, and
> he was given acceptable housing in Nairobi.

These terms were agreed verbally, but never in writing. The Director and Elliott met in December 1978 with Court Parfet present. Meetings continued in early February

1979. The Director knew that Rodney's presidential backing was manifest through the police rather than the Minister and Ministry of Tourism and Wildlife. Yet, as will become apparent, he had less to fear from the police than Rodney at that stage believed. Although it was agreed that outwardly Rodney would be the Director's second-in-command, the terms he laid down reduced the top man's status to a mere figurehead. The power Rodney sought[104] was –

> *"that all my recommendations in relation to the reorganisation of the Department and all general executive decisions issued by me will be accepted and implemented promptly by the Director. My office to be situated immediately adjacent to that of the Director."*

To ensure that there was no misunderstanding he then bolstered this overall control with fifteen clauses on specific topics, the most important being that Elliott would be responsible for all hiring and firing of staff. He also wanted the Department reorganised by 1st July 1979 on the lines he recommended. New staffing grades had to be created immediately. No officers or men could be transferred without Elliott's approval in writing. He wanted a guarantee that funds would be available to supplement salaries and purchase new equipment. Further he demanded that no licences or permits or purchases of vehicles, aircraft, firearms and ammunition should be made without his assent.

Because he had never been a bureaucrat and had a poor grasp of how the bureaucracy worked, Rodney pasted himself into a cul-de-sac with these demands. Even with the best will in the world and a real wish to accede to the presidential wishes, many of these requests were beyond the Director's powers to either accept or implement. Creating new grades in the civil service was the Public Service Commission's prerogative and not a departmental head's. Providing funds for additional salaries or purchasing equipment was a Treasury responsibility. And unless there was a Presidential directive to bypass both Public Service Commission and Treasury, which there was not, he could only approach them through his Ministry which, at that point was in the dark over Elliott's position. Nevertheless the Director did not air these points at his meetings with Rodney, though he must have been aware of them. Very intelligent, he knew both how the Civil Service worked and how to use it, while Rodney's demands displayed his own lack of grasp in this realm.

The Director's command appears when two letters are compared. The first was Rodney's draft of a letter he wanted the Director to sign written on 12th February 1978. Its opening sentence was –

> *"Major R. T. Elliott has been appointed as my Personal Executive Assistant, and I have instructed him to take such measures as may be necessary through the respective Wardens ..."*

The version the Director signed[105] on 14th February read –

---

[104] Elliott R.T. *In litt.*

[105] Letter WCMD 4/2 of 14th February 1979 Director WCMD to Whom it May Concern.

> "*Major R. T. Elliott who is employed by Mr C. Parfet of Solio Ranch Limited, Box 2, Naro Moru has very kindly agreed to assist in putting forward suggestions ...*"

Between the two is a wealth of difference and the Director never conceded that Elliott was once more a member of his Department. On the one hand Elliott, the direct military man, assumed that orders from the President were all that was required for everyone to fall into line. On the other was an astute strategist who saw that, as issued, the presidential order could be endlessly and legitimately delayed.

Nowhere in the correspondence on the issue or the reports that Rodney wrote for the Commissioner of Police is it recognised that the root of the problems facing conservation in Kenya lay in basic conservation policy. By omission it was accepted that nothing was wrong and that to get things back on an even keel one only needed a Department governed by probity and discipline to put policy into effect. As already pointed out, the Game Department had never been truly successful. With a flawed basic policy – which Elliott did not at that time appreciate – nothing he might do would achieve much in the long run. This notwithstanding, it was strange that the Director was given any latitude to 'fail'. Elliott's rôle determined in the verbal Presidential order to the Commissioner of Police should have been passed on to the Director by the Commissioner as an order and not relayed to him by Elliott. At the same time, the Commissioner should also have passed on the President's order to the Director's superiors in the Ministry of Tourism & Wildlife so they, too, had to comply. The most basic flaw was assuming that an agreement over lunch could be transmuted into effect through a single command. By analogy, what was going wrong in Kenya's Government and its game policy was akin to an engine in which all moving parts needed replacing and was running on an inappropriate fuel. Inserting a new set of sparkplugs alone could not rectify such a general malaise. Parfet clearly underestimated what was required when he offered the new set of plugs in the form of Elliott.

It should never have been left to Rodney, a civilian with no official status, to negotiate what he wanted with the Director of WCMD. He was in an impossible position from the outset. This is so obvious that one must assume the purpose of the project was no more than to placate prominent and critical conservationists. The word would go out, 'Old Rodney is on the job,' and slacken off the criticism. Recall that Woodley was returned to Tsavo in 1978 and for the same reason.

Elliott's deadline of 1st March came and went without being met. Rodney had no contract of employment, as had been promised, no Range Rover, no driver, no office, no secretary, no rangers and no house. On 29th March 1979 he wrote the Director confirming his requirements, but by 6th April he had had a meeting with the Attorney-General, Charles Njonjo, at which Rodney stated that he could not work with the Director. His notes[106] sum his feelings –

> "*. . this arrangement is unworkable as it is being deliberately turned into a black/white confrontation by certain elements loyal* [to the Director] *and fearing my return . .*"

---

[106] Rodney Elliott's notes in the author's possession.

The connection with Njonjo was unofficial, but because he was one of the most powerful men in the land at that time, a friend of Court Parfet's and in addition personally interested in conservation, Rodney enlisted his support directly. At this point Rodney's brief to reorganise the Department was dropped. Instead he was more realistically appointed a Superintendent of Police in the Kenya Police Reserve and asked to investigate the WCMD from an office in CID headquarters. In that respect the Director had won again.

Yet Government had at long last run out of patience. Embarrassing conservation matters arose all too often over the Game Department. Without his Kenyatta era protectors, it was politic to dump the Director so his last victory over Elliott was Pyrrhic. Sent on indefinite leave in May 1979, he never came back to WCMD. If anyone brought this about, no matter how indirectly, it was Rodney Elliott.

* * * *

Within the CID where he worked, Rodney's immediate superior was Superintendent Musa. Above Musa was Assistant Commissioner of Police Joginder Singh Sokhi, who answered to Assistant Commissioner Jack Irwin, the Deputy Director of Criminal Investigation; above all was Ignatius Nderi, the Director of Criminal Investigation and head of the CID. Rodney's mentor in police procedure was Jack Irwin, a long-standing friend. Although placed five levels down from the top of the CID, through the presidential edict Rodney still officially retained direct personal access to the Commissioner of Police, Ben Gethi.

In the WCMD Rodney retained close contact with Daniel Sindiyo, the new Director whom he had known and liked since the 1960s. With some justification Rodney believed that his welfare depended on limiting the number of people aware of what he was doing and what he knew, so he was extremely secretive. It is not possible within the scope of this book to detail all evidence Rodney accumulated against the WCMD and a few cases must suffice to illustrate its extent.

First he went through the CID's earlier case files. Chief Game Warden's Permits (which changed their name to Director's Special Permits with the amalgamation of Game Department and National Parks in 1976) were worth examining. In fifteen months of 1970-1971 Chief Game Warden's Permits[107] had been given to thirty-one wardens to shoot eighty-two elephants in the highland forests only. The police established that this rule had been broken with elephants being shot far from the mountain forests. One warden's meticulously kept vehicle log showed he had been in western Kenya at the time that he claimed to have shot his two elephants in central Kenya. The evidence had been irrefutable but the police had not prosecuted him.

An executive officer – Mr Auko Daudi Mabinda – and a clerk – Mr Peter Benson McOdoyo – both purchased Chief Game Warden's Permits for forest elephants. They handed them to an Asian who shot the elephants for them, taking a commission for his services when the tusks were sold. Subsequently this hunter was caught by the Game Warden Malindi, David McCabe, having shot four elephants in the Coast Province. He produced Mabinda's and McOdoyo's forest elephant licences claiming that they covered

---

[107] Report INQ/33/71/3 of 7th April 1972. *Ibid.*

two of them. When it came to Mabinda's notice that a second elephant shot two hundred miles from the mountain forests was claimed on his licence, he went to the Chief Game Warden for guidance. Later he was to tell the police –

> "..I reported the matter to the Chief Game Warden. The Chief Game Warden's ruling was that license [sic] No 892 would remain valid and that I should take out a "Bush Elephant" [i.e. normal] license to cover the elephant that had been shot [by the Asian at the coast] ... The Chief Game Warden also ruled that this new bush elephant licence be issued in my name but paid for by the Asian."[108]

Considering what had gone wrong with the Department, this cluster of events was minor. Yet the Police had written confessions from both Mabinda and McOdoyo that they had not been present when an elephant was shot in their names. The case against them was straightforward. The Asian could have been charged with killing two elephants without a licence in the Coast Province – an offence punishable by six years in jail and/or fines of £1,250. The Chief Game Warden could have been charged with being an accessory after the fact contrary to Section 396 of the Penal Code. The evidence was incontrovertible but the CID had not prosecuted.

Elephant hunting was supposedly stopped in 1974. Proof that it continued was obtained when a German at the coast was arrested for having two tusks under his bed! At the time of his arrest he could not produce evidence that they were lawfully his. Subsequently he produced a permit of legal possession[109], on which it was stated that the elephant had been shot on a Chief Game Warden's permit of 17th March 1975 issued for one Bush Elephant[110]; both issued after elephant hunting had been banned.

The hunting ban in 1977 and the revocation of all dealer's permits in 1978 were supposed to have ended all hunting, capturing and trading wild animals or their parts and derivatives. Yet under the Director's special authorisation the Mount Kenya Game Ranch was given permits to capture 1,043 animals between 19th July 1977 and May 1979 which, at prevailing international zoo prices were worth £338,235. The list contained 20 black rhino, 30 bongo and 10 Hunter's antelope, species for which there was sufficient conservation concern to have made any capture or hunting a matter of intense public interest. There is no evidence that the animals allowed were in fact caught. There is no evidence of any impropriety in issuing the permissions, other than they occurred after the public had been informed that there was a ban on hunting and dealing in game animals and trophies. Where black rhino were concerned, international trade in them or their parts was banned under the Convention on International Trade in Endangered Species of Fauna and Flora (CITES) to which Kenya formally acceded in 1979, but to whose rules the country had paid lip-service since its inception.

It was publicised during the 1970s that the recipient of these permits had exported animals to Nigeria at the Kenya Government's request as part of an inter-governmental arrangement. It was assumed initially that the permission for him to trap after the ban was

---

108 Statement to Police in author's possession.
109 No. 899332 dated 1st April 1975.
110 Letter APU/HQ/CONF/0/21/50 of 9th November 1979 from E.C. Goss to Director WCMD.

part and parcel of this standing inter-government arrangement. Asked to detail shipments made on this order, the permit holder wrote on 16th June 1979[111] that there had been three: in 1973, '74 and '75. There was seemingly no connection between permits issued after the hunting ban and the Nigerian/Kenyan arrangement.

However, permit of possession 275756 issued on 30th June 1978, stamped 'export allowed' and endorsed on 4th August by the Central Bank of Kenya, relates to 100 crowned cranes and 25 colobus monkeys caught on a Capture Permit dated 5th October 1977 – i.e. issued after the hunting/capturing ban. These were destined for the United States, not Nigeria. There was other similar evidence and contrary to public perceptions and the Government's announcements, certain elements in the game capture business operated after the bans were announced. I say 'certain elements' because some of the best-known of the animal trappers had been categorically denied any permission to continue 'because of the presidential ban on hunting and capture'.

On the face of it, Director's Special Permits were still being used in the late 1970s just as Chief Game Warden's permits had been earlier in the decade. Elliott urgently needed to establish whether the Diretor had been acting on his own initiative or with sanction from higher authority. When Rodney sent CID men to seize the Director's Special Permit books to examine the folios of what had been issued, four books were inexplicably missing. The new Director could not explain the absence of documents that were supposed to be under his personal control, eroding confidence in him.

A new aspect of permits arose when the Game Department and National Parks merged. Until that point, permission to prospect for minerals in National Parks had been jealously guarded. In the late 1960s it was discovered that southern Tsavo West National Park was rich in semi-precious and precious gemstones. Permission to look for and mine them had to be obtained not only from the Mines and Geological Department but also from the Director of the WCMD. A spate of concessions was granted, among them one to the more prominent of those awarded Collector's Permits to recover ivory and rhino horn. Any normal entrant to a national park is bound to stay in his or her vehicle other than at designated places. Someone possessing a prospector's permit can of course leave the roads and move about on foot. Such licence would coincide neatly with the interests of someone looking for ivory and rhino horn or who might have a rendezvous with poachers.

On 26th July 1973 at Sagante in Marsabit District, police officers stopped vehicle registered KNH 947, found four elephant tusks in it and arrested the four men in the truck. It was an open and shut case in which all four initially admitted their guilt. If convicted they were all liable to be jailed for up to six years, and or be fined up to £1,250 each, and in addition, the vehicle should have been confiscated. However, on 1st August, the accused and their vehicle were released. This was because a woman – Field Marshal Muthoni – arrived, produced her Collector's Permit, claimed that it was valid all over Kenya and that the accused were in fact her agents[112]. As if it was necessary, here was proof that the Collector's Permits were being used as Ian Grimwood had feared and that the trophies involved had nothing to do with Mau Mau days in the highland forests of central Kenya.

---

[111] Letter dated June 16 1979 from Don Hunt to The Licensing Officer, WCMD.

[112] Letter dated 1st September 1973 from Junior Assistant Warden Stephen Kubai, Marsabit National Reserve, to Director of National Parks, Nairobi.

The value of these miscellaneous records was not so much as cases to prosecute, as in sketching in the detail of what had been going on and what Rodney was up against. The public perceived the WCMD as a Department, which had many good men in it, but which was tainted by a few 'bad eggs'. What he found was an organisation far more deeply corrupted than the most sanguine of the public had imagined. Wardens had stolen petty cash from their imprest accounts, stolen National Park gate receipts, obtained goods on false pretences, embezzled their scouts' and rangers' salaries, misappropriated public funds to buy private houses in Nairobi, and sold a Departmental vehicle, as well as poached and trafficked in trophies. It was as though if a scam was possible, it had been done. Much was so brazen that one can only laugh.

Despite lipservice to the principle of not trading rhino products, Kenya's Wildlife Conservation & Management Department nevertheless undertook such trade.
(Photo Peter Davey)

Such an instance concerned a United Nations vehicle. UN's Food and Agricultural Organisation (FAO) had spent millions of dollars on a programme[113] to develop legitimate wildlife use. As always, the project was run by imported experts equipped with expensive vehicles displaying red, diplomatic number plates. When such a programme was over, the vehicles were usually given to the host Government Department most closely connected with the project. In this case its vehicles had been given to the Game Department which should then have taken them on its Government register and replaced the red diplomatic registration plates with Kenya Government plates. Thus Toyota Land Cruiser 40UN292K should have become GK267B. Lax, the local UN authorities never asked for the return of their red plates and the GK plates were simply held in storage. For the next three or so years the Game Department men had their own 'diplomatic' vehicle, ran it on government fuel and because it had red plates could pass anywhere without being searched. It would have been particularly useful in carrying rhino horn and small quantities of ivory, or as an escort for trucks with larger shipments.

Three other cases further illustrate the situation. The first I shall call the Semboi file. Charles Harris, a Laikipia farmer, was surveying a dam site in an area of thick bush where several rhinos lived. They were a danger to the survey party, so Harris asked for and was given Game Scout Corporal Semboi Nairangai as an escort. As things turned out, it was the rhinos that needed protection from Semboi, for he shot one for its horn, was caught red-handed by Charles Harris, prosecuted and went to jail. Harris knew that there had been more to this case than ever came out and consequently Rodney made a point of looking for the man who had, by then, served his sentence. With little love for the wardens under whom he had served, Semboi willingly made a statement on his activities when in the Department.

While serving at the Kinna outpost in Meru he shot elephants as well as zebra for the Meru warden. On this warden's instruction he guided Indians who also came to take zebras in greater numbers than they had licences for and was ordered not to make them sign the

---

[113] UNDP/FAO Project Ken:71/526.

hunters' register when they entered and exited the area. While at Kinna he also shot zebra for the Divisional Warden at Nyeri who sent his official Land Rover and driver to take Semboi hunting for them. Transferred to Muringato, closer to Nyeri, he was detailed to shoot animals for his new warden and told not to worry as this warden was related to President Kenyatta. He shot three rhino, a leopard and four elephants on this basis – two of the latter for the President, so he was told.

When Semboi voiced doubts about what he was doing, his commanding warden threatened him with dismissal if he talked about it. Later he withheld Semboi's salary for six months, the better to control him. At the end of the six months this warden told Semboi to go and shoot rhino and to keep quiet and, if he did so, he would start paying Semboi's salary once more. The rhino he shot on Harris's land was such a rhino. Ironically for Semboi, because he got caught he could not receive his withheld salary. Semboi gave signed statements[114] and the names of others that could testify against the wardens who had made them hunt illegally. They were contacted and corroborated Semboi's story. The Dol Dol Anti-Poaching Safari was clearly not an exception but an example of what had been going on routinely.

The second case commenced when Rodney Elliott read *Daily Nation* number 5,491, Saturday June 10th 1978, which had an article on a major seizure of ivory and game trophies in Nairobi's Parklands suburb where Asians own most of the houses. A photograph showed a parcel ready for dispatch when it was seized. It was addressed to K. N. Deitrich, Berlwerstrasse 10, Postfach 349, 6750 Kaiserslautern, West Germany. Rodney got Jack Irwin to ask the German Police for information on the address. They complied and seized a parcel arriving at the address in Germany, which contained rhino horns. They returned the packing to the CID in Nairobi so that the Kenya Police could determine the post office of origin. At this point a penny dropped: the post was an obvious way to get contraband out of Kenya.

A record of all parcels sent overseas was kept centrally in Nairobi regardless of which of the City's many post offices they are despatched from. This was seized by the CID and Rodney struck gold. K. N. Deitrich of Kaiserslautern was a regular destination for heavy airmail parcels, most of which came from Nairobi's Desai Road Post Office. The senders always gave one of two post office box numbers – 30529 and 40135 – in Nairobi. The former was rented by one of the biggest game trophy dealers in town. The second box renter was not identified. Most of the parcels were delivered to the post office by one man – Joseph Jackson Muasya Kingele. Interestingly, he had once been a postal worker himself, and no doubt not only knew the system, but also people behind the counters and was well placed to dampen curiosity about such regular business.

Between 12th January 1978 and 31st May 1978, 636 air parcels weighing 6,320·15 kilos and averaging 9·9 kilos had been sent to K. N. Deitrich. Two hundred and eighty-two of the parcels had been sent from Nairobi's Desai Road Post Office. The rest had been spread about the post offices elsewhere in town and it is a reasonable assumption that this was done to reduce the risks of attracting attention. Once the pattern was established the central parcel register was examined for evidence of similar systems and several emerged. Of these the largest was always addressed to Transatlantic Speditions GmbH & Co. KG,

---

[114] In the author's possession.

This photograph in the Daily Nation of June 10th 1978, led to finding that the principal means of moving rhino horn internationally was by post, of which none of conservation's experts knew.

Neptunestrasse 18-22, Postfach 100905, 2800 Bremen, West Germany and was always sent by Hassaco Limited, Kampala via Nairobi. The name was fictitious, the senders obviously hiding their identity. Between 2nd January and 24th March 1979 Speditions received 180 parcels from Hassaco weighing 1,605 kilos and averaging 8·9 kilos.

The parcels characteristically were close to the maximum acceptable weight for parcels (10 kilos) and their contents must have been valuable enough to warrant the high airmail costs. Curio shops regularly airmailed ebony carvings to tourists' home addresses in this manner, but such wooden carvings were characterised by variable weights. Further with tourist orders it would have been rare for the same address to appear twice. Consistently high weights and the parcels' common German destinations were in keeping with the contents being ivory, rhino horn and other contraband.

This information was forwar-ded to Germany which resulted in several successful hits, not only where ivory and rhino horn were concerned, but other illicit commodities too. The *Daily Nation* photograph of 10th June 1978 paid handsome dividends.

The third case again started fortuitously. Rodney happened to go up to the WCMD offices where trophies were stored and saw officials busy packing tusks for export. This seemed irregular and he took action. From his CID offices he rang the Attorney General and told him of what he had just seen. Charles Njonjo ordered that the ivory be seized, taken out of WCMD hands entirely and stored at the Police General Service Unit headquarters pending completion of Elliott's enquiries.

At this point the CID called me as an expert in ivory and the trade, asking if I would value the seized tusks. I examined the whole consignment, tusk by tusk and rhino horn by

The Wildlife Conservation & Management Department started issuing itself permits to export ivory and rhino horn. I valued the ivory and rhino horn covered by this permit, for the police and estimated that it was being exported for not more than 30% of its true value.

rhino horn. As suspected, the ivory and rhino horn in the GSU store had been sold for 32% of its worth. The five lots of ivory (and some rhino horns) had been sold for $63,523 by 'secret auction' arranged by the Minister of Tourism and Wildlife, Mr Mathews Ogutu. Their real worth had been at least $197,283. We knew that both Minister and game wardens concerned were well informed about international ivory prices, so could not claim ignorance as ground for having accepted only 32% of value. We assumed that the difference between what was paid and the ivory's real value ($133,760.57) had gone into private accounts.

Elliott had more than enough ground to cancel the ivory and rhino horn sales. Expressions on WCMD faces were gloomy when we left the GSU headquarters. Certain that the ivory was safe at the GSU, Rodney set about further investigation. Several days later he saw a Ministry of Tourism & Wildlife official whom he knew coming out of the Director of the CID's office. The man looked rather pleased and Rodney wondered why. Two days later a junior CID Officer quietly tipped him off, at that moment the ivory held at the GSU base was being repacked for immediate shipment out of Kenya.

Rodney went to Musa's office; he was out. He went to Sokhi's office but he, too, was out. Jack Irwin was away at the coast on leave. Finally he went to the boss of the CID's office to find he, too, was not on the premises. Having climbed the CID chain of command without success, Rodney telephoned the Attorney General. Njonjo gave a curt order: "Get out there and stop it." Suggesting that, in the circumstances, his sole presence might not be effective, Rodney was told to go and that while he was on his way Njonjo would arrange back-up.

Arriving at the GSU store Rodney found several WCMD men sewing the tusks back into the sacking from which I had removed them some days earlier. Also present were two of Musa's CID officers who had worked with Rodney. He knew immediately then that the CID itself was colluding with the WCMD to bring his investigation to nought. Nevertheless he told the GSU officer in charge of the stores that the process should stop as an order was en route to reverse it. Within five minutes a message was delivered from the officer commanding the GSU ordering that all the ivory be returned into the store and all WCMD and CID personnel were to leave GSU premises.

Next morning Elliott was called to the Director of CID's office. Arriving he found him and Sokhi waiting him for him, the former very angry indeed. He accused Rodney of acting outside his authority and going over his head. The charge was rebutted and Rodney asked the Director to confirm with his Secretary that he had tried to contact each of the men under whom he served – in turn Musa, Sokhi, Irwin and Nderi himself. This only increased Nderi's rage and he ordered Elliott to remove his belongings from CID premises and be gone within twenty-four hours. He obeyed the order, removing all his files except those locked in Jack Irwin's safe and was gone. Reasons why the CID had not prosecuted so many clear-cut cases now seemed self-evident.

Five days later Jack Irwin sent a message, "Come back, all is forgiven." Rodney returned to see his erstwhile colleagues and agreed to go on working with them, but declined to reoccupy his office in the CID headquarters. Given what he now knew, he felt it would be healthier to keep a substantial distance between himself and the CID. Understandably, his hopes and ambitions were much diminished.

One challenge remained: to establish the degree to which the Minister of Tourism & Wildlife himself was party to what was going on in the WCMD. This did not take long. The Secretary General of CITES notified the Kenya Government of anomalies concerning 4,068 kilos of ivory imported into Germany[115]. Following protocol, the query was sent through Kenya's Ministry of Foreign Affairs to the Ministry of Tourism and Wildlife, but was also well aired in both press and conservation circles.

As a first step Rodney wanted to inspect the relevant export permit books. These should have been in the hands the WCMD Licensing Officer. He did not have them and had given them to the headquarters' telephone switchboard operator. Cleaners excepted, this gentleman was the Department's most junior employee. Amazed, Elliott interviewed the telephonist, who had nothing to hide. The Minister, the Honourable Mathews Ogutu, had called the switchboard man and asked him to bring the relevant books to him personally at the Ministry. This he had done and had kept a slip of paper on which he had written down the numbers of all permits individually, which he then handed to Rodney. Why the telephone operator should have been selected was never apparent, though if he came from the same village as the Minister it would have made sense in an African context. Nevertheless Rodney now had clear proof of the Minister's personal involvement. Yet, despite having the duplicates of the export folios in his personal possession Ogutu told the Press[116] on 17th June 1979 –

> *". . Our records show that the consignment which is being held in Bremen was not exported by this ministry."*

---

[115] Letter SV 0832 B/KF-2 of 25th April 1979 from Secretary General, CITES to Hon. M. J. Ogutu, Minister for Tourism & Wildlife, Nairobi.
[116] *Nairobi Times* of Sunday 17th June 1979.

With his evidence of the Minister's involvement, Rodney went to the Commissioner of Police. He had earlier been promised that the Government would prosecute anyone shown to have broken the law, irrespective of how senior he might be. Now he felt he should let those above him be aware that he had a Minister in his sights. Gethi heard him out and said, "That is good evidence; don't lose it," yet gave no directions about what to do next. This was left for Elliott to arrange with his CID colleagues.

The ivory held in Germany was not the only export in which the Minister had taken a direct hand. Warden Noor Abdi Ogle of the WCMD testified in writing that he had been ordered to supervise the export of 3,128 kilos of ivory through Jomo Kenyatta International Airport. It had belonged to a Mr Sardeny, but had been carried for him in WCMD transport from Mombasa. Ogle personally saw the –

> *"letter of authority conveying the minister's permission, allowing the firm to export the trophies".*

At CID headquarters it was decided[117]–

> *"... to obtain a full explanation from the Minister the Hon. Mathews Ogutu as to his activities in the matter of the West German ivory and his apparent intent in removing documents in connection with it and failing to place the matter in the hands of the C.I.D. at the time."*

Ogutu had told the press that the matter of the illegal tusks in Germany was in the CID's hands. In fact neither he nor anyone in his Ministry had contacted the CID on the matter, or passed any information to it or requested police help. Ogutu did this to divert the intense press interest away from himself onto the CID who were not amused. This explains the somewhat hostile attitude so at odds with the obvious collusion between Ministry and police over the GSU ivory. An official interview would help them get their own back. Rodney assisted with a list of questions for the Minister who was then interviewed by Sokhi and Musa and asked to make a signed statement[118], which he did. In paragraph three the Minister said –

> *". . there is a clause under Section 45 [of the new 1976 game laws], the Minister has absolute power to permit export of game trophies within [sic] assigning any reason therefore, in accordance with the condition of the permit issued by the Minister authorising such export. It also empowers me, as the Minister to refuse any export of game trophies within my discretion without assigning any reason."*

Minister Mathews Ogutu was putting over the same point as that which the Chief Game Warden had handled so effectively over the years: 'as Minister I can empower anyone to

---

[117] Internal CID loose minute dated 12th September 1979 by J. D. Irwin D/D. C. I.
[118] Written statement dated 19th September 1979, 10.25 a.m. at CID HQ Nairobi signed by the Minister for Tourism and Wildlife, the Hon. Mathews Ogutu M.P.

do anything where wildlife is concerned, therefore what I have allowed cannot be illegal.' Nevertheless Sohki and Musa guided him into admitting that no ivory could leave Kenya as per Legal Notice 57 of 23rd March 1976. He had therefore contravened this in the shipment witnessed by Warden Ogle and in the consignment seized in Germany in 1979.

One can read between the lines of his statement that as the interview proceeded and the focus of the CID questions became sharper, he decided to deny everything. He denied receiving the crucial documents delivered to him by the WCMD telephonist. On the contrary he said that they had been seized by the police. On 5th October 1979 Ogutu wrote to the CID[119] sending legal notices, a statement that there had been no trapping or export of live rhinos and copies of his New York monthly bank statements from 3rd January 1979 to 5th September 1979 to show no unexplained payments had been made into his account. Under Kenya's exchange control laws of the time, holding overseas accounts was only possible with special dispensation. It was not something that politicians readily admitted and Ogutu's actions suggested panic.

The statement led Elliott to draw up a further list of questions to ask Minister Ogutu, casting the net much wider. They were never asked, however, for the issue was becoming too hot politically. Once it may have looked as though reshaping the WCMD was all that was required. Now, with the CID questioning a Minister, it was obvious that far more was involved than originally anticipated during that visit to Solio in October 1978. The President had to count the political cost of letting a Minister be prosecuted. While it might make a big splash with international conservationists, they did not vote in Kenya. Of greater import was the possibility that Ogutu's Jaluo people in western Kenya would see his prosecution as an attack upon themselves and influence their relationship to the regime.

By this stage Rodney Elliott knew that his chances of getting anywhere were nil. The Director of WCMD had been changed, he had proved CID collusion in an illegal attempt to export ivory and got the CID to interview the Minister of Tourism and Wildlife. To have got this far was extraordinary in itself. He had gone as far as he could go and called it quits. So he drafted a final report to the Commissioner submitting it on 26th November 1979[120].

Rodney concluded that the WCMD was beyond repair. The only solution was a new organisation with a new staff. In less than a year, of seventy-seven wardens in the Department he had put no less than forty-seven (61%) under investigation for suspected offences. The balance of thirty were not necessarily 'clean' either; he had just not got round to looking into their activities. Appendix A of the final report headed 'Details of Direct Sales and Undersale *[sic]* of Ivory and Rhino Horn' listed a sample of WCMD sales, showing that ivory was sold, on average, for 35% of its market value and rhino horn for 13%.

It also presented some trophy buyers' names together with the relevant official receipt numbers. These included the boss of the Central Bank of Kenya who bought tusks for 54% of market price; senior officers of the CID who respectively got tusks at 7% and 8% of market price, the Minister of Tourism & Wildlife who paid 36% of what the tusks were worth, four judges who respectively paid between 4% and 36% of value. Another

---

[119] Letter MJO/MTW/S/1 of 5th October 1979 headed Secret to the Director of CID from M. J. Ogutu and copied to the Attorney-General.
[120] Elliott R.T., 1979. *Report on my Special Confidential Enquiries Carried out in Accordance with CP/MISC/74-89/VOL. 11/71: to the Commissioner of Police.* Typescript.

purchaser was the Chief Justice himself. The Deputy Public Prosecutor also got ivory for 36% of value. All these purchases were made in 1978. The sales to the members of the CID were made on 11th March 1978, two days before the revocation of dealer's permits. The Minister, the three Justices and the Deputy Public Prosecutor bought on 20th April 1978 after trading in game trophies was banned.

These purchases were of course quite legal because they were sanctioned by the Director of the WCMD, but given the stature of the men concerned, many minds would, rightly or wrongly, suspect their status influenced both the decision to sell and the prices. Purchases by senior members of the judiciary after a ban on sales was, to put it delicately, injudicious.

Finally, Rodney observed that the pattern of illegal trade in game trophies had not changed much from colonial times. The primary producers were Africans. The marketing was overwhelmingly, but not exclusively, in the hands of Asians. And as in colonial times, there were many cases of illegal dealing, but few ended in court, let alone in convictions.

Elliott went to the Commissioner of Police, told him that the project had ended and returned to CID headquarters out of courtesy to thank the various officers who had helped him. ACP Sokhi said that he had 'never thought it (the project) would work' but everyone was very friendly (and clearly relieved).

No other person in conservation in Kenya ever went to the lengths Rodney Elliott did in trying to redress the mess that overtook the erstwhile Game Department and WCMD, or had taken the risks that he took in doing so. Some laughed at his secretiveness and one wag nicknamed him '006 $^{7}/_{8}$'. Others held that he tilted at windmills, which is probably true. Yet no-one did more to establish just what was happening or enquired so deeply into corruption in the country's power structure. It was epic.

Kenya's Chief Justice Millar obtained substantial discounts for trophies he acquired from the Wildlife Conservation & Management Department.

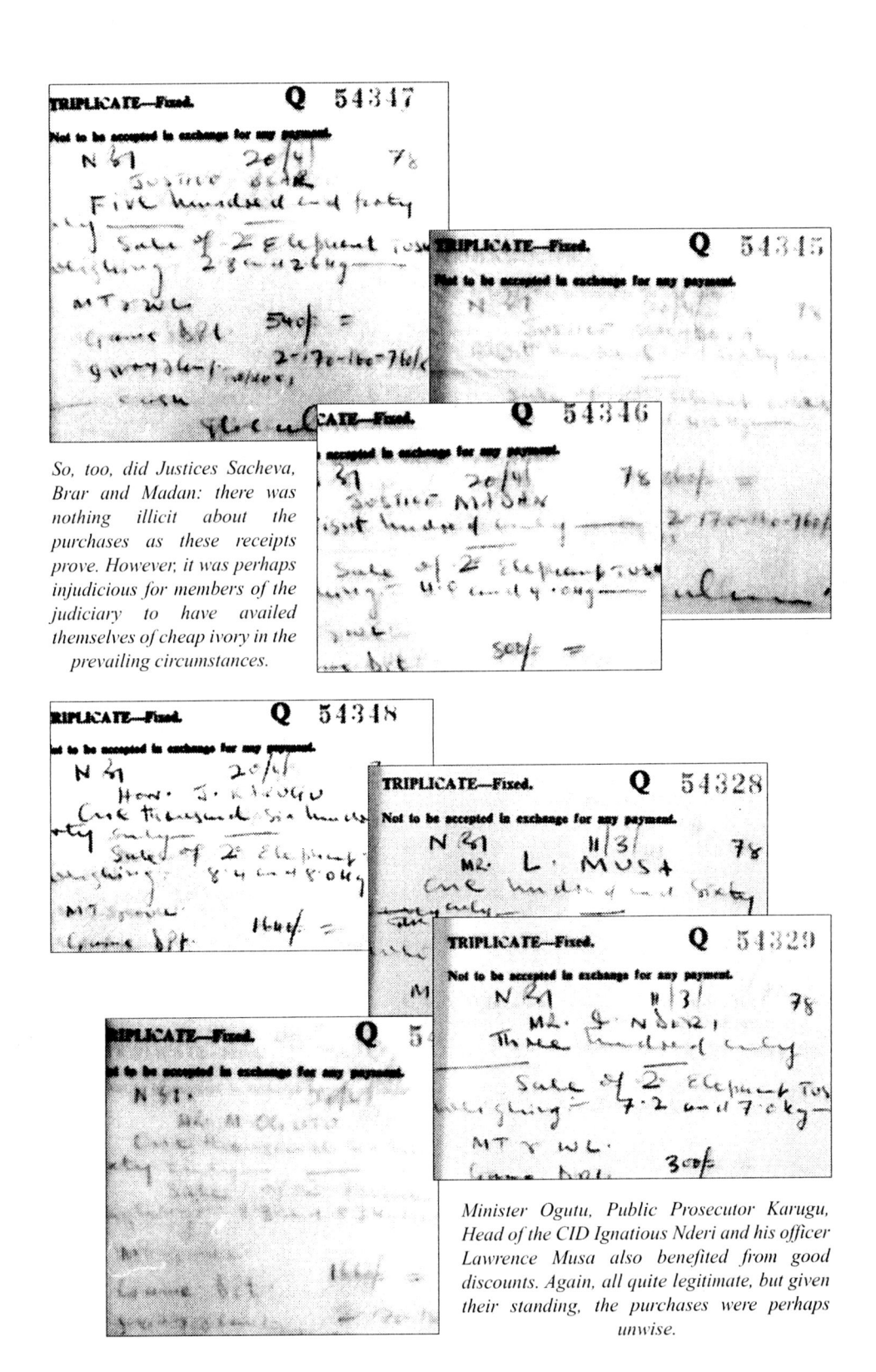

So, too, did Justices Sacheva, Brar and Madan: there was nothing illicit about the purchases as these receipts prove. However, it was perhaps injudicious for members of the judiciary to have availed themselves of cheap ivory in the prevailing circumstances.

Minister Ogutu, Public Prosecutor Karugu, Head of the CID Ignatious Nderi and his officer Lawrence Musa also benefited from good discounts. Again, all quite legitimate, but given their standing, the purchases were perhaps unwise.

Black-backed jackal: in the Kalahari winter these animals develop a heavy pelage that was in high demand for karosses (a blanket of skins stitched together)
(Photo Peter Davey)

Genet skin karosses were an item in consistently high demand and a source of substantial income for the Bushmen hunters who trapped them.
(Photo Peter Davey)

CHAPTER EIGHTEEN

THE FIRST IVORY CONSULTANCY

The events in Kenya that so drastically influenced the lives of men in conservation like Woodley, Sheldrick and Elliott were not so critical for me. Our private firm was able to operate internationally and rough times in Kenya could, to a degree, be offset by taking work in other countries. This also gave us a continental perspective as the imperial skin was sloughed off Africa through the 1960s. Over the years I undertook consultancies in fourteen African states. Among the more interesting of them were those to do with my personal albatross – ivory.

* * * *

Peter Becker was a professional hunter and a farmer when I first met him in 1957. Later he brought clients to the Galana area and over the years we built a firm friendship. He foresaw difficulties living on in Kenya after independence and in 1965 moved with his family to Botswana, then the least developed of all Anglophone countries in Africa. It offered ample scope for hunting and wildlife based tourism. Many hunters from Kenya made the same move. Indeed they not only took themselves thence, but their Kenya safari crews as well. And they went in such force that as Batswana were drawn into the industry many learned to speak Kiswahili. To this day there are men in Botswana who can speak this East African language.

Becker saw the flourishing safari industry would need a taxidermy service. Converting an old creamery into a tannery and engaging a taxidermist, he started Botswana Game Industries (Pty) Ltd (known from the outset as BGI). In no time he was also buying curios and handicrafts from the local people. I visited him in 1968, became the company's consultant on conservation matters, was a shareholder until 1978 when I had to sell my shares to avoid a conflict of interests, and often visited Botswana over the next thirty years.

Unlike in Kenya, where Africans could not hunt, the Batswana were allowed to and could sell trophies from any animals lawfully taken. BGI bought them and was soon running a substantial game skin tannery. Initially the quality of hides and skins brought to the Company was poor with over seventy per cent proving worthless once tanned. Peter sent teams to the four corners of Botswana to instruct local hunters in trophy preparation and it was not long before rejects fell to less than five per cent of material presented for sale. Every trophy had to be accompanied by the relevant licence upon which it was obtained and the policy of 'no licence, no sale' soon reflected greater adherence to the game laws.

By the early 1980s BGI was the biggest wildlife-products company in Africa and possibly anywhere. It was Botswana's biggest company outside the cattle and mining industries, had a turnover of around $13 million annually, bringing substantial income to some 5,000 rural Batswana, mainly through buying trophies and curios from them.

As the Batswana were allowed to shoot elephants Peter also became an ivory buyer. He knew a little about the commodity from Kenya days because, as had so many of us, he had

The sheer quantity of tusks moving in trade, each year exceeding the total number of elephants conservationists thought existed, on its own, proved that ideas of how many elephants there were in Africa had to be revised upward radically.

shot elephants on licence and sold their tusks. Yet he did not know enough to trade effectively and turned to me. Would I describe the East African ivory trade for him? It was my first consultancy to do solely with ivory.

* * * *

Of the East African Game Departments, only Uganda's had published annual records of its ivory production in any detail and even that stopped after independence. In Tanganyika details could have been extracted from such records as they had kept, but it would have entailed several months of archival research. Those records of Kenya Game Department's that had not been destroyed were abysmal. There was some information on elephants taken on licence, there was a little on what the Department's own control work had produced, though even this was incomplete, but there was nothing in the Department (and never had been, even before the archives were burned) which indicated production by district.

I had assumed that the only available facts would be in the Game Departments and that I would have to dig for them. This was in 1970 and, aware that all was not well in the Kenya Game Department, I anticipated reluctance to have me going through the records. In the Ivory Room in Mombasa I collected complete auction records for 1959 to 1969. Yet while these produced information on the ivory offered for sale, they gave little on the trade overall. Warden Robin Pereira in charge of the Ivory Room thought that the Customs authorities might be able to help.

Taking up the suggestion I struck gold. Year in and year out the Customs Department had published the volume and value of all ivory imported into and exported from Kenya, with amounts, origins, destinations and prices. The system was not peculiar to Kenya, or East Africa for that matter. Wherever I looked, Customs records gave all the basic information required to describe the international ivory trade. No other wildlife product was so thoroughly documented except, perhaps, for the Hudson Bay Company's fur business in Canada. As no other evidence, this revealed the commercial importance of ivory in African history. Funds limited what I could do so I concentrated on the East African Customs records between 1925 and 1970.

More ivory was being exported in 1970 than in 1925 and, except during the war years, the upward trend had persisted across this entire period. Every game or park warden to whom I mentioned this was as astonished as I had been, illustrating just how out of touch with reality we were. All of us were aware how much of Kenya in the 1920s had been occupied by elephants and how much less they occupied in 1970. There had to be far fewer elephants in 1970 than there had been in 1920 and from this indisputable fact we assumed that ivory production would have followed suit and fallen. That the reverse was true amazed us. The volume of ivory produced annually from Kenya had risen 105%, from Uganda by 133% and from Tanganyika by 923%. Incontrovertibly, the rate at which elephant were being killed had been rising between 1925 and 1970. The eventual outcome of more and more elephant taken annually off less and less land was self-evident.

A second point was that while the rise in the volume of ivory produced over the forty-five years had been more or less constant from year to year, the price of ivory had fallen 68% in the Great Depression and had only recovered its 1925 value in the late 1960s. In real terms the 1970 value was still far below the 1925 value. A rising price had certainly not driven increasing ivory production. Possibly the reverse was true: the lower prices had driven a rise in production and more elephants were having to be killed to generate the same income. This was, however, unlikely as while government Treasuries had always welcomed ivory revenue, no government ever set annual targets for how much ivory should be produced.

An obvious possibility was that rising ivory production was the incidental product of rising conflict between humans and elephants. People were increasing steadily and moving into elephant ranges with ever increasing interaction between them. Obviously there would come a point at which there would be no more elephants and in my report[121] I wrote –

> *"There have been marked contractions of elephant ranges generally, a decrease in overall numbers and a high rate of ivory production. This continues today and the state of the populations as evidenced by the type of ivory produced as well as the other biological evidence suggest that elephants and ivory production will decrease rapidly in the coming decade."*

Events since 1970 bore this out in both East Africa and the continent generally. At the time though, no notice was taken of it. In the same report I estimated that the number of elephants in Kenya as 'at least 100,000,' in Uganda as between '40,000 and 50,000, but might be lower,' and for Tanganyika as 'possibly five times as many [as Kenya], giving projections of 200,000 and 500,000.' There had to have been at least this many elephants to have produced and be producing the volumes of ivory that the Customs records showed exported. From the auction records in Mombasa and Dar-es-Salaam we knew average tusk weights and numbers, so arriving at the number of elephants represented by the Customs figures was simple. Con-servationist reaction to the indisputable evidence was oddly

---

[121] Parker I.S.C., 1970. *The East African Ivory Trade 1925-1969*. Mimeographed report for Botswana Game Industries (Pty) Ltd.

hostile. For all that the evidence was published statistics, they asserted that my estimates of elephant numbers were wildly overstated.

Subsequently I took my data to Murray Watson, who, though controversial, was an original thinker and a competent statistician. Combining our various aerial census results from Kenya and Uganda we postulated relationships between elephant densities, rainfall and human densities from which we suggested that Kenya's elephant population at the end of 1969 was 169,000, Uganda's of the order of 40,000 and Tanganyika's as around one million. Again the evidence was ridiculed when submitted for publication. While the first to admit that our estimates were speculative to a degree, they were the best we (or anyone else for that matter) could do with the limited data to hand. Yet when a benchmark was later needed from which to measure decline in Kenya's elephant numbers, our estimate of 169,000 for Kenya was accepted without question. The volume of ivory exported over the next twenty years proved that, if anything, we had underestimated the number of elephants that had been alive in all three East African countries.

Whether game wardens and conservationists believed the results was of little consequence at the time. My client, BGI, was satisfied that they gave a comprehensive insight into the ivory trade. Ironically, however, the company did not act on my advice. Predicting that in the short term both ivory production and price would rise (an accurate prediction!) and knowing that ivory was available cheaply in Southern Africa (mainly because the sellers were ignorant of the market), I recommended BGI sold only enough to cover its cash outlays and stockpiled the balance. After a decade, whatever had been hoarded would be worth many times what it had cost and the profit run to millions of dollars. Instead my clients pursued a hand-to-mouth policy which, while still profitable, threw away the grand commercial opportunity.

* * * *

In 1972 the UN's Food and Agriculture Organisation contracted me as a consultant to comment on an elephant and hippo culling programme they had been managing in Zambia's Luangwa Valley. I was very critical. Elephants were supposed to have been taken as we had taken them in Uganda: whole herds shot in seconds (our average was forty seconds for a herd averaging fifteen). FAO said that whole herds had been taken, but the evidence suggested otherwise. Dropping a herd in seconds called for elephants to be shot through the brain. In the Luangwa Valley boneyard the number of elephant pelvises displaying bullet holes told of animals being shot running away and not when clumped. I found this when enquiring about the very low average tusk weight produced by the programme. Our East African culls had averaged over three kilos per tusk; in Luangwa it was half this weight. This needed explaining as the research biologist attached to the project, John Hanks, had reported that while tusk size and growth in the Luangwa Valley[122] was less than that reported from East Africa by Dick Laws[123], the difference was marginal.

---

[122] Hanks J., 1972. *Growth of the African elephant (Loxodonta africana).* East African Wildlife Journal 10 No. 4.

[123] Laws R.M., 1966. *Age criteria for the African elephant (Loxodonta africana.)* East African Wildlife Journal 4, 1-37.

Rather than cut tusks out of elephant skulls while fresh, the managers had opted for the lazier course of dumping skulls and ivory in the boneyard and drawing tusks from their sockets after the retaining tissues had decomposed days or weeks later. Unsurprisingly, given the value of ivory, when workers went to collect tusks they routinely found that others had beaten them to it. The abattoir manager conceded that many tusks had been stolen. This smelt like rat. If tusks had been stolen once, surely measures should have been taken to either remove the tusks earlier or to place a guard on the boneyard?

Zambian wildlife authorities and FAO had publicised the Luangwa Valley cropping as successful. This was wrong. The rationale for culling was confused. One school said it was necessary to protect the habitat, another that wildlife should be used to benefit local people and, in this case, to relieve the poor valley people of chronic protein deficiency. With hindsight it seems that it started for no better reason than that it was fashionable to cull. Initially meat from culled animals was smoked and dried and made available for the local Bisa people at a nominal price. Surprisingly, they did not avail themselves of this wonderfully cheap meat.

Again with hindsight this is easily explained. First the Bisa had little cash, for theirs was a subsistence economy. Second, the average person would have had to walk between ten and fifty kilometres from their homes to buy the smoked Game Department meat. Third, even when smoked, the meat had a short 'shelf life'. Put these factors together: was it reasonable to expect rural peasants to spend money – of which they had little – to make round trips of between twenty and 100 kilometres to buy a perishable product? If they bought sufficient to warrant the long walk – had they been able to afford it – they would then have had to store what they could not eat immediately. Without refrigerators and the modern conveniences that we take for granted, food storage is a problem. Even dried meat is vulnerable to insect pests. Simple economics would have been reason enough for the locals not to buy what the Zambian Game Department offered them.

There was, however, a more compelling reason. The Bisa were not protein deficient. An American anthropologist, Stuart Marks, lived among them and published a book[124] on his work. He found that they ate nearly as much meat annually as the Argentinians who, arguably, are the world's greatest meat eaters. The Bisa hunted for themselves and, in the Luangwa Valley with its abundant wildlife, were never short of meat. There was no market for Game Department produce.

When FAO arrived to help the Zambian Government make better use of its wildlife, the rationale for the Luangwa culling went unchallenged. Accepting that there was no market in the valley itself, the meat should go to urban centres: specifically the Copper Belt where there were surely big meat-hungry mining populations? To sell the meat in these centres it had to be fresh, hygienic and frozen. The outcome was an abattoir on the banks of the Luangwa River which looked like a whaling station that by some incongruous magic had ended deep in the African interior. It was expensive to build and operate. The meat, once frozen, had to be trucked over 1,120 km in refrigerated vehicles. Nevertheless FAO said the project was a success.

The facts did not bear this out. To break even the abattoir had to process a minimum of 500 tons of meat a season. In not one year had it processed more than 170 tons, so it could

---

[124] Marks S., 1975. *Large Mammals and a Brave People*. University of Washington Press.

not have been profitable as claimed. Yet more telling, early in the whaling station's operation, an FAO economist pointed out that it could *never* break even because its annual capacity was only 350 tons; that is 150 tons less than the break-even volume. There was a planning fault!

That was not all. While much was said about the benefits of getting good red meat to the miners, the miners would not buy it. Here the planners overlooked a cultural problem. While rural Zambians ate elephants and hippos readily, those in the towns did not. Such meat, they said, was for primitive and backward people. In the towns civilised people ate beef and the flesh of domestic animals. In the end, the bulk of the meat was sold as dog food, subsidised to the tune of some £250 a ton.

FAO did not like these facts. While the men in the field were pleased with the report (one even said that it was the best consultant report he had ever read), FAO's experts in Rome declined to comment on it, let alone publish it. Perhaps they had in mind my remark to the Head of the Zambian Game Department on my way back to Kenya. He had asked for my overall impression of the project – off the record. He wanted to know, not what I would write with consultant diplomacy (and an eye on future consultancies), but what I really felt. I had replied that I thought the whole business was a matter for the police. Criminal, to my mind, was the enormous waste of money. Closer to my own speciality, where had the ivory gone?

Several years later I did find out what had happened to the ivory. An Asian ivory buyer from Livingstone told me that he had bought both the 'legal' tusks entered in the records *and* the 'illegal' tusks stolen from the boneyard. The former were bought legitimately from the project. He got the stolen tusks, not from Africans, but from three white men: the abattoir manager, a staff member of the Cold Storage Board (which was responsible for selling meat, hides and skins etc.) and a senior member of the Game Department. In turn, he had sold the ivory to Hong Kong.

Any doubts about his veracity were dispelled when I examined the Hong Kong import records. If the Luangwa Valley elephant had produced a similar volume of ivory to their East African counterparts, which according to Hanks should have been the case, then given the number of elephants taken in the culling I knew approximately how much ivory should have been added to Zambia's normal ivory exports in the years of the cropping. The Hong Kong imports given below reflected not what the official Luangwa project recorded, but the greater amount that should have been reported –

Year	Tons Imported from Zambia
1965	1. 6
1966	6. 5
1967	4. 0
1968	**17. 6**
1969	**15. 7**
1970	**16. 0**
1971	**16. 7**
1972	2. 1
1973	8. 8
1974	1. 7

(Years in bold are the culling years)

Supposed to provide meat for protein-starved Africans, hippos culled in the Luangwa Valley mostly ended up in Zambia's copper-belt towns as dog meat, subsidised to the tune of £250 a ton.
(Photo Peter Davey)

The cropping years were 1968-71, averaging 16.5 tons of ivory annually. The pre- and post-cropping years 1965-7 and 1972-4 averaged 4 tons annually. The difference was what I would have expected had the Luangwa elephants grown tusks of the sizes that Hanks said they grew. The Zambian tale reflected not only Game Department incompetence, but events that were similar, though on a lesser scale, to what was happening in East Africa in the early 1970s. Those responsible, however, were white expatriates, proving that neither Africans nor Asians had a monopoly over the ivory fiddles.

Stimulated by contact with revolutionary France and Imperial Britain, Egypt's ruler Mohamed Ali set out to make the Nile Valleys part of his dominion.

CHAPTER NINETEEN

SUDAN

I predicted rising ivory prices, but trends in 1971 and 1972 were far greater than I had forecast. They were increasing at a rate that, paradoxically, was driving the established ivory traders crazy. Traditionally, ivory trading was a conservative business in which annual price movements had been much less than ten per cent and were seldom dramatic. Many of the big dealers held their assets in ivory rather than money, buying and selling only a portion of their stock annually. When value started to rocket it placed them in a quandary. If the price kept rising or rose and then held steady, their traditional strategy would be viable. Yet their Newtonian suspicion was what went up would come down again equally precipitously. If that happened they could face a repeat of the Great Depression's drop in ivory prices. Being so conservative they were not the sort of people to risk the huge losses that might happen if the trend reversed. If they sat on their stocks and stopped buying, they would open the field for newcomers to get at their suppliers. Their quandary was stay out of the market and risk losing their business positions, or buy at the rising prices to stay in the trade and risk disaster if the trend reversed. They would only sleep well if they no longer held stocks and merely bought and sold new material as quickly as possible, changing the whole nature of their businesses. No-one seemed to know what was happening or why.

* * * *

Again BGI asked if I would investigate further. If the traders outside Africa were nonplussed by the combination of rising prices and rising production, the answer had to lie inside the continent. The only way to get it was to see the business from the inside. So, with BGI backing, I started my research by becoming an ivory buyer in the Sudan.

First, a recapitulation on the political background to the Sudan in 1973. While almost all states south of the Sahara were designed on European drawing boards and bore little similarity to the ethnic reality of European nations, the Sudan is an extreme case of the consequences. After Napoleon had shaken Egypt awake at the end of the eighteenth century, that country developed imperial ambitions and in 1820 the ruler, Mohammed Ali, sent a force under Ismail, his third son, to subjugate the Sudan. By 1823 he had succeeded and in 1824 the seat of Egyptian authority was established at El Khartoum (ironically, meaning 'the elephant' and deriving from the shape of the peninsula at the confluence of the Blue and White Niles). In 1839 the Egyptians moved up the White Nile and reached Bor. Subsequent expeditions in 1840 and 1841 got as far south as Gondokoro near present day Juba. From then until 1869 the Egyptians exploited the area for slaves, ivory, and gold with no pretence of a civilising mission. In 1869 and under pressure from Britain, Khedive Ismail of Egypt appointed Sir Samuel Baker a Pasha, Lord of the Ottoman Empire and a major-general, commissioning him to suppress the slave trade and bring civilised rule to the Sudan.

Compelled to control Egypt so that, in turn, it could control the Suez Canal, which was believed essential to the defence of its Indian Empire, Britain intervened in Egyptian affairs. Initially the British influence was manifest through Egyptian employees like Baker and then General Gordon. Paid by the Khedive (Turkish Viceroy of Egypt: nominally Egypt was part of the Ottoman Empire), they nonetheless imposed very British policies. Suppressing the slave trade automatically disrupted the ivory trade, drawing the linchpins of Egypt's Sudanese economy.

More formal control of Egypt appeared when in June 1879 Britain engineered the Khedive Ismail's departure and replacement by his son, Tewfik. From then on until 1907, however, the real ruler of Egypt was Britain's Proconsul Sir Evelyn Baring (later Lord Cromer and father of the anti-poaching Sir Evelyn Baring who was Governor of Kenya in the 1950s). This position was consolidated with military force in 1882 when Britain invaded the country to put down an Egyptian Army revolt that had, in turn, unleashed widespread Islamic opposition to European domination.

In the Sudan resurgent Islam was turned against anything and everything Egyptian by the Sudanese Mahdi – Mohammed Ahmed Ibn el-Sayyid Abdulla. In 1883 he defeated the Anglo-Egyptian army, besieged and captured Khartoum killing General Gordon on January 26th 1885, but himself died later in 1885. From then until 1898 the Sudan was controlled by its local Arab people under the Mahdi's successor, the Khalifa Abdulla until they were defeated by a British Army under Kitchener at Omdurman.

The French made a brief but unsuccessful attempt to secure territory on the Nile, but apart from this the Sudan was effectively ruled by Britain until 1956 when the country was given independence. Yet the Sudan was never treated as a colony or protectorate in the conventional sense. Indeed Britain never challenged Egypt's claim to ownership. Even at the Battle of Omdurman Kitchener fought under Egypt's, not Britain's, banner and the Sudan was ruled as a condominium and in name at least, jointly by Britain and Egypt. If the Egyptians had been kinder, or had their break southward not happened as European anti-slaving sentiment rose to a frenzy, it is possible that events might have taken a different turn. But because their outlook was so different to that of the British, the Egyptians were very much the junior partners in the condominium.

Neither Egyptian nor British rule ever addressed the ecological predicament of both Egyptians and northern Sudanese trapped in the Nile valley by the Sahara. For centuries the people in this trap have been under pressure to look south beyond the desert to fertile land where there is space. Whatever governance prevails must accommodate the ecological reality of people from the north wanting to move south or, at very least, have access to its resources.

Britain blocked such accommo-dation. In the wake of the slave-raiding in which Northerners devastated the south between 1846 and 1869, it protected the uncivilised Southerners from northern exploitation and kept north apart from south. Integral to this policy was converting the pagan and animist Southerners to Christianity and creating a religious divide between north and south that entrenched their separation.

If, during the years of the condo-minium, Britain had encouraged the mass conversion of the southern people to Islam and blocked the spread of Christianity, north and south would have melded more easily into a single country when the Sudan became independent. As happened along the East Coast of Africa, Arabs under Islam blended and merged with the native peoples to form the Swahili culture. They did this more effectively than white Christians have ever merged with other cultures. Yet by injecting Christianity into the

Khedive Ismail's hopes of profiting from ivory and slaves were dashed by British determination to stop the slave trade and control the Nile Valleys.

Northern merchants attending the Juba ivory auction. (Photo John Heminway)

Sudanese equation, Britain virtually ensured that north and south would clash when left to their own devices. Britain should either never have Christianised the south or, having done so, should have established the south as a separate country when she dumped her Empire. That she did neither is all water under history's bridge, but it was only weeks after independence before north and south were at war. Seventeen years later this ended and ostensibly north and south were reconciled.

Two events illustrated just how paper-thin the goodwill was. I had not been there long when Juba was visited by Egypt's Ambassador to the Sudan. This man earlier gained notoriety while serving as Ambassador to Rome. At one of Rome's airports the Italians had opened a coffin *en route* to Cairo, purportedly containing a body being sent home for burial. The body proved to be of an alive but drugged Egyptian dissident being shipped home unwillingly to face the authorities in Cairo. The Ambassador was responsible and had so upset Italian sensibilities that he was recalled to Egypt as *persona non grata*. His lack of diplomacy was still apparent. Arriving in Juba and seeing the vast empty spaces, he had turned enthusiastically to his hosts and said, "But we could put a million fellaheen here." The Southerners seethed with anger as such northern sentiment was the very nub of the seventeen year strife just ended.

* * * *

My visits to the Sudan were described in *Ivory Crisis*[125] and need only brief recapitulation. I went at the invitation of Isaiah Kulang Mabor, head of the regional government's conservation department. The contact went back to when he was a student at the Mweka College of Wildlife Management in Moshi, Tanzania. That pan-African institution trained future game and park wardens and was sponsored by governments, FAO and international conservation bodies. Our company, Wildlife Services, had given the College a standing invitation to visit any of our field operations so students could get hands-on experience alongside our staff. Also, I occasionally lectured at Mweka.

Students remembered these contacts when they went back to their countries, as had Isaiah when he needed help to auction the Regional Government's stock of tusks. This was to be his Department's contribution to the celebrations marking the end of the first year of peace between north and south. The successful auction gave me an entrée into the Sudan and freedom to go wherever I wanted in the Southern Region. My request for permission to buy ivory was also approved.

It was generally known that ivory had been a currency throughout the civil war and was still widely held as such. The Southern Regional Government declared an amnesty whereby this stock could be converted into money and eliminated. Today, with the foresight of hindsight, I would point out that if people hold ivory as currency it is because they do not wish to hold conventional money. Then I merely thought the proposal made sense as an aspect of the return of organised governance. Provided that the amnesty was too short to give hunters time to shoot large numbers of elephants, I supported the idea. Ideally it should not have lasted more than a month. Yet the amnesty's success would depend on all ivory holders hearing of it in time to dispose of their tusks. In the conditions prevailing, a month was too short a time for the news to disseminate throughout the region.

The vastness of the Sudan only came home to me when flying over the its featureless western tracts. Flat or gently undulating, the endless woodlands were more like stretches of ocean than a terrestrial environment. It was dangerous country for light aircraft with limited endurance. With just a magnetic compass and no functional radio beacons, pilots who were not meticulous in their navigation could disappear all too easily. Today's ubiquitous GPS navigational systems, that use satellite transmissions for very accurate fixes anywhere on the planet, had yet to become available.

Driven by war to hide in far reaches of the wilderness, small communities had tenuous contact with the Regional Government. News could take literally months to get to them and a similar time to be verified before people holding tusks might be expected to bring them out of hiding. The Government had to trade a reasonable time for the amnesty to be effective against the number of elephants that would be shot by people in the towns who got news of it by telephone on the day it was declared. Thus while a month might be ideal, six months was more practical.

Though I had expected ivory to be generally available, nothing had prepared me for the volume that came on offer. It was far more than BGI's funds could have bought, even if I

---

[125] Parker & Amin. *Ibid.*

had had the time and inclination to go after all tusks of which I heard. Consequently I bought only a sample of what was available.

Seventeen years of fighting had never stopped ivory moving. It may not have gone from the Southern Sudan north to Khartoum as it once did – though even this route never closed completely – but it had flowed southwards into Zaire and Uganda, eastward into Ethiopia, west to the Central African Republic. The amnesty never achieved what the Southern Government had hoped for. A lot of tusks came forward, but the project was compromised by ivory owners having to accept Sudanese money.

Tusks were an international currency but the Sudanese pound was valueless outside the country (and all but valueless inside it). Who, in their right minds, would swap ivory for such worthless paper? Tusks were the region's *de facto* currency. And the proof of their general negotiability was that seemingly everyone had tusks to sell. As I told the regional government when finally reporting on my buying project, I had bought tusks from government ministers, a judge, policemen, army officers, and administrators. Perhaps that might have been expected as they were in positions of power to which a valuable item like ivory would gravitate. Yet I had also got it from waiters in the Juba Hotel, peasant farmers and people in the rural villages. Everywhere I went ivory was available.

In the year after the civil war ended, at least 200 tons of ivory worth more than $4,000,000 had left the region. The revenue had naturally not returned directly to the Sudan, and much was held in accounts outside the country. Understandably, having acquired an internationally negotiable asset, no one wanted to convert it into non-negotiable money and my biggest problem buying ivory had been persuading sellers to accept Sudanese pounds. Had I paid in dollars or any hard currency, which would have been easy, I would have contravened the country's exchange control regulations and risked seeing Juba jail's high walls from the inside.

While in Juba I became friendly with Gaafar Elias Busaid – a trader from Khartoum who had come south to buy ivory. Mutual curiosity drew us together. I was keen to hear a Khartoum ivory trader's views, and he to find out just what was behind the incursion from the south that posed such a potential problem for northern interests. While I never made notes of the dialogue in those long mosquito-plagued evenings in the Juba Hotel, the essence of what Gaafar said was as follows –

> *"My company, my history, my family, and my life: these are all tied to the Nile: with this area we now call the Southern Region. I have as much right here as anyone. For 200 years maybe, this is where we made our living. How much longer do you have to use a place for it to be accepted that it is yours? I think maybe we have been coming here and using this region even longer than some of the tribes themselves have been here. So I will go on using it. So will my sons. So will my sons' sons."*

That I could recall from the literature, two hundred years was stretching it, but only slightly. The first Arab venturers up the Nile this far south since ancient times arrived slightly before the first white men. Nevertheless, Gaafar's claim portrayed the northern attitude. We had talked of slavery in which his forebears had been involved (as had mine) –

> *"True, we took slaves – as did your own people until recently. The Koran does not say that slavery is a sin. Only those of Islam may not be enslaved. The people we took were infidels and slavery brought them closer to the Faith. If we of Islam had not been stopped from continuing here, by now all these people would have been of our faith. Slavery would have declined as the people joined Islam and gradually we would have become one people with no separation between the North and the South of Sudan."*

'If ifs and ands were pots and pans' ... but he might have been right. As already pointed out, it had happened on Africa's Swahili coast. And Gaafar had been direct and friendly in his advice –

> *"You are one man on your own. You have influence over some men in the Juba Government. But you play politics as a beginner. You talk of ivory but you will find that in Africa ivory is power. If you want to know about it ask me. If you want to trade in it, then why not make a partnership with me? Send your ivory through Khartoum."*

I kept in contact with Gafaar for some years after meeting in Juba. The last time that we met was in London when he and Peter Becker discussed an ivory sale. It was in a bare basement flat in Paddington which Gafaar kept as a 'gymnasium' where, so he said, he brought girls for exercise. We had sat on hard, straight-backed chairs facing one another while he plied us with neat Black Label whisky in thick glass tumblers in the middle of the day. Eventually we escaped, much the worse for wear. Meeting far away up the Nile had been less stressful.

* * * *

There is a sequel to the foregoing tale. In 1985, while on a consultancy for CITES, Zimbabwean ecologist and my good friend, Rowan Martin, met Gafaar and wrote the following memoir –

> *"I had been languishing in the Hilton Hotel in Khartoum for several days having rather worthless meetings with the wildlife department officials, all of whom were trying to avoid me. I eventually had to drag that waster Hasaballah out of his feigned sick-bed to get any sort of satisfaction. A feeling of being ineffectual was creeping over me combined with a feeling that there was a hell of a lot more going on in the ivory trade in Khartoum than I was getting close to. So I made an amateurish effort at being James Bond (003). I chatted with the man behind the counter in the curio shop in the Hilton where various ivory trinkets were on sale and got it across to him that I was not really interested in trinkets but wanted to buy raw ivory. He directed me to a relative of his in downtown Khartoum near the Sahara Hotel. I went there at eight o'clock in the evening and a less savoury part of any city I hope never to have to enter again. I was shunted from one contact to another in various shops and finally taken to*

a green door in a wall and led inside. I ascended a set of stairs with carved banisters and wall-to-wall carpeting with a luxurious pile. At the top of the stairs telex machines were chattering away in well-appointed secretarial offices. I was ushered into a palatial office with Spanish Barcelona designer chairs and Persian carpets. Various specimens of tusks graced the walls and I was later to learn that each type of tusk had a name describing it. A heavy-set man in Arab garb rose from behind a huge desk, stepped forward effusively and said "Hello, Mr Martin – we were expecting you to visit." You can imagine my confusion – I had not given my name to anyone and therefore had to draw the conclusion that some very rapid investigations had been taking place during the time I was being decoyed from one place to another.

"In order to cover my confusion, with total panache I greeted him equally effusively and blurted out that I brought greetings from Ian Parker in Nairobi. This was a total flyer – I had no idea who the man was or whether he had ever heard of you. The reaction was remarkable. He stopped in his tracks and began to question me about your health, activities and whereabouts – clearly he had the highest respect for you and bore no ill-will for any perceived frustrations you may have caused him! The remainder of the meeting was very profitable. Much of what you recount he also told me."

Rowan found out, as I had done, that many involved in the ivory trade were honourable, decent people. They spoke openly of their businesses, of ivory, elephants, politics and life in general. Coming, as they did, from very different backgrounds, it is not surprising that their views were different to ours. That this was never appreciated in the world of conservation and the media was largely due to the hostility both exhibited so openly towards anything to do with ivory: but that is to digress.

* * * *

The desire to control ivory, the south's only viable export, crystallised issues. Gafaar's reference to playing politics was a gentle warning. With hindsight he had also been aware that the Southerners had not just asked me to grade and auction their ivory and given me freedom to buy tusks through the amnesty. Isaiah Kulang and his Minister had asked me to draw up a policy and draft laws to regulate the ivory trade and, specifically, to move control of it from Khartoum to Juba, the regional capital. Their rationale was straightforward. Virtually all the Sudan's elephants were in the Southern Region and ivory was unequivocally a southern resource. The game laws dating from the condominium days placed the seat of the Game Department in Khartoum. Under these rules the export of ivory came under Khartoum's control and the trade pattern was of district and regional flows heading first to the capital from which exports were made. All foreign exchange therefore went to Khartoum and the Southern Region received payment in Sudanese currency. The Southern Region wanted all the foreign exchange that it could get as, understandably, the more it received, the more real its autonomy from the north.

The Southerners' request that I grade and auction their ivory and their permission for me to buy it were never as naïve as it might at first have seemed. They revolved about a major political issue. In 1973 ivory was the only product of any value that the Southern Sudan had for export – just as it had been a century earlier (if one discounts the slaves that were taken to carry it, for they, too, were an export of sorts). The men in Khartoum were as keen to monopolise this resource in 1972 as they had been when Egypt laid claim to the Nile valley in 1820. They looked with extreme disfavour upon my arrival from Nairobi, diverting 'their' ivory away to the south. If I succeeded others would follow. As Nairobi was closer to Juba than Khartoum and the big seaport of Mombasa a lot closer than Port Sudan – the nearest Sudanese port to Juba – it was geographically logical for trade to flow from Juba southward and not north down the Nile. While Khartoum understandably wanted to nip this prospect in the bud, the Southerners were equally as interested that it should develop. The request that I draft new law and policy thus had ramifications beyond ivory. Contrary to most public perceptions, the ivory issue was more important politically than it was commercially.

The draft law that I drew up required that any ivory leaving the Southern Region had to be authorised by permits issued in Juba – the Southern Region's capital. Similarly, no elephants could be shot or ivory traded except on permits issued in Juba. I also recommended that an Ivory Room based on the Mombasa precedent, be established at Juba, to which all government ivory from whatever source – whether seized from poachers, found or from animals killed in defence of life or property – should be sent, graded and auctioned at regular intervals. Traders from overseas should be invited to these auctions directly and the catalogues circulated widely in Europe and the Far East. With such laws in place and with Juba accessible from Kenya and Uganda, Khartoum's rôle as an entrepôt would be obsolete.

There were other aspects of policy that I proposed. One concerned ivory that became available from natural mortality. Throughout Africa, this must always have been the largest and most consistent source of tusks. Even the Wata of Kenya's coastal hinterlands obtained more tusks from it than from elephants they actually killed. Because of the lack of risk in collecting ivory from natural mortality, it was open to incompetent hunters, as well as women and children. Anyone with business in the bush or forests could get it. This would apply in the Sudan as much as elsewhere. Further, given the general distribution of elephants and the minuscule force at the wildlife authorities' disposal, there was no possibility of Government itself, recovering other than a small fraction of the tusks becoming available.

The British Government had paid portage for found tusks brought in and surrendered to it throughout its African territories. The rate of portage in East Africa averaged about 10% of the ivory's commercial value. The system was really quite silly. It guaranteed all tusk finders a basic price upon which they could fall back, but being as small as it was, it was at the same time strong incentive to look for someone who was prepared to pay more. To obtain ivory an illicit dealer only had to pay 20% of the prevailing legitimate price thus ensuring the finders received 100% more than they would from Government. *That* simple logic was ignored. The argument against Government paying a higher price was that it would encourage poaching. It might, but it was incorrect to assume that if Government paid less or nothing there would be no poaching. Incentive to hunt had little to do with what Government paid for ivory. If government raised its ante to 50%, then the illicit buyer only had to offer 60% to get it and still show good profit. *Only* if Government had offered close to prevailing market rates would it have had reasonable prospects of receiving the bulk of ivory available from natural mortality. Any discount it might have offered would relate to the seller's risk of being caught and punished for selling to an illegal buyer. As this, across Africa, had been minimal, discounts should also have been minimal only. Failure to perceive this and failure to understand that while the presence of a market might create incentive to take elephants, it was a non sequitur to then argue that offering a low price would reduce the problem. The system I proposed sought to eliminate this.

I knew that no-one finding ivory in the bush would leave it to rot. It would, one way or another, legally or illegally, eventually be marketed. The only debate was how. What Government paid for found ivory would determine how much of that found was offered to it. I thus recommended that the Regional Government offer close to market prices for ivory brought in from natural mortality. It might not make a cash profit, but the policy would be worthwhile because it would monopolise the resource and receive foreign exchange for it.

I appreciated that such a system would give cover under which tusks taken illegally would be presented as having been 'found' from natural mortality. Yet it was not as if there would be no illegal take if Government refused to make handing in found ivory worthwhile. As I had been finding wherever I went in Africa, governments lacked ability to stop people taking ivory. It was taken regardless of game laws. As a brake of sorts on the number of elephants killed, I recommended that the Regional Government determine an annual quota of ivory that could be exported. Production in excess of this limit would have to be stockpiled as part of subsequent years' exports. I was aware of the proposal's imperfections, but it was more realistic and open than any approach to the vexed problem of found ivory elsewhere.

Separately, I offered BGI's services to put an Ivory Room in place and to manage it on contract. Yet nothing came of these proposals for the documents containing them found their way into the Khartoum authorities' hands and displeased them exceedingly. The Northerners' problem was to stop me without making the issue a north-south confrontation; mine was to anticipate and get round their moves.

The first step to hobble what I was doing was insistence by the Northerners that I have approval from the Central Bank in Khartoum to export any commodity which, in turn, called for submission of invoices, proof of payment in foreign exchange, receipts, etc. There may have been an amnesty but there was no likelihood of obtaining invoices and receipts from the people who were selling tusks, least of all from the senior civil servants concerned. This attempt to 'play it by the book' was not in keeping with reality. The Regional Government itself did not, however, need Central Bank approval, so the ivory I bought was surrendered to the regional authorities who exported it in their name. They had the evidence of payment having been made in Sudanese pounds converted from dollars that I brought into the country and changed in the bank at Juba. Round one to the South.

Air traffic clearances (I flew myself to and from the Sudan) were controlled through Khartoum. To ensure there were no hold-ups over this, every clearance I put in was accompanied by a request for it from a Minister in the Regional Government. The Northerners could not refuse it without irritating the Southerners.

My ivory purchases were flown out of the Sudan in East African Airways' DC3s whose handling agent was Sudan Airways based in Khartoum. At the last moment there was invariably a hitch and the agent was unable to load the DC3 or prepare manifests etc. So I did the loading and prepared load sheets, manifests, etc. myself, helped by the East African Airways crew. Round two to the South.

When these hindrances failed, stronger measures were called for and I was turned out of bed around midnight by the police who took away my passport. First thing next morning I was at the office of the Minister under whom wildlife affairs fell and by midday my passport was returned with apologies. Round three to the South.

Khartoum decided on stronger measures. The next East African Airways aircraft would be permitted to get to Juba, but regardless of any letters from the authorities in Juba, permission for me to follow would be refused on a technicality. I would not be in Juba to load the cargo and therefore face a big bill for demurrage while the aircraft waited for me to arrive. Hopefully it might not wait until I got my clearances and I would have to pay for a fruitless trip. Finally, if the aircraft was loaded, it was to be seized by bank officials on such pretext as they could trump up and held as long as possible which, again, would be expensive.

The northern plan was passed to me by a Southerner in Sudan's Nairobi Embassy. At the time that he gave this forewarning, he pointed out a loophole. I should not apply for clearance to fly into the Sudan until a certain date when the Ambassador and his deputies were out of town and the Southerner would be acting Head of Mission. I did as instructed obtained my clearance and flew to Juba. Knowing the authorities at the airport – all Northerners – were primed, I played along. It was usual for those needing favours to make small gifts to those whose goodwill they needed. Until now I had never done so. This time, however, I took a dozen kilo packets of sugar (sugar being difficult to get in Juba) and several cheap watches.

Sure enough, there were Northerners on hand to welcome me. Arabs are past masters at being really nice when mischief is afoot and they were nicer than I had ever known them. On a wave of bonhomie we talked of weather, what it must be like in Khartoum, and much more of no consequence. In due course as evidence of my esteem for them I gave each man a kilo packet of sugar. After more gossip, I produced the cheap watches. Their eyes betrayed their delight: the rat was in their trap; he must need a serious favour and they knew what that favour was. He had no clearances. After letting them savour the approaching dénouement for a minute I had stood up and said that I supposed they wanted to see my clearances, and produced them. Their looks were magic and I could not but laugh out loud and a couple who bore no personal ill-will laughed too. They knew then their plan had been rumbled. Round four to the South.

The southern authorities provided an armed guard of ten men from the Wildlife Department specifically to keep any Central Bank people away from the aircraft. So we won round five, but that was the last load I brought out of the Sudan. Sooner or later the Northerners were bound to win a round, so I called it a day. In any case the research results were beyond my wildest dreams and while the business aspects still had potential and I am sure that I could have bought a lot more ivory, trading had never been my primary objective. I had gone far enough.

Ivory illustrated the fundamental political situation starkly and I had a clearer view of this than was available to most at that time. The war may have stopped, but the Arab north still wanted the south's resources. The ethnic divide between Arab and southerner was as clear as ever. The Egyptian Ambassador's 'we could put a million fellaheen here' had

Within a decade of the Battle of Omdurman in 1898, Alexandria was connected to Mombasa by a chain of railways, river and lake steamers, enabling travel between the two almost as quickly as by sea. The chain broke in 1956 and by 1976 along much of its length, all that remained were reminders such as the once proud Robert Coryndon, that sank at anchor at Butiaba, its home port on Lake Albert in Uganda. So much for progress!

summed the ecological problem succinctly. Pragmatically it is not whether the north will gain access to the south, but when and how.

Looking back over the past two centuries, strife has been the norm across the Southern Sudan. Fifty-eight years of British rule between 1898 and 1956 brought a pause, but only a pause. There seems no end in sight, and while ivory *may* not be contributing now, it was at the centre of events for most of the past 200 years, providing excellent perspectives on the political scene.

* * * *

The sheer volume of ivory coming out of the Sudan revealed an international trade far greater than understood in any conservation circles. It was not just the Sudan's home production. Many tusks were coming by truck out of north-eastern Zaire to Juba, then going down the Nile by steamer to Khartoum and by air from Khartoum to Antwerp. This trade was handled by a cartel of Senegalese traders who answered questions readily as they had no ground to be cautious with another who was also buying ivory. Buying and bartering widely across northern Zaire, the tusks accumulated and were shipped down-Nile. These Senegalese were no newcomers to the business. Peace in the Sudan made Juba and the Nile route easier than it had been, but even during the civil war they had still occasionally used it. In 1972-3 it was a particularly good route, but if any problems arose the flow would go towards another country – Uganda or the Central African Republic, for example.

Another flow of ivory was from the Central African Republic into the western Sudan town of Wau. This was not managed by Senegalese, but Sudanese Arabs, who railed it to Khartoum. My most useful informant was an old trader who had lived in Wau all his life. My stepfather, Hugh Woodman, was for years a doctor in Wau, and when I asked if the old man recalled 'Nkoronkoro' (Hugh's Zande nickname), he not only did, but recalled being treated by him in the 1930s. Learning that I was the doctor's stepson he could not help me enough. He made the point that while the flow in 1973 was from the Central African Republic to Wau and the Sudan, it could just as easily have been the other way around. If trade in the Sudan was hard, then the flow would reverse direction. It had always been thus.

Was the Sudan representative of Africa more generally? No other country had had a civil war of the same order, but chaos was growing across the continent. To the south Idi Amin was magnifying the destruction that Milton Obote had started back in 1966. And while other countries may not have been at war, they were making disastrous political experiments that, in economic terms, produced similar results. The Customs statistics to which, at the time, I had access (principally Hong Kong's) suggested that the Sudan was not exceptional.

Many people thought my work in the Sudan must have been illegal. They simply could not believe that one could lawfully buy ivory in bulk in the second half of the twentieth century. I never corrected them, particularly other traders, as they would not have been so forthcoming as they were, had they thought I was not one of them.

The madness was not entirely without its method. It opened doors and an opportunity to look at what was going on in Kenya where the trade, if technically legal, was in spirit unlawful. When the Deputy Chief Game Warden seized my last Sudan consignment as it lay in transit at Nairobi airport, it was known to other traders – almost within minutes. The same would have applied when it was released and it was naturally presumed that I had made an appropriate payment, further enhancing the sense that I was one of them. Several Asian buyers keen to get on my Sudanese bandwagon readily told of what was going on in Kenya. It was a free-for-all. Ivory was being traded by people who had never been in it before. Not only were tusks from within Kenya being bought and sold, but much from neighbouring countries was being imported and re-exported. The transactions were of course overwhelmingly legitimate as they were sanctioned under Chief Game Warden's Permits. What Elliott, Sheldrick and Woodley had been seeing from the inside, I was now seeing from the outside.

Combining the Kenya evidence with what I knew from the Sudan, and with reference to the Customs statistics, the volume of tusks in trade represented at least tens of thousands of elephant deaths annually. Further, this was not a sudden phenomenon, but had been building up for decades. Alone, this indicated that elephants must be present in their millions across Africa; indeed it could *not* be otherwise. For the first time I had a glimpse of the ivory economy's size and the sense in Gaafar Busaid's words, "… In Africa you will find that ivory *is* power."

CHAPTER TWENTY

HOISTING HUNTERS WITH THEIR OWN PETARD

Zambia and the Sudan greatly expanded the backcloth against which what was going on in Kenya could be viewed. It was ever more obvious that the game laws and the policies they reflected were drawn up in ignorance of African reality. One could see why the Woodleys, Sheldricks and Elliotts had made no ground dashing themselves quixotically against ivory's economic might. Clearly more than game laws were necessary to influence the political forces it involved and I doubted whether they could be controlled at all.

* * * *

For all that Kenya's rising corruption through the 1970s was no secret and its impact upon wildlife was widely talked about, few individuals were prepared to stand up and publicly condemn what was happening. As always there were exceptions to the rule and Peter Jarman was one. A quiet-spoken biologist, he was acting head of the Game Department's Fauna Research Unit in 1973 where he had worked for the previous two years. As a sop to public outcry about the rate at which elephants were disappearing the Chief Game Warden asked Jarman to report on what was happening and submit recommendations to stop it. This was tricky. Peter knew there was much for which the Game Department could be criticised. Further, his employment contract would be up for renewal shortly after the report was to be submitted and a negative report would block further contracts. Nevertheless, writing confidentially, he hoped honesty and the evidence would not offend.

The essence of his findings[126] were that the country's elephants had dropped from ±160 thousand in 1968/69 to 118 thousand in 1973; a loss of 42,000 (26%) in five years. Note, though Jarman did not make the point clearly, at least 15,000 had died from drought. The rate of decline in 1973 had reached ten per cent annually, elephants had been *legally* over-hunted for at least the past six months, but possibly for the preceding two and a half years.

Jarman pointed out that between 1958 and 1965 licences to hunt elephants had fluctuated between 200 and 250 a year. In 1970 this rose to 444, in 1971 it was 582, in 1972 829 and in the first half of 1973 1,358. He referred to the 'notorious Collectors' letters [permits]' that had originally been granted on presidential instruction, which allowed the holders to 'collect' ivory as and where they wanted and said –

"They typify the kind of abuse of power which has spread corruption-by-example ..."

This implicitly criticised whoever had ordered Collector's Permits to be issued. Peter noted that ivory, rhino horn and other trophies were being disposed of other than through the established channel, pointing out that in the 1960s over ninety per cent of all Kenya's Government and National Park tusks were sold through the Ivory Room. This had dropped

---

[126] Jarman P. J., 25th October 1973. Confidential Report to the Chief Game Warden Concerning Elephants and the Ivory Trade in Kenya.

to 80% in 1971, to 44% in '72 and only twenty-three per cent in January to July '73. Sensitive information indeed as the trend was still outside public awareness; Jarman went on –

> *"Leading personalities, including both Assistant Ministers in our own Ministry, were among those buying from HQ,"*

ameliorating the force of this observation by saying –

> *"Without knowledge of the prices paid there is nothing to show that this was reprehensible; but it is a practice that should be avoided because it is so obviously open to public criticism."*

Obviously he knew members of the Ministry of Tourism and Wildlife were, with the Chief Game Warden's collusion, taking ivory and rhino horn worth millions of shillings for which they were paying less than market rates. The report may have been confidential, but doubtless Jarman could pass the information to others. Hardly surprisingly, Peter did not have his work permit renewed and he left Kenya not long after his report was received.

Ellis Monks, a retired chemist with no interest in fame or publicity, was devoting his time to conservation matters through the East African Wildlife Society and Joy Adamson's Elsa Fund. With Jack Block, a wealthy businessman and Charles Njonjo, the Attorney General, he had set up a local chapter of the World Wildlife Fund (WWF) as it was then known. He was a Kenya citizen and both an honorary national parks warden and an active honorary game warden with all the powers that went with such status. In both rôles he embarrassed those in authority. The 'Nairobi Orphanage' was a small zoo attached to the Nairobi National Park, originally set up to care for orphaned or injured wild animals prior to their release back into the wild. Ellis was critical of its management which he felt left much to be desired. Consequently Ellis's honorary national parks wardenship was revoked: elementary logic of course, if you don't like criticism, remove the critic.

A little while later, he seized an illegal trophy (an immature caracal skin) in a Nairobi ivory and curio dealer's shop. The proprietors should have been prosecuted, but were let free. Instead, on the personal order of Minister of Wildlife & Tourism – the selfsame Mathews Ogutu whom Rodney Elliott had investigated – Ellis's honorary game warden status was cancelled and again the critic silenced.

Monks was interested in the information I had been accumulating both in and outside Kenya and in October 1973 asked me to address a small meeting at his house. One of those who attended was Jack Block who asked me to commit the facts to paper for him – which I did. Some of the conclusions I drew have stood the test of time.

So far as I am aware, the twenty-eight page report was the first overview of the international ivory trade since Kunz[127] had produced his seminal review in 1916. I concluded that ivory was not so much a commodity as a currency whose value to Africa ran to millions of dollars annually. The accumulated quantities outside Africa represented investments worth many hundreds of millions of dollars: both conclusions were later substantiated. There was

---

[127] Kunz F., 1916. *Ivory and the Elephant in Art, in Archæology and in Science.* Doubleday & Page, New York.

no effective management of the ivory resource, the conservation departments seemed ignorant and ill-equipped to understand or manage the ivory trade. Banking and finance ministries seemed more appropriate.

In the 1970s strict exchange control laws in all the East African countries made it difficult to transfer capital internationally. The degree of their economic mismanagement was incentive to break these laws: the greater the mismanagement the greater the incentive. As ivory did not bear the seal or stamp of any one nation, tusks were difficult for the layman to ascribe to source and were thus readily negotiable. The Southern Sudanese Government, for example, had asked me to buy new Land Rovers for it in Nairobi, payment in ivory.

Ellis Monks had his honorary game warden status cancelled for seizing an immature caracal or African lynx skin that was held illegally by a firm that, unfortunately, had 'protection' (photo Peter Davey).

The principal ivory buyers in Kenya were people originating in India and Pakistan. They bought tusks in local money, exported them, received payment in hard currency, of which they repatriated only some or none to Kenya and evaded exchange control rules. In the first six months of 1973 I calculated that the sum thereby 'lost' to Kenya was of the order of $2.3 million (equivalent value in 2003 would be over $100 million). The only way to stop this was to stop Indians and Pakistanis trading ivory. I was of course wrong. Someone else would have stepped into the gap.

No less than 214 tons of ivory had left Kenya in the first six months of 1973 of which I calculated seventy-six per cent (163 tons) was accounted for by 'collector's permits'. Much ivory was not of Kenyan origin, but imported from neighbouring states. Further, important people were involved in the ivory trade. Naïvely, I suggested that a continental market for ivory be established which could regulate sales as De Beers do diamonds and thereby obtain the greatest benefits for Africa. This, to my knowledge, was the first time that an ivory cartel was proposed. Where the Kenya Game Department was concerned I wrote –

"Likewise the Game Department is clearly the most inappropriate organisation to handle any aspect of ivory trading. By demonstration it has proved inept and any re-organisation of the current situation that still included Game Department control would create no public confidence either locally or internationally."

Naturally the report was confidential because Government would not have liked the criticism. Unbeknown to me, Jack Block did show it to WWF and IUCN, both of whose headquarters were in Switzerland. Indeed the only copy I now have of my original document was stolen off an IUCN desk in that organisation's Swiss headquarters (though not by me!) The salient point is, however, that both these bodies had hard evidence on the dimensions of the ivory traffic in 1974, but did nothing.

Reggie Destro, Glen Cottar and Dave Williams; three professional hunters who were outspoken in their complaints about the Wildlife Department's plunder of wildlife and lost their livelihoods as a result.
(Photo Peter Davey)

* * * *

No group was more directly influenced by the events concerning elephants and rhinos during the 1970s in Kenya than the professional hunters. In the system inherited from the colonial era, the country was divided into more than eighty hunting blocks or 'controlled areas'. Any party intending to hunt had to book blocks in advance. Most blocks were limited to having one overseas hunting party and one resident hunting party at a time. This system collapsed under the surge of licensed hunting which accompanied the rising price of ivory after 1970 – as illustrated by Peter Jarman. Professionals taking their clients into areas they expected were exclusively reserved, found numerous other parties hunting without bookings. Of these, many were unlicensed, and some were being helped by both game wardens and scouts.

Coming across ever more evidence of what was going on and because it affected them directly, the hunters spoke out. So did their clients. To placate the international clamour and show that something was being done, in mid-1973 Government stopped all elephant shooting, deftly wrong-footing the professional hunters. Many had taken deposits from clients who wanted to shoot elephants, and they cried out in alarm. The ban was temporary and lifted after several months. On June 28th 1974, however, the sale of normal elephant licences was stopped permanently. The immediate impact on the safari industry is apparent in the following extract from the East African Professional Hunters' Association records[128] –

> "Figures obtained from a portion of our Members showed that at least 215 clients had deposit-paid Bookings to hunt Elephants in Kenya in this half of 1974. These Bookings totalled US$768,271 in direct Fees and would employ 603 Citizens for 18,090 man days. The total loss of Foreign Revenue to Kenya would be over one and three-quarter million Dollars."

By attacking the hunters' pockets, the Government turned them from the offensive demanding action on poaching, to the defensive arguing against a hunting ban. This drove a wedge between the hunters and a western public that could not equate opposition to the

---

[128] East African Professional Hunters' Association memo dated July 15th 1974 to all members from the Executive Officer D. H. M. Dempster, being the text of an address by the Chairman at an informal meeting of members.

hunting ban with opposition to poaching. The media never corrected a public assumption that hunting was virtually the same as poaching. It was a public relations coup by the Kenya Government from which the hunters never recovered and which the media should have picked up but neither did nor wanted to. And because of it the professional hunters were denied both the recognition and kudos for being the first to protest publicly about what was happening in Kenya's wildlife affairs.

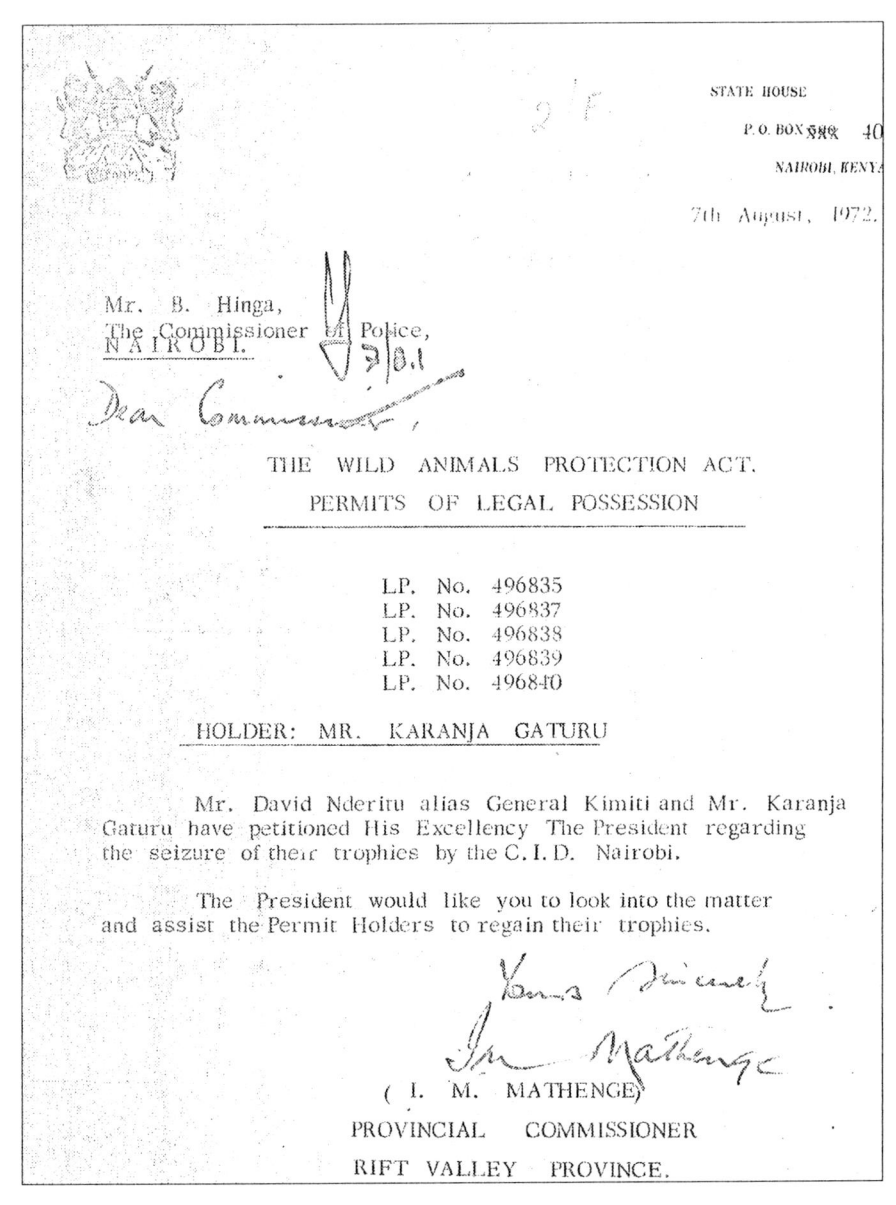

STATE HOUSE
P.O. BOX 40
NAIROBI, KENYA

7th August, 1972.

Mr. B. Hinga,
The Commissioner of Police,
NAIROBI.

Dear Commissioner,

THE WILD ANIMALS PROTECTION ACT.
PERMITS OF LEGAL POSSESSION

LP. No. 496835
LP. No. 496837
LP. No. 496838
LP. No. 496839
LP. No. 496840

HOLDER: MR. KARANJA GATURU

Mr. David Nderitu alias General Kimiti and Mr. Karanja Gaturu have petitioned His Excellency The President regarding the seizure of their trophies by the C.I.D. Nairobi.

The President would like you to look into the matter and assist the Permit Holders to regain their trophies.

Yours sincerely,

(I. M. MATHENGE)
PROVINCIAL COMMISSIONER
RIFT VALLEY PROVINCE.

Evidence of protection in high places. Justified as a reward, some Mau Mau fighters were allowed to recover ivory they had supposedly hidden in the highland forests. In short order they were collecting ivory all over the country, in areas far removed from the former conflict zone, and no pretence of connections with Mau Mau.

A thoroughly bamboozled western public praised the steps Kenya had taken. What it did not appreciate was that while regular elephant hunting licences may no longer have been issued, Chief Game Warden's permits went on being issued in greater numbers than ever before.

One has only to glance at how many such permits were issued to appreciate the point:

1970 = *215*, 1971 = *217*, 1972 = *156*, 1973 = *163*, and 1974 = **2,407**[129].

Those in favour with the Game Department went on hunting elephants and stopping normal elephant licences was a cosmetic exercise to simultaneously teach the professional hunters a lesson and placate a critical western public. The hunters were, however, poor pupils as they did not take the lesson to heart and went on complaining about what was happening in the field. Consequently all hunting was banned in 1977 and again the world applauded what, at the very least, was a brilliant public relations coup that hoist hunters with their own petard.

* * * *

In 1974 the American NBC television network wanted to make a documentary film on what was happening in the realm of ivory. I agreed to help. It was the first of several television films on the subject with which I became involved. They all suffered from the media's driving desire to please the public and from what I call the 'rose syndrome' that arises through abbreviating information. Need to shorten maybe unavoidable: a film has to fit a given time or a news article into so many pages or so much space on a sheet of paper. Thus, for example, many of the world's television stories have to fit into half hour or fifty minute slots. Yet forcing the facts into a predetermined mould distorts them. Let me illustrate: I say my wife smells like a rose. It is a compliment. The journalist shortens this to fit space and correctly but incompletely reports that I said my wife smells, and my praise becomes an insult involving me in calamity. That is the rose syndrome: reduction to the point of inversion.

In the search for sensation film makers concentrate on public figures rather than nonentities. Where Kenya and ivory were concerned they focused on the Kenyattas. It was widely rumoured that the President was involved with ivory and that his wife, Mama Ngina, was the country's biggest exporter. The film producer was keen to get evidence of this. To my knowledge, then and now, President Kenyatta never personally traded ivory. He gave some tusks to prominent people: for example his erstwhile Minister for Agriculture, Bruce McKenzie, was given a pair of large ornamental tusks from Tsavo (many years later I valued them). There was nothing improper in that gift. Kenyatta was said to have ordered the collector's permits and we have the evidence from Field Marshal Muthoni[130] and Ian Grimwood that he did. I do not think, at least *initially* with so

---

[129] Casebeer R.L., 1975. *Summaries of Statistics & Regulations Pertaining to Wildlife, Parks & Reserves in Kenya.* UNDP/FAO Project KEN:71/526 Project Working Document 8. FAO, Rome.

[130] Njagi. *Ibid*

contentious an issue, that anyone would have pretended that they spoke in the President's name – no matter how highly placed. Later things may have changed. David Nderitu alias 'General Kimiti' and Karanja Gaturu were definitely among the original permit holders along with Field Marshall Muthoni. In 1972 the CID had seized ivory belonging to them. It resulted in the following letter –

The letter was on red-printed State House stationery. Whether it did in fact reflect a Presidential edict is moot. It was widely known that Isaiah Mathenge was close to President Kenyatta, and many people in Kenya would not have risked questioning its provenance, particularly on a matter that did not concern their personal fortunes. The CID had taken it at face value as an order from the President; the two men had their trophies returned and enquiries into how they acquired them terminated. The evidence is, however, conclusive proof that where Nderitu and Gaturu were involved they were fireproof. Their original licence came from the President and protection subsequently was, if not directly from him, then very close to him in the hierarchy.

It is assumed that Kenyatta's motives for granting these concessions were political, that he created bonds that were worth the risks of local and international criticism. Yet whether he had anything whatever to do with the later issuance of collector's permits is questionable. A list of the collectors who had been operating in Kitui District[131] gives the names of Field Marshal Muthoni and of six local Kamba and a Jaluo. Other than Muthoni, the others were all seemingly petty traders operating in small trading centres in a crescent round the eastern rim of settlement in Kitui, abutting onto either the Tsavo National Park or the rangelands used nomadically by Kamba and Orma graziers, and where there were many elephants. They may have been given collector's permits as part of Muthoni's or Karanja Gaturu's networks. They may have been awarded permits independently. What the precise situation was may now never be known, but intuitively I suspect that President Kenyatta created a situation that was then developed by others lower down the hierarchy and that some of the later permits were issued without his authorisation or knowledge.

I never saw evidence that his wife Mama Ngina traded ivory though there was a lot of hearsay to that effect. It was a different matter, however, where his daughter Margaret was concerned. Her company, the United African Corporation (UAC) which at the time was managed by a man named Pusey, an ex-banker so it was said, certainly exported a lot of ivory during the 1970s. Mr Pusey admitted this, saying that it was all bought legally and I am sure he was correct. Proof of legality was a bit of cheap paper. Nevertheless he refused press interviews and was very reluctant indeed to talk about the UAC's ivory dealings. When I did talk with him, even though the day was cool he sweated and kept running a finger around inside his collar. Talking about ivory seemed to make him uncomfortable.

There would have been an uproar if a film crew had walked into the airport hangars and pointed a camera at Margaret Kenyatta's ivory where it was held prior to shipment. Nevertheless, merely walking past it caused no stir. The NBC cameraman produced a camera eye-piece on a long, jointed stalk which could be twisted in different directions. With the camera under his arm like a bagpipe sack and the lens facing backwards, he could

---

[131] Letter CID/SEC/4/4/73/74 of 25th January 1974 from the District Criminal Investigation Officer, Kitui to Director of Criminal Investigation, Nairobi.

see behind himself. Facing away from the tusks he filmed them for as long as he wanted without raising anyone's suspicions.

We paid a freight agent's employee £20 per copy for airway bills covering ivory consignments. I learned subsequently that he did not use his own Company records, but the Customs authorities'. Slipping into their airport offices when the officials were out, he picked up any ivory airwaybill available, copied it on the Customs photocopier, returned the Customs' copy and emerged with his own. Inevitably he was caught. Two Customs officers returned unexpectedly and found him at work. They were angry, not because of what my informant was doing, but because they had not been cut in on the business. As *their* airwaybills were being copied on *their* photocopier, they felt that *they* had been entitled to most of the £20 fees.

Airwaybills illustrated the nature of the UAC's business. In March 1974 the Company made eight shipments weighing 6,052.6 kilos made up of 521 tusks that averaged 11.62 kilos[132]. All were carried by Pakistan International Airlines and all were addressed to the China National Light Industrial Products Import & Export Corporation, Peking Arts & Crafts Branch, Peking, China. If this month was representative of 1974, UAC averaged exports of just over 200 kilos of ivory a day (about 73 tonnes or somewhat over 3,000 elephants a year).

We interviewed General Kamiti – one of the more prominent holders of a collector's permit. A braggart in fancy cowboy hat and boots, he loved the idea of being filmed and assumed that because the film crew were with me, fellow dealer, he had nothing to fear. Buying in the Sudan had its bonuses! We went to his 'factory' – a two-roomed house in Nairobi's Eastleigh slum where his 'craftsmen' were cutting tusks and making bangles. There were tusks under the beds and thrown higgledy-piggledy about the room in a strange amalgam of squalor and opulence. When asked about legality, the cowboy general laughed and with a meaningful look said, "Some of us are always legal."

They say that one has to have money to make money and in curious parallel, now that I had a certain volume of knowledge about the ivory trade, more kept arriving almost of its own accord. The flow of information seemed to have acquired a momentum that kept it going and helping with the NBC ivory film produced yet further insight into what was going on. At no point had I overestimated what was happening. To the contrary, the ramifications became ever wider.

---

[132] Pakistan International Airlines Corporation Airwaybills # 62683725 and 62683740 of 8th March; 62683751, 62683761 & 62683762 of 15th March; 6268351 x 2 (there were two awbs with the same number but different amounts of ivory) of 18th March; 62683621 of 22nd March and 62685343 of 29th March 1974.

CHAPTER TWENTY-ONE

SPADES ARE SPADES

If I had taken no interest in ivory and confined my work to aspects of conservation involving animals other than elephants, perhaps life would have been easier. The issue is moot. Yet even when matters concerned other animals or research, it was never easy for Wildlife Services Ltd to work in Kenya. At the best of times the Game Department and the Ministry of Tourism & Wildlife always seemed hostile. No doubt my personal attitudes played some rôle in this as they were out of step with the placatory post-colonial manner in which westerners generally treated with Africans. I have never believed there is a case for lowering academic standards or expecting less efficiency in government or business management because Africans are involved. Nor have I bought the other side of that coin, that Africans have come to expect special treatment – though it is very human for them to have taken it when offered. Yet, as will be explored in this chapter, there was more to the hostility we experienced than can be accounted for by any personal idiosyncrasy.

* * * *

It has been said that I am not diplomatic – which may be true. In 1972 I attended the Third World Conference on Animal Production in Melbourne. While there a Melbourne paper printed an article I wrote that ended in criticism of Papal resistance to birth control. It followed a series of brushes with the Australian media arising from a seminar in Sydney University I had been inveigled to give by my good friend the late Graeme Caughley. It was his idea of humour to invite me to the university for a seminar and only when there announce that I was conducting it. My unoriginal hypothesis had been that as people compete with other animals for space, human increase creates the primary need for conservation. Failure to recognise this made much conservation a waste of time. Following up on this in an interview with a journalist – also set up by Caul – she had interrupted my spiel by asking if professional hunters slept with their clients' wives. Being neither a professional hunter nor a client's wife I opined that I was unqualified to answer that, but thereafter the interview deteriorated and she implied my presence as a white in Africa must in some way be to the blacks' disadvantage. I answered (rather poetically), "But for the dice of fate your forebears might have gone to Africa and mine to Australia, but they didn't so I suppose by definition I am the imperialist and colonialist." The drift of the resulting article was 'imperialist and colonialist says conservation is a waste of time.' The train of events which followed included an interview on television into which I had been lured on the pretence that it was an intellectual programme. Hoping to retrieve some gravitas I was mortified to find that it was a variety programme and I appeared as a comedy turn to answer inane questions about condoms for elephants, between interludes of (quite lovely) high-kicking girls. The incident left Graeme Caughley and his then wife Jude smirking in high good humour.

There was, however, a serious side to this chain of trivial incidents. The Kenya High Commission in Canberra read the papers and sent clippings back to Nairobi. Undiplomatic

Alistair Graham's widely respected study of crocodile population biology on Lake Turkana was a seminal point in our understanding of these saurians. Yet the Kenya Game Department, for whom it was undertaken, never passed comment on it, other than to suggest the whole project was a scam to make money from crocodile skins. (Photo Peter Davey)

as my comments on the Pope may have been, they were within the limits of free expression. Further, I had been in Australia as a private individual and not representing Kenya in any way. Nevertheless three months later, but for a friend in the right place at the right time tipping me off, I would have been deported from Kenya, because my utterances in Australia projected a bad image of Kenya. Forewarned, I forestalled the event. Yet in the process I learned that both the Chief Game Warden, and the Director of National Parks had been consulted and recommended the move. The real reasons had nothing to do with anything I had said or written in Australia.

* * * *

As a result of work done in the Kenya National Parks in 1966, my company Wildlife Services owed them £300. At the same time the Game Department owed Wildlife Services £300. As both Parks and Game Department came under the same government ministry, we asked its accountants to pay to Parks the money that Game Department owed us and thereby square both debts. The Ministry never paid. We had completed a three year study of the Lake Turkana crocodile population for the Game Department, and needed to know government reaction to the findings and management proposals before formally terminating the project. We had offered Government a shareholding in a new company we would form to undertake game culling projects in Kenya. We had proposed financing a student selected by Government for a course at the College of Wildlife Management at Mweka in Tanzania. These three examples of programmes held in limbo were typical. The Ministry of Tourism

& Wildlife simply refused to answer any correspondence from us on any subject. Effectively, it blocked our ability to operate in Kenya. Via the grapevine from within the Ministry, source of this policy was always the Permanent Secretary – Alois Achieng.

Matters came to a head over the £300. A determined foray into the Ministry of Tourism & Wildlife's Accounts Department was more than the accountants could handle, and we learned that the money had been taken by the Permanent Secretary himself. It was a small amount but at last we had a handle on him.

One of my fellow Wildlife Services Directors, Tony Archer, and I went to see the country's top civil servant, Geoffrey Kariithi, Permanent Secretary in the President's Office and Secretary to President Kenyatta's Cabinet. We laid complaints against Alois Achieng, accusing him of deliberate obstruction and, where the £300 was concerned, of apparent corruption. Kariithi said he would look into the matter, later confirming that he had done so[133] and Alois Achieng would meet with Tony Archer (but not me). A meeting took place, Achieng confirmed that he had nothing against us or Wildlife Services Limited. He expressed interest in our proposals and set up meetings with his subordinates to go into the details. And the National Parks received their £300. Seemingly, we had faced him down and won. Our aggressive reaction surprised Achieng as well as many friends, because it was so contrary to the prevailing reluctance to gainsay anyone in power.

"You shouldn't have done that ... you'll only upset them," was the common reaction. Yet it was just such acquiescence to the early stages of inefficiency and corruption that brought about the rapidity with which these ailments afflicted Kenya. Had we colonials risked a few sacrificial deportations and made a fuss every time we came across corruption and theft, it is possible that the process might have been stayed.

Where Wildlife Services Ltd was concerned generally and Archer and Parker in particular, reporting Achieng meant it was only a matter of time before he sought revenge. It was not long in coming. In the 1960s and '70s I wrote frequently on both the philosophical and practical aspects of conservation. This, together with the fact that our company was one of the few concerns in East Africa involved in the field – putting our money where our mouths were, so to speak – gave us a certain prominence and we were invited to contribute to international conferences and symposia.

As a matter of common sense when it came to using wildlife as a source of meat and related commodities, I held that marketing was best supervised by a government arm specifically set up to handle agrarian commodities – such as an Agricultural Department. This was an issue of principle, applicable worldwide and was not made with reference to Kenya in particular.

In 1969 I was asked by the FAO to write a discussion paper for a working party on Wildlife Management that was held at Lomé, in Togo on 15th-18th January 1969[134]. Its harmless gist is best given in my conclusions –

---

133 Letter GEN. 329/230/002A of 15th May 1968. G. K. Kariithi, Office of the President to I. S. C. Parker.
134 Parker I.S.C.,1969. *The Development of Wildlife Use as a Food Source.* Document FO:AFC/WL – 69/5 dated 8th November 1968 & presented at the African Forestry Commission Working Party on Wildlife Management, Third Session. Lomé, Togo 15th-18th January 1969.

"The reason for the current slow development of wildlife use lies primarily in the orientation of the organisations charged with conservation. The use of fauna for basic human nutrition is contrary to the founding precepts for preservation. As a result, Game Departments and Park Authorities are not geared to exploiting animal productivity. These fundamental aspects influencing wildlife policy must be appreciated if further progress is to be made.

"To permit wildlife use, considerable re-organisation of Departments involved must take place. They must be augmented, or current staff replaced, by technically competent and trained staff. In many instances game utilisation would be better undertaken by Agricultural, or sub-departments of Agricultural Ministries."

FAO responded[135] –

"This is exactly what we wanted, and it has been processed for distribution to all participants".

The East African Wildlife Society obviously thought the ideas sufficiently interesting to publish them in its journal[136], which produced an astounding official response. Written by Perez Olindo, then Director of Kenya National Parks, dated 8th April 1969 and addressed to M. J. Sawyer, Chief Executive of the East African Wildlife Society it read –

"I am directed by my Ministry to write to you for a clarification of the stand of the East African Wildlife Society concerning the article, and I quote, "Encouraging development is being held up" appearing on pages 10-11 of Volume 3, no. 9 1969 of Africana.

"It is the official view of my Ministry that by virtue of the fact that this article was published in the Society magazine, an article in which the writer advocates the re-organisation of Ministries currently constituted under the Presidential Prerogative, the Society lends support to it. The article advocates the dissolution of institutes and authorities constituted under the laws of this land, and it is felt that an official reaction from you should be recorded as to whether the interpretation given above is correct.

"We consider this article to be a direct attempt to interfere in the organisation of the Government of Kenya, and following discussion within the Ministry of Tourism & Wildlife, it appears that the Game Department and Kenya National Parks might be instructed to cease forthwith their association with the East African Wildlife Society unless a course of action to dispel the misgivings created by the publication is implemented immediately.

---

[135] Letter FO 19/4,III. FOm of November 13th 1968 from A. de Vos, Forestry and Forest Industries Division, FAO Rome to I. S. C. Parker.

[136] Africana Volume 3, #9 of 1969. *Encouraging Development is Being Held Up.* East African Wildlife Society, Nairobi.

> *"You may be interested to know that the Kenya Government has already protested to the FAO for having given this same article publicity amongst papers being presented to a Conference held in West Africa, and that our Government cancelled within 48 hours participation by a high-powered 5-man delegation from Kenya because of this. FAO has since formally dissociated itself completely from that paper, and their apologies have been accepted by the Republic of Kenya.*
> *"We hope to hear from you at your earliest convenience."*

Inter alia, this letter shows why white men tended to keep quiet and not comment, let alone criticise the Government. It is a classic, albeit minor illustration of how ideas that did not accord absolutely with Government's were avoided. Ever fearful of deportation and government wrath, aliens opened the door for petty tyrants like Alois Achieng to throw their weight around. What was far more distasteful was not the Kenya Government line, but that taken by FAO. The Organisation never told me of the Kenya Government reaction and never mentioned that it had apologised or formally disassociated itself from the views expressed. FAO's craven performance illustrated typical post-independence syndrome in which aid organisations and western governments bent over backwards towards the independent states, going to astonishing lengths to avoid giving offence. Even when no wrong had been done or offence committed, they would abase themselves and apologise when no apology could reasonably be demanded. In this instance, petty as it was, FAO overtly agreed that I had been in the wrong, whereas it believed nothing of the sort.

Perez Olindo wasn't quite finished with me. Between 1968 and 1970 I had developed interests in Botswana and wanted to establish a fishing company to take advantage of the huge but erratic production of fish when Lake Ngami occasionally filled. Good relations with the Botswana Government inexplicably turned sour and it took several years to establish that Perez Olindo had authored a letter to the Botswana authorities stating that I was *persona non grata* with the Uganda and Tanzanian authorities. This was untrue. Where the Ugandans were concerned our relationship had always been cordial and the only problem we encountered in Tanzania was trying to work as Kenya-based capitalists in Nyerere's socialist paradise. By the time that I had discovered this and had rectified the situation with letters of commendation from both Tanzania and Uganda, Olindo's assertions had done their damage. Years later when I asked him why he had written as he had (which was easy as we have known one another since he was a rookie game warden in the Ivory Room and got on well personally), he said quite simply that he had been ordered by Achieng to sign both the letter concerning the FAO paper and the letter to Botswana.

On the evidence presented up to this point in early 1970, the reader would be justified in assuming that the campaign against me was driven by Achieng's personal animosity. His star was waning, however. Arrested for embezzling funds intended for fishermen, convicted and jailed, he disappeared from public view. Rumour had it that his real crime was too close a friendship with Tom Mboya, the assassinated Luo leader who had posed a credible political alternative to the Kikuyu-dominated Kenyatta regime.

Achieng disappeared off the scene just as Wildlife Services was again taking legal advice on action against the Ministry of Tourism & Wildlife for this harassment and the new Permanent Secretary, G. M. Matheka, quickly brought developments to a head. He met with us, told us that he had personally been through the records and that all the allegations against Wildlife Services and its Directors collectively or individually were baseless. It looked as though, at last, we would have co-operation from Government.

* * * *

Ironically, this was when I took the consultancy to analyse the East African ivory trade. As the reader is now aware, by 1970 the Game Department was collapsing and, in particular, that all government ivory was no longer being sold through the Ivory Room auctions. For those responsible, it was not an auspicious time to have an independent investigator sniffing through the ivory records. I knew about Assistant Ministers in the Ministry of Tourism & Wildlife taking ivory for themselves and about the collector's permits and the poaching that was developing as a result. Yet at the time I did not appreciate just how badly things were going wrong. If I had, I might not have made the formal approach for permission to examine the records that I did. After a month's silence the Ministry referred me to the Chief Game Warden. I talked with him and he gave me permission to approach the Ivory Room staff for information.

With hindsight this must have been a fraught decision. By choice, and given what was going on, I am sure that the Chief's inclination was to refuse permission. Yet he knew that as a one-time warden I knew a great deal about how the Ivory Room worked and that I also knew more than most about ivory and the ivory buyers. If he assumed that I would have got my information anyway, regardless of permission or no permission, he was right. What he did not know was how the Customs records had already provided what I wanted. Giving permission established some goodwill, but at the same time, there can only have been apprehension over my research.

In 1971 I was engaged as an expert witness to examine the jaws of four elephants shot in the Coast Province, two of which were falsely claimed on the forest elephant licences taken out by Messrs Mabinda and McOdoyo in Game Department Headquarters (referred to in Chapter 17). In respect of these cases the police already had evidence of the Chief Game Warden's improper conduct[137]. As explained earlier, in 1972 I had extended my research into the ivory trade by actually going into the Sudan and buying tusks. Mentioned, too, was that in early 1973 the third and final shipment of my Sudan ivory was in transit through Nairobi's Jomo Kenyatta International Airport when it was seized by the Acting Deputy Chief Game Warden.

I met him and asked what the problem was. In his view there was no problem: the ivory belonged to the Game Department and he, as its representative, was entitled to do what he wanted with it. He did not justify the seizure, merely said that he had the power to seize it and had done so. This was a head-on blatant, malevolent corruption. The Deputy Chief Game Warden's actions reflected a general disregard for the law, particularly where ivory was concerned. He felt that there was only one way that I could get the ivory going on its way, and that was by paying. And as the Game Department now owned the ivory, the price would be high. Yet again it was the sort of situation brought about in the first instance by acquiescence. Faced with this sort of behaviour the reaction of most then

---

[137] Letter INQ/33/71/3 of 7th April 1972. *Ibid.*

As has been known for centuries with domestic animals, elephants can be aged from their dentition. A collection of elephant jaws such as this provides a great deal of information on the population from which it was collected. (Photo Peter Beard)

trafficking ivory would have been to negotiate and pay if only because so much ivory business was on the fringes of legality.

Further discussion with the Deputy was pointless, so I went after the Chief Game Warden. After several unkept appointments I physically hunted for him and ran him to earth in a flying training lecture at Wilson Airport – the second of Nairobi's two civil airports. Joining the class I sat at the back of the room. As soon as it was over and while the other students were collecting their papers together, I was at the Chief Warden's side and speaking loudly enough for those close by to hear about the ivory he had 'nicked'. The tactic worked for he immediately asked me to see him at his office. There we had a blunt, one-sided talk. My points were simple. He had seized ivory belonging to a Botswana Company in transit from the Southern Sudan to Hong Kong. The tusks were obtained lawfully, had all necessary documentation to prove their legality, had been seized illegally and that he was asking for a ransom to release them. If they were not released forthwith then there would be an international incident. The Botswana Government would be asked to intervene, the press would be called and I would make sure that every detail was well and truly aired.

The Chief Game Warden was conciliatory. He had not, he said, known of the details or that I was responsible for the ivory moving through Nairobi. His deputy had made a mistake and he would get the matter sorted out immediately, which he did. The ivory was on its way within twenty-four hours. Again, it proved that head-on confrontation when the rules were being bent paid off.

The foregoing sets out a chain of events, all trivial in themselves, but which cumulatively made my presence a nuisance – not to Kenya – but to some individuals. They wanted me gone. It was this that made my published thoughts about the Pope and birth control something of a godsend – so to speak. If they were ground for my deportation, the real causes of embarrassment need not be revealed.

CHAPTER TWENTY-TWO

EBUR

The more I learned of ivory, the more obviously it was entangled with politics. Many conservationists refused to acknowledge this, insisting, that conservation should not be politicised and kept separate in the maelstrom of human affairs.

* * * *

Individuals who felt differently were a small minority. Among them were Tony Dyer, Chairman of the East African Professional Hunters' Association, and Jack Block. Tony realised that the hunting fraternity faced a crisis. The ban on elephant hunting in 1974 was the tocsin for worse to follow. One of few hunters aware of how unpopular the professionals were with the Government, he foresaw the difficulties of keeping his association going and wanted to really know what Government's attitude towards conservation was and to have its involvement in poaching objectively assessed.

Jack Block had a similar interest, though from a somewhat broader perspective. As a businessman he wanted a general measure of the degree to which the Kenya Government was corrupt, and given the Block holdings in tourism and the hotel industry, had a particular interest in what was happening in both Game Department and the National Parks. He also felt that as a member of WWF he should be *au fait* with what was going on.

Tony Dyer and Jack Block jointly commissioned me to report on the ivory trade and its ramifications. The result would be confidential and for their eyes only. We were aware that it would touch on matters that the Government preferred to keep hidden and would have wide political implications. Ten copies of my report were produced in October 1974 and I called it simply *Ebur* – the Latin for ivory. The opening statement illustrates my mindset –

> *"This report arises from a growing awareness of the inadequacy and ineffectiveness of Western involvement in Africa. It is incomplete in that it concerns itself mainly with ivory. However it is presented as a first step to a wider and more detailed discussion."*

Comparing African Customs export statistics with Hong Kong's and the United Kingdom's import records showed big discrepancies in both volumes and values, the general trend being understatement at the African end. Taking Tanzania's discrepancies between its stated ivory exports and other countries' stated imports as an example, I suggested that they correlated with the fortunes of that country's Asian community. The worse off they were, the greater the difference between exports and imports with the latter always being larger.

These Customs statistics merely confirmed what the Indians themselves had said. As faith in their future in Tanzania waned, their desire to move capital and disinvest rose. Ivory was an excellent but relatively cryptic medium for achieving this and was used intensely. The incentive to buy and export ivory was sharply stimulated by the product's rising value through the early 1970s. Yet this was not really new other than in scale, the following extract summing my view –

> "The salient point is that in East Africa the ivory trade and official corruption have gone together for a long time. In view of the commodity's value, this is not surprising. That Indians were prime movers and that corrupt white men sanctioned them is now history. Today it is Africans who dominate the trade in ivory producing countries (through their positions of authority). The only differences are the record volumes involved and the unprecedented blatancy of corruption."

I presented thirty-nine 'cases' of corruption to illustrate both what had happened in the past and was going on in the present, and went to some lengths to get information on the subject that concerned issues other than ivory. The nub of my findings was summed –

> "This disregard for law would be a minor issue if it concerned ivory alone: but it does not. Corruption extends to all walks of national life, in business, land purchases, acquisition of citizenship to specify but few instances. It is blatant in that senior Government officials such as the country's Provincial Commissioners, many Permanent Secretaries, and Ministers themselves are wealthy beyond any savings of salary, and beyond the most optimistic interest rates upon their salaries if the latter were saved in toto.
>
> "In this situation of general corruption, there is little hope that the ivory 'racket' can be tackled or contained on its own. Even though it has grown to unprecedented proportions under African 'management', it is only a small part of the national malaise. To approach the problem from the ivory angle alone is analogous to treating the patient's in-growing toe-nails before considering his generalised affliction of leprosy."

I felt strongly then and now that controlling the ivory trade could only be achieved by going to the very roots of the problem: a problem that had nothing to do with either ivory or elephants and to which the last chapter of the report was devoted.

Tony Dyer received *Ebur* calmly, paid his share of the consultancy fee and, in keeping with, the original agreement, used the evidence to orient his thinking, but kept his information confidential. After hunting was closed down completely in 1977 the Professional Hunters' Association went into liquidation.

Jack Block reacted differently. Having read the report he said to this effect –

> "Just what we have really wanted all along. Now we can get something done. We must get it to both the American Ambassador and to the British High Commissioner."

Quite what Jack meant by getting something done was never spelled out. With hindsight, he knew that matters had gone badly awry in conservation and that corruption was now severely influencing Kenya's commercial life. Yet *Ebur* was the first time that he

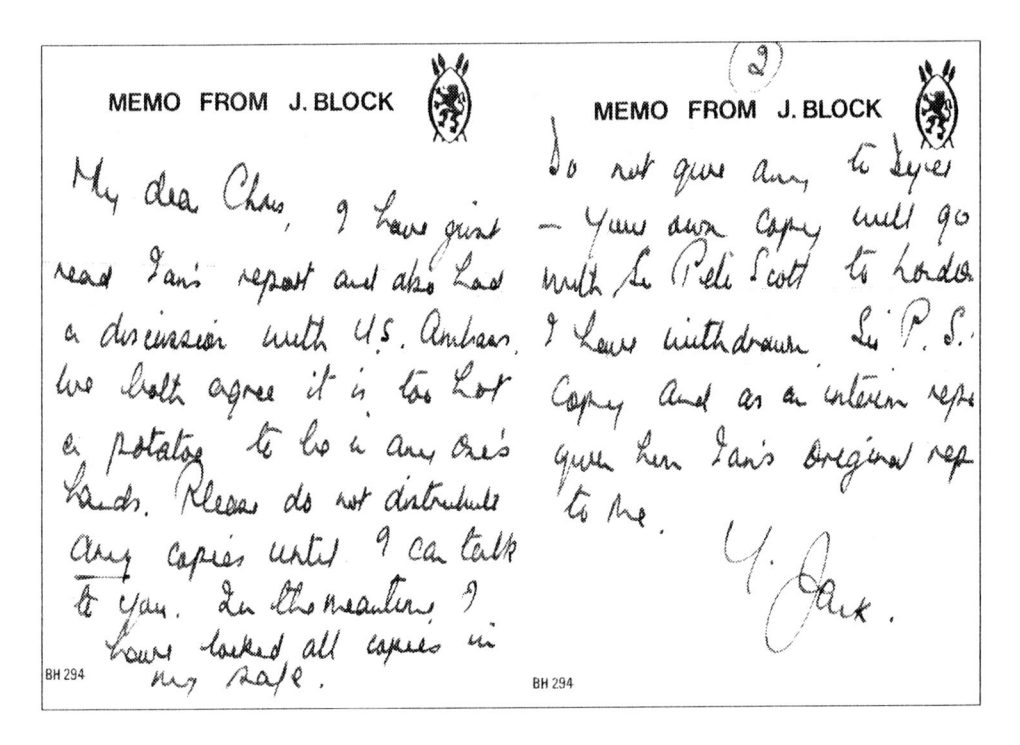

> **MEMO FROM J. BLOCK**
>
> My dear Chris, I have just read Ian's report and also had a discussion with U.S. Ambass. We both agree it is too hot a potato to be in any one's hands. Please do not distribute any copies until I can talk to you. In the meantime I have locked all copies in my safe.
>
> **MEMO FROM J. BLOCK** (2)
>
> Do not give any to Dyer — your own copy will go with Sir Peter Scott to London. I have withdrawn Sir P.S. copy and as an interim rept given him Ian's original rep to me.
>
> Y. Jack.

had seen facts laid out coherently on paper. It wasn't at all obvious to me what either the British High Commissioner or US Ambassador could or would do. Both had professionals gathering all manner of intelligence on what was going on in the country and were likely to be far better informed on both corruption and its ramifications than I was. *Ebur* was never meant for even restricted circulation. Although Jack founded the local chapter of WWF, and I knew he would pass on information to the organisation's international headquarters in Switzerland, I thought that, as agreed, the report was for his eyes only and that it would be used as Tony Dyer did. What he passed on would be distilled through his own knowledge and senses of diplomacy. Taking it as a document to British and American representatives was outside our original understanding. Nevertheless, bowled along by Jack's enthusiasm, we took copies, first to the American Ambassador then to the British High Commissioner – Tony Duff.

Early on the day after Jack and I had delivered these copies of my report, I flew out of the country and was not present when Jack Block 'phoned my wife in panic. Apparently both the High Commissioner and the Ambassador had returned their copies of *Ebur* to Jack saying that as far as they were concerned they had never seen it and didn't know it existed. They thought the Kenyans would think it seditious and even though both knew it was not, they did not want to appear involved with it in any way. They warned Jack not to risk being connected to the report. He asked Chris to remove the copies from his office. He was out when she got there to find a letter in Jack's hand.

The second paragraph was somewhat incoherent. Tony Dyer already had his copy which could not be withdrawn. Neither Tony nor I were aware that Jack intended making a copy for Sir Peter as it contravened our agreement. We never knew what he meant by referring to 'your own copy', but assume that he tangled it mentally with the report completed

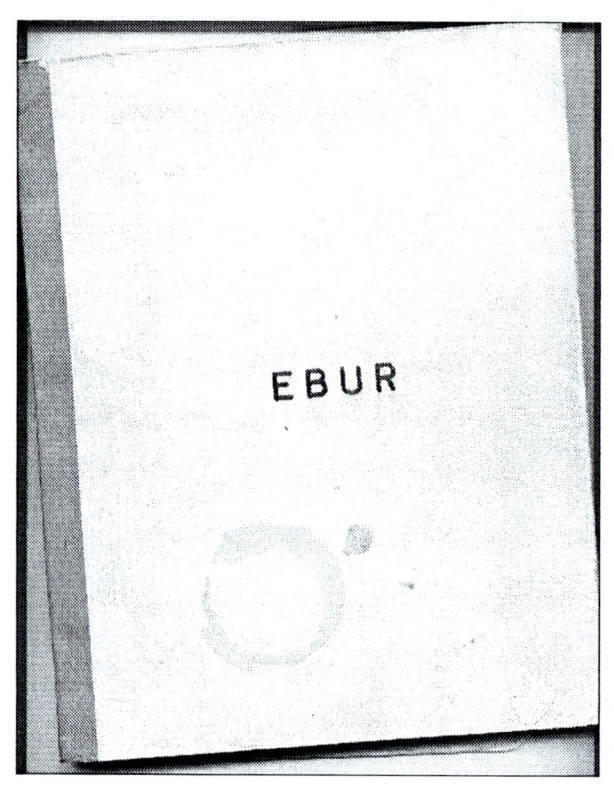

"The report that caused so many ripples.

in October 1973, which (unbeknown to me) he had copied and circulated already. The fact is that Jack had copied both reports and circulated them. Christine removed *Ebur* from Jack's office. When I returned to Kenya a few days later I met him and sought explanation. He felt that I should never have put all the evidence on paper, contradicting the brief that he and Tony Dyer had originally given me. We had agreed at the outset that the report had never been intended for any but their eyes. Fortunately Tony could confirm this.

Jack was a good friend and remained as such, but his reactions showed that playing in the big league of commerce and behind-the-scenes international politics, as he did, he appreciated how pursuing conservation objectives had to be constrained by the wider issues of *realpolitik*. The same reasons also restricted Charles Njonjo's support for Rodney Elliott's final investigations. While it is easy with hindsight to pinpoint issues where stands of principle should have been taken, 'the greater good' at the time may seem different later in history.

The message from my report was that redressing the ivory debacle without addressing the broader issue of a general respect for law was pointless. Despondently Jack had conceded that, to control ivory, one had to go further into raw politics than his commitment warranted. Going as deep as necessary would be akin trying to catch a butterfly in the midst of a pride of lions.

We agreed *Ebur* should be forgotten. Fate would not let us forget, however. One afternoon in January 1975 when I was again out of the country, there was a knock on Jack's office door and Kenya's Director of Intelligence walked in. After pleasantries he had been direct: "Jack, I want a copy of that ivory report." As Jack described it to me later, he had nearly fallen out of his chair. Rummaging in his desk he had produced a copy of my October 1973 report (it became known as the 'white' report from the colour of its cover). The Director thanked him and departed. Twenty minutes later he was back and with fewer pleasantries had said, "Jack ... I want *the* report. Not this one," and handed back the white report.

What Jack did next was disappointing. Knowing I was out of the country he could have said so, delaying things until I my return. Instead he told the Director to ask my wife. Within a short while two Special Branch officers – Ian Barratt and John Arkle – arrived at Chris's office, demanding a copy of the report. She stalled them saying that they would have to speak to me as she didn't know where the report was, but managed the interview so that it was a two-way affair and she gathered from what they unwittingly said that Special Branch clearly knew a lot about the report. As soon as she was free to, she removed all evidence of *Ebur* from our house. Two days later on my return from Rwanda, Chris met me, hugely relieved that her Special Branch friends were not waiting at the airport too. They knew that I was flying in, but unaware that it would be in our own aircraft to Nairobi's Wilson airport, they had gone to the city's main Jomo Kenyatta airport to meet a scheduled flight from Rwanda that, coincidentally, landed at about the same time.

Forewarned of the official interest, I placed copies of *Ebur* in the hands of three journalists: Martin Meredith of the *Sunday Times* in London, Dial Torgensen of the *Los Angeles Times*, and James Pringle of *Newsweek*. All three agreed not to publish the contents unless I got into trouble with the Kenya authorities. If the Kenyans were unpleasant, then these journalists would publicise the evidence as widely as possible. I then called Special Branch and invited Barratt and Arkle to tea, taking the initiative in preference to letting them make the running.

I knew neither Ian Barratt nor John Arkle well, but had met Ian in the past and had heard of John. Both were very pleasant and had a simple message: President Kenyatta, personally wanted a copy of the report. In fact, as I recall, one of them used the words 'climbing around the ceiling' with impatience. It really was a matter of considerable urgency. They took pains to assure me that, irrespective of what the report contained, there would be no retribution. Naturally I wanted to talk to Jack Block and Tony Dyer before making any move so told the two policemen that (i) I did not keep the report on the premises and (ii) even if I did, I would not give them a copy until it was sanctioned by my clients. After all, it was a confidential document that, as far as I was concerned, was clients' property. They countered by saying that they knew that Jack Block was my client and that he had already told them to get it from me. They were unaware of Tony Dyer's involvement. Nevertheless, it would do no harm if they thought there had been several clients, which is what I led them to believe.

We talked a bit about *Ebur's* content, its potential to embarrass the Government not only with the western media and conservation circles, but politically. How would it appear if the information was made public that a prominent member of the Kenyatta family – Margaret, Mayor of Nairobi – was trading extensively with Red China and obtaining her ivory at least in part from Somalis? At the time both Red China and Somalia were not in favour with Kenya and the Kenyatta's political opponents would surely see this as consorting with the enemy? Ian and John left without the report but an assurance that they would be contacted once I had spoken with my clients.

They returned around midnight waking us from deep sleep and were not quite as friendly. The President, so they said, was really angry and wanted the report immediately. Further, if they didn't get it, then Special Branch would get rough. Sending two white men like them was the gentle approach. If they failed then others, meaning Africans, would come and pull the place apart and there was no knowing what else they might do, including roughing up Chris. Behind the threat their discomfort was evident. Both had trained as British policemen and strong-arm tactics ill became them.

I was angry for two reasons: the greater was at being threatened, the lesser because I was shivering. I shivered because standing in my nightwear of a *kikoi* (loin-cloth) it was cold, but angry because they might interpret it as fear. Trying to keep calm I offered both a whisky and stated my position as clearly as I could. I had lodged the report with the media in Europe and the USA. If there was any harassment, indeed if the milkman on his 'pushy' (Kenya slang for a bicycle) had the misfortune to bump into me, I would take it that they had sent him and the report with all its contents would be splashed across the headlines. The journalists holding the report, would be checking twice a day to see if all was well. If I was detained or inconvenienced in any way, they would go public with it. Whether they did or not was thus in the Kenya authorities' hands. They went away with my assurance that on the morrow I would come to Special Branch offices and discuss the matter with their boss, the Director of Intelligence. And that is what happened.

Clearly someone wanted the report badly and would make life uncomfortable until he got it. The assurance that we would be subject to no retribution notwithstanding, was there anything that would make him go back on its word? Chris and I reasoned that, if there was, it would be reference to Kenyatta himself. Knowing that Special Branch obviously knew quite a bit about the report, we could not doctor *Ebur* too much. Eventually we removed one sentence. Chris needed a day to replace the page bearing it.

The Director was urbane and easy going. He confirmed that the President wanted a copy of the report and we had talked a bit about its contents and political ramifications. He assured me that there would be no repercussions. For my part, I said that I was sure the report's sponsors had no interest in publicising the document. Its value to the regime's opponents was recognised, but as we rated them as worse prospects than the party in power (for which he wryly thanked me), we saw no point in acting against the incumbent government's interests.

I asked that, if my clients agreed to hand over a copy of *Ebur*, and once the President had read it, it would be possible to discuss it with him. I was keen to hear direct from one of Africa's leading political intellects just how he saw the whole issue of conservation. I had several meetings with the Director, the last being when I handed over a copy of *Ebur. Inter alia,* in the course of these meetings I got a strong message to the effect that white folk were welcome to come to Kenya, welcome to do research on wildlife or simply to enjoy it. But they had to understand that it gave them no right to interfere in African politics and that, in the last resort, what Africans did with wildlife in their country was their affair: it *was*, after all, *their* country. The Director was sure, however, that the President wanted the ivory trade controlled. Finally, he would try to arrange a meeting for Jack Block and myself to discuss the report with the President. My attempt to find out how Special Branch came to know about *Ebur's* existence was turned aside with a laugh, though with an inference that it came from the British High Commission.

No meeting with Kenyatta took place. It seemed that issues of graver urgency always got in the way. The last of these was the disappearance and death of politician J.M. Kariuki and I never got to speak about ivory with Jomo Kenyatta. In truth I never had proof, other than what the Special Branch men said, that the President knew anything about the ivory report. 'The President wants it, or has ordered it' was one of the most compelling and coercive arguments any civil servant could muster then and now in Kenya. It would have been logical for Special Branch to have used the ruse in getting me to part with *Ebur*. Yet while it may have carried weight with a nonentity such as myself, would they have risked such a threat with someone like Block who did occasionally meet the President face to face and could have raised the subject directly? Now, all these years later, I wonder if they were speaking the truth. At the time I certainly believed them.

* * * *

Jesse M. Kariuki, known to everyone as 'J.M.', was one of Kenya's more charismatic politicians in the late 1960s and early 1970s. During Mau Mau he had been arrested and detained as an adherent and subsequently written a book[138] on this experience. A Kikuyu who represented the Rift Valley Gilgil constituency, his charisma marked him from the outset. President Kenyatta appointed him an Assistant Minister for Tourism & Wildlife in the 1960s and he quickly used this position for personal ends. He was among the earliest beneficiaries of Government ivory deflected from public auction to private purchases at favourable prices. He had used his position to get zebras caught for his ranch at Gilgil on favourable terms and it was almost certainly his influence that got witnesses to deny their evidence and block the Dol Dol Anti-Poaching Safari prosecutions. J.M. was ambitious and clearly wanted a higher post than Assistant Minister. When it was not forthcoming his loyalty to Kenyatta wavered. Choosing the poor and the ex-Mau Mau fighters as his particular constituents, he agitated against the Government and, tongue-in-cheek, spoke out against corruption among the country's most highly placed people.

The hierarchy kept tabs on all wrongdoings by such as J.M., down to the smallest technicalities, filing them for use when an offender got out of line. Alois Achieng's (the Permanent Secretary who had given me such strife) prosecution was such an instance. His association with the assassinated Tom Mboya was sufficiently threatening for him to be prosecuted and jailed for an offence that, on the scale of official embezzlement then prevalent, was almost petty. As it was virtually unheard-of for the wife of someone so elevated as an Assistant Minister to be prosecuted for anything, it was unusual when Esther Mwikari Kariuki, J.M.'s wife, was prosecuted in 1973 for possessing ivory unlawfully. Relative to the value of the ivory involved, the £450 fine imposed mocked justice and an Appeal Judge reviewing the case complained that it was ridiculously low, but he could not increase it as the prosecution had made no request for this to be done. However, I think the object of the fine was not so much to punish Mrs Kariuki as a warning shot across J. M.'s bows which he chose to ignore.

J.M. spoke out ever more strongly: a tactic which found great favour with the disaffected and particularly among the university students. More and more he was challenging Kenyatta and, most seriously, turning some of his own Kikuyu people against the President. In early 1975 there were bombings in Nairobi. The last of them was placed in a briefcase in a Nairobi cinema. The police were warned and evacuated the building, but never found the bomb because a thief had stolen it. Putting distance between himself and the scene of his crime he had boarded a bus without opening his prize. The bomb went off blowing him, several other passengers and the bus to glory, greatly perturbing the Government. While Kenya is unparalleled as a hot-bed of

---

[138] Kariuki J.M, 1963. *Mau Mau Detainee*. Oxford University Press.

imaginative rumour, such was his anti-establishment invective that many thought J.M. might in some way be behind these bombings.

I took no particular interest in J.M.'s pronouncements or doings and was only aware of them from what I read in the papers or heard as passing comment. This changed abruptly on a Tuesday afternoon an agitated Jack Block called me. The gist of what he said was –

> "Ian, ten minutes ago I was in the foyer of the Stanley Hotel having had lunch. J.M. Kariuki came up to me, clapped a hand on my shoulder and said in my ear – Jack, I'm right behind you on that ivory report. How the hell did he know of it? What should we do?"

I was shocked, because this was exactly the sort of situation in which *Ebur* could be used damagingly, and Jack had gone on –

> "Do you think it was this that he referred to in the House the other day?"

Jack meant Parliament a day or two previously when J.M. had ranted against corruption and thought he had said something to the effect that –

> "Next Monday I shall place before this Honourable House a report which documents corruption by the highest in the land."

Ebur was such a report. It was reasonable assumption that if he knew about it, it might be what he intended tabling in Parliament. And if that happened then it was equally reasonable to assume that either Jack Block or myself, separately or jointly, would be blamed for having given it to him. If government concluded this, they would be unfriendly. In our own interests we had to assume that J.M. had been referring to *Ebur* and was going to table it. I would not allow this to happen by default and asked Jack to leave matters with me.

As soon as he had put his 'phone down I called the Director of Intelligence and told him what had happened. Stressing that we had even less of a clue as to how J.M. knew of the report than how Special Branch had found out about it, I wanted Government to be aware he had not received *Ebur* from either Jack or myself.

We will never know what report J.M. was referring to as two days later he was taken away from a function at the Hilton Hotel in Nairobi and, other than by the policemen who took him away, was never seen alive again. A day or so later his body was found behind the Ngong Hills south-west of Nairobi where he had been taken in the dead of night and shot: the incident being heard by some nearby Masai.

That J.M.'s murder was sanctioned by high authority seems reasonable assumption. That it happened because of his growing challenge to the regime in power is equally logical. Yet both Jack and I wondered if *Ebur* had influenced the cascade of events: J.M.'s statement to Jack Block on Tuesday that he was behind him on the ivory report, the relay of this to Special Branch; his arrest on Thursday and murder on Friday made him conveniently unavailable to put a report on corruption before Parliament. It is the sort of query that tends to stick!

I had always assumed that Jack got the idea that J.M. was going to put documents before Parliament from what had been reported in the local newspapers. Writing this book, I felt I should verify J.M.'s purported statement from Kenya's Parliamentary *Hansard* record. With some unease, I have been unable to find reference to it in any of his last speeches to the House. We will now never know the truth and with hindsight it all seems a tad melodramatic. In all probability our tiny influences played no rôle in J.M.'s demise. Yet at the time it seemed real enough for both Jack and I to wonder whether *Ebur* may have been at least a straw on the camel's back.

* * * *

Conserving elephants (or anything else) depended so absolutely on both local and international politics that I was losing focus on conservation in favour of

J.M. – we never figured out how he learned of Ebur.
(Photo Camerapix)

the wider world. What went on with tusks was not peculiar, but an aspect of a general situation. Others must have been aware of what was happening and none more so than the British and American diplomats stationed in Kenya. The Cold War still dominated political agendas. If the price of keeping Kenya within the western camp was a blind eye to corruption, then the exigencies of the moment must prevail. Just over ten years after independence, Kenya's Africans were still sensitive to being manipulated by white men: just how sensitive is now slipping from the memories of all involved. In the 1970s the illusion that African states were *independent* was fostered assiduously – at least in public. For as long as conservationists deluded themselves that their cause was apolitical, it ensured that they would not rock any strategic boats.

Showing that conservation called for political solutions and redress of so widespread and deeply entrenched a phenomenon as corruption in Kenya could disrupt the *status quo*. I had stated that if western conservationists wished to stop what was happening to elephants they would have to involve themselves in managing African affairs directly. Such a view ran counter to the post-colonial liberalism masking the west's anti-Communist strategies.

Ironically a quarter of a century later and in the wake of the Soviet empire's collapse, the intervention in African affairs that I had said was necessary is now as overtly routine as was its denial in the 1960s and 1970s. The World Bank and International Monetary Fund lay down conditions on how governments shall run their economies, aid donors demand 'transparency' – a synonym for honesty – and all demand adherence to plural political systems. The United States now threatens all manner of trade sanctions upon those who buck its will in CITES – a conservation trade treaty. But that is all by the by: in 1975 I understood why the British High Commissioner and his American counterpart did not want to be associated with *Ebur*.

The report brought me into contact with Tony Duff, the High Commissioner, who later became involved in intelligence work in London. I recall one occasion in particular. We had talked of the political implications of corruption, where it would lead and where one's loyalties should lie. Feeling an emotional loyalty to Kenya and things Kenyan, I wanted to contribute to its development. Yet I had retained my British citizenship and while I didn't sing *Rule Britannia* before going to bed at night, I asked Tony Duff what was expected of a patriotic Briton.

He had laughed. "I'm not walking into that trap: you must do what you have to do, but I'll tell you what my job is. Britain has £460 million invested in Kenya. I am here to look after that investment. If anything gets in the way of that, I have to sort it out. It is that simple."

"So if I annoy the powers-that-be and stir up feeling that impinges on Britain's investment in Kenya, your duty would be to shut me up?"

"You said that," Duff had replied with a smile. Of course I did not record our exact words at the time, but that was their gist. Duff had not made a threat, but simply stated facts. Commenting on both my 'white' report and *Ebur* he wrote[139]

> ". . it seemed to me that the conclusion you came to in your last paper [the white report] was wise, & this second one [Ebur] leads me in the same direction. It seems right to me."

Referring to Margaret Kenyatta's ivory going to China –

> "the only thing I am inclined to doubt is the political significance of the destination of the product."

His doubt referred to the political significance of Margaret Kenyatta's ivory going to China. As stated earlier, in the mid-1970s Kenya's relations with China were frosty and there was a contradiction between this outward position and one of the President's family trading with China that I felt could be exploited politically.

Tony Duff agreed the central point: conservation was tertiary in the hierarchy of issues that govern human affairs. The primary problem of corruption could not be solved by going after the Kenya Government for mismanaging ivory and elephants; but if corruption in general was brought under control then the ivory problem would be resolved incidentally. Publicising what was going wrong with ivory would merely annoy those involved and stimulate action not to save elephants, but to muzzle the publicists – as the professional hunters had found out to their cost. It would generate ill will that could spread beyond conservation and into more important fields: hence the warning to Jack Block that he was playing with fire – which he heeded. And when the greater issue of bringing corruption under control was considered in the context of the Cold War, then clearly both the British and Americans felt it less important than keeping Kenya friendly and in the western camp.

Block never paid his share of the *Ebur* consultancy fee. He never said why, but uncharacteristically ducked and weaved whenever the subject came up. I assume he felt

---

[139] Duff A., 1975. *In litt.* to I. Parker, in the author's possession.

that paying would be proof of his rôle in its genesis which, given his many responsibilities, was too big a risk to take. Without paying he could always say that I had exceeded my brief and avoid blame. When the project was originally planned, it had been for Jack Block and Tony Dyer. Subsequently Jack behaved as if WWF was the client. His note to Chris indicated that even if he had kept WWF (Kenya) ignorant of *Ebur*, he had at least discussed it with Peter Scott and, through him, WWF (International) was aware of its contents. Reading between the lines of that note, Jack had broken confidence and blabbed to Sir Peter. Just how far had he gone with Peter Scott and WWF (International)? Had he copied the report to them? There was one way to find out: ask WWF to pay for it. If WWF denied knowledge of *Ebur* or played dumb, then the ball was still in Jack's court. If WWF paid it would prove it must have the report.

On 14th September 1976 I wrote on the matter directly to Prince Bernhard of the Netherlands – the past President of WWF – and on 18th September received a telegram –

"Many thanks letter payment being arranged regards Bernhard."

and I was paid. This establishes another seminal point. From 1975 WWF possessed factual evidence of the scale on which ivory was being produced, knowledge of where that evidence was most easily available (the importers' Customs statistics) and some idea of ivory's deep political ramifications. All this in addition to the lesser report received via Jack Block a year earlier.

* * * *

Ebur made me take stock. Progressively over the years I had lost sympathy with 'conservationists' and their organisations such as the East African Wildlife Society, the World Wildlife Fund (WWF) and the International Union for the Conservation of Nature (IUCN). The more I had had to do with them, the more I doubted their sincerity. It made me look at my own motives for involvement in conservation. An interest in natural history continued together with a wish that, in as far as it seemed reasonable, fauna and flora were preserved. Yet I had no passion or priestly devotion to it as a cause, and my attitude had no more spiritual content than my interest in farming.

In this frame of mind I looked back at recent personal involvements with ivory. There had been a succession of interconnected events. First was the insight Customs records gave into elephant mortality and the tusk trade, and the unexpected finding that far more elephants were being exported as ivory in the 1960s than in the 1920s. This led to ventures in the Sudan and discovery that ivory was used more generally and in far greater volume than understood in conservation circles. It was becoming ever more apparent just how politically important ivory was. Reactions from the United States Ambassador, the British High Commissioner and Kenya's Police Special Branch, confirmed better than words, that ivory was more than a conservation issue. WWF possessed some of the evidence, but from subsequent actions chose to ignore it. Over the next twenty years it went on behaving as though poaching for ivory was a simple, apolitical, cops and robbers business without wider ramification. From then on it was difficult to take seriously either IUCN or it seriously.

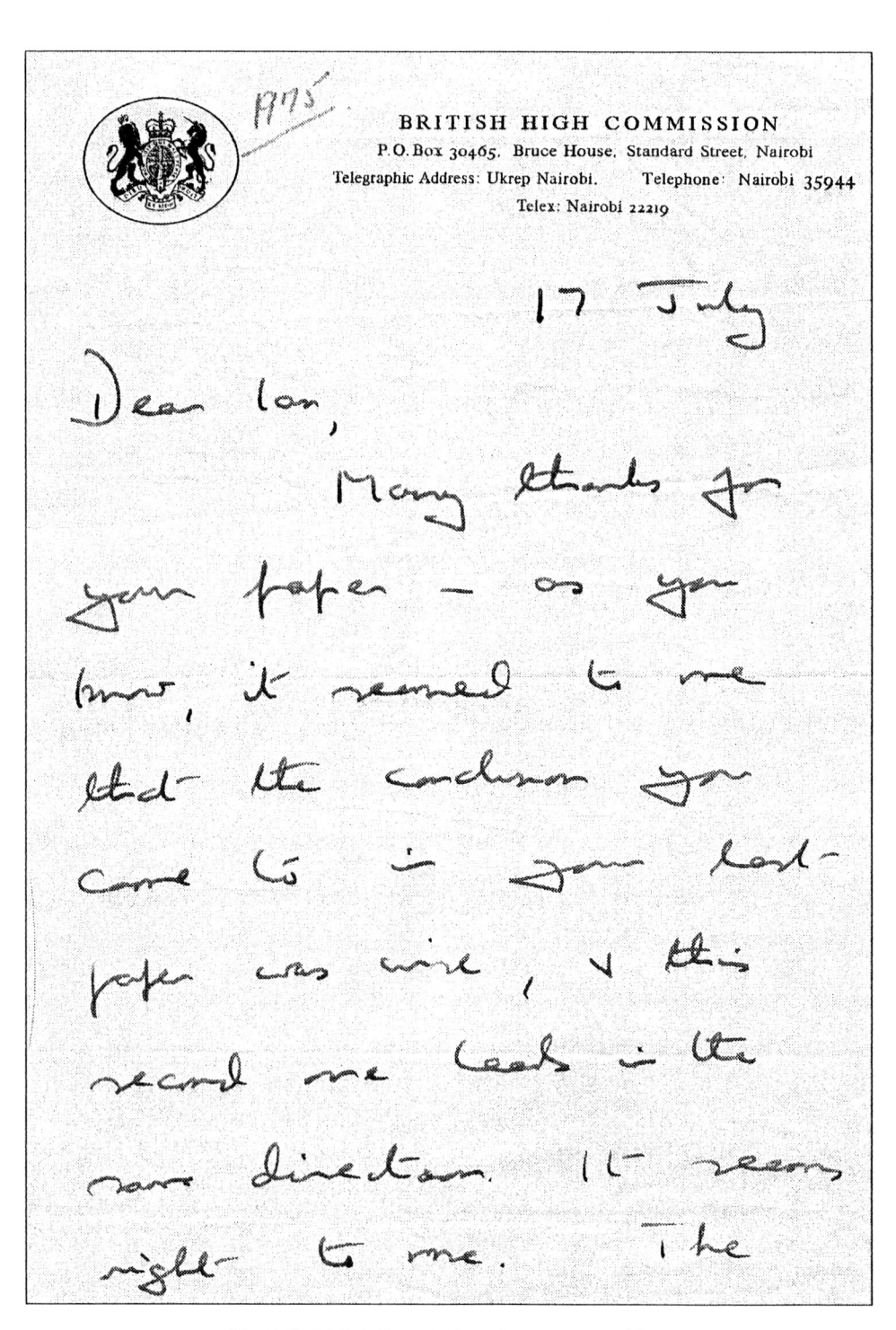

BRITISH HIGH COMMISSION
P.O. Box 30465, Bruce House, Standard Street, Nairobi
Telegraphic Address: Ukrep Nairobi. Telephone: Nairobi 35944
Telex: Nairobi 22219

17 July

Dear Ian,

Many thanks for your paper — as you know, it seemed to me that the conclusion you came to in your last paper was wise, & this second one leads in the same direction. It seems right to me. The

The British High Commissioner's comment on Ebur.

*   *   *   *

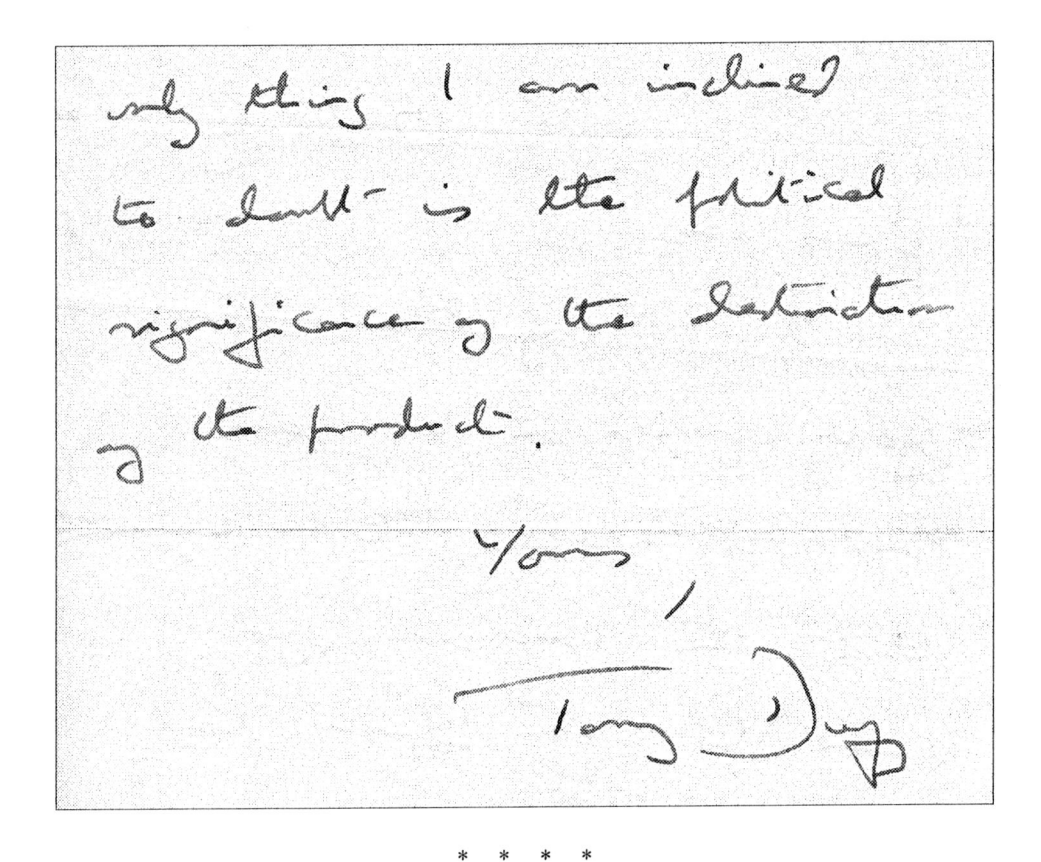

When Rodney Elliott's data were combined with my own, the picture of conservation in Kenya was a shambles. In the seven years 1962-8, when white wardens were still dominant, the auctions sold 163.4 tons of ivory[140]; yet the country's own Customs records[141] indicate that 283.3 tons were in fact exported, 1.7 times as much as was auctioned. This discrepancy is not made up by imports from other countries, for they were accounted for separately in the Customs records. The ivory sold by licence holders for the same period was of the order of only twelve tons annually. What we witnessed in the 1970s was development of a trend that was running through the early 1960s when white officials were still in control. Comparing Customs recorded exports with Ivory Room auctions indicates that the discrepancies go back even before the 1960s. I find them inexplicable and suspicious. Just what had been going on in the 'good old colonial days'?

The gap between Customs and auction totals increased in the 1970s: in 1975, for example, Customs at 106.2 tons was 3.8 times the 27.6 tons auctioned. From overseas Customs records[142] we know that Kenya's exports were yet greater than the local Customs

---

[140] Ivory Room auction records.
[141] Annual Reports of the Commissioner for Customs & Excise assembled in I. Parker 1979, the Ivory Trade ms for the US Fish & Wildlife Service.
[142] I. Parker. *Ibid.*

records, they continued at high levels after the cessation of public auctions and there is no accounting for Margaret Kenyatta's United Africa Corporation ivory. According to her manager, Mr Pusey, it had come mainly from government sources, but for which I never came across any accounting in either the Nairobi or Mombasa ivory registers. If the volume of the licit trade was so much greater than either conservationists or the media were able to accept, then what was illicit raised it even further beyond the believable. Seemingly so much so, that they simply ignored it and, in so doing, themselves became part of the chaos that prevailed.

CHAPTER TWENTY-THREE

A TRIP TO WASHINGTON

When Britain granted independence to Africa, most in conservation were pessimistic about the future. We knew game law was unpopular among Africans: only tax law was disliked more. Some felt national parks might be kept, but that game departments would be abolished altogether and hunting thrown open to everyone. We thought white wardens would surely be replaced quickly as their posts called for no academic qualification. We were wrong. Game departments were retained and some white wardens stayed in service with Kenya's new Government longer than most other categories of civil servant. Inaccurate though our perceptions may have been, they were nonetheless strongly held. We still believed that, few as we were, our presence was the reason why any game existed (as I have pointed out, we were wrong in that).

We reasoned that collapse could be prevented by keeping at least some whites – any whites – in the conservation organisations. The loophole through which to achieve this was science. Few Africans were trained biologists and it would be years before many were. Conservation would have to rely on white scientists for a long time to come.

White wardens started to ask for 'scientific advice' where they had never asked for it before. Suddenly it was fashionable to back management with scientific authority wherever possible. The fashion caught and was self-sustaining as fashions often are and classically illustrated in the mid 1970s when the World Bank agreed to provide Kenya $17 million for conservation. The decision coincided with growing public awareness of what was going on in the Wildlife Conservation & Management Department. The bank's dilemma was ensuring funds were used as intended – among other things for improved law enforcement.

Assuming scientists' 'monitoring' of elephants would deter poaching and ivory trafficking and observe whatever was going on, the bank funded a biological research programme into elephant numbers and distributions. Monitoring elephants alone would be a trifle disingenuous, so for camouflage the survey would take in two other large species: black rhino and buffalo.

Out of this was born the very large herbivore study – or VLH programme – which the World Bank funded. Of course the VLH study no more rode shotgun on the bank contributions to Kenyan conservation than pigs fly at Mach two, but the bank could and did say, "Well, we covered our asses!"

It did not take long to appreciate that scientists would not fulfil the surrogate rôle expected of them and enthusiasm for having them around soon waned among the old-style law-enforcing park wardens. Doyen of them was Myles Turner of the Serengeti. As did most of us, he initially welcomed biologists. Yet like many, he reversed this attitude. In his autobiography[143] Myles made the following summation –

> *"Out of many hundreds of thousands of dollars spent on research in East Africa during the 'fashionable' decade of the 1960's, little if anything has been*

---

143 Turner M., 1987. *My Serengeti Years*. Elm Tree Books/Hamish Hamilton. London.

The pastoral nomad's upbringing and environment fitted him far better to being a field ranger than anything that he could be taught behind a desk. (Photo Peter Davey)

achieved to my knowledge. Far better the money had been spent on antipoaching and education. How much was spent on research in East Africa during those heady years? I have heard the figure of $10,000,000 quoted by a man in a position to know. He may be right. One thing is sure: it was a great confidence trick, and virtually nothing has ever come out of it to help the hard-pressed animals of East Africa."

What Myles Turner did not know, or never mentioned in public if he did, was the opinion of John Owen, Director of Tanzania's National Parks, in the 1960s and early 1970s. John, son of the missionary, Bishop Owen, had been one of the 'Heaven-born' selected for the Sudan's administrative service in the days of the Anglo-Egyptian condominium. He established the Serengeti Research Institute (SRI) as East Africa's biggest wildlife research institute and was brutally frank about his reason for doing so. He felt research was marginally relevant to conservation, but scientists' attracted the western attention essential to raise money for conserving parks. There you had it. The man responsible for foisting the SRI on his wardens did so because research helped him raise funds.

* * * *

Judge Russell Train, American jurist and wildlife enthusiast, knew East Africa from safaris he had made through the region. Foreseeing the rapid changes that independence from Britain would entail he, too, worried that conservation would collapse following the departure of white wardens. His logical solution was not to install white scientists in lieu, but to create a cadre of educated Africans to fill the gap as quickly as possible. More sensible than the 'put-in-scientists' ploy, Judge Train's theorem was nevertheless deficient. He did not appreciate how uneducated wardens of yore really were. Among our numbers there were several who had been to university, but if the majority of us had a common characteristic, it was that we had left school as early as we could.

Such administrative abilities we had were not formally taught and in large measure must have been acquired by cultural osmosis from our backgrounds. They were the same attributes that, during the Mau Mau years, resulted in uneducated youngsters of 19 being appointed DOKGs and given administrative charge of locations that might contain more than 10,000 people. They had nothing to do with a formal academic education and everything to do with leadership.

As apparent from Elliott's work in northern Kenya, Woodley's and Sheldrick's in and around Tsavo, the conservation organisations did have men with leadership qualities. Like us colonial wardens, they were unlettered. They were at home in the bush, good with weapons, self-confident and not averse to a scrap. Unfortunately, in the eyes of the incoming rulers, such qualities were inferior to formal learning. They wanted men with degrees and diplomas: an attitude in keeping with Russ Train's American outlook.

Myles Turner, whose succinct assessment of the value of research was held widely by many other game and park wardens.
(Photo Michael Turner)

To avert what he perceived as looming disaster Russ Train established the African Wildlife Leadership Foundation (AWLF) in Washington with an African headquarters in Nairobi. While the ambition was to produce African leaders in the field of conservation, the title implied that, Pied Piper-like, the Foundation would lead wildlife: where or why was left to the imagination. Many years later its name changed to African Wildlife Foundation (AWF) eliminating the problem. In due course AWF helped produce educated and lettered wardens, but they were young, inexperienced and not leaders of older, bushwise men. AWF's stress on 'education' was as big a failure as was our hope that researchers would plug the gap.

The first head of the Foundation's Africa office was Francis (Frank) Minot, a wildlife enthusiast keen to conserve Africa's fauna. Dry humoured, outwardly cynical but above all pragmatic, he stressed the Foundation's leaning toward education, hoping that African conservation leaders would emerge from it. Unusually for a conservation organisation, AWLF under Frank shunned the limelight and quietly got on with what it said it would do.

The destruction of Rwanda's elephants was held crucial to the conservation of its gorillas.
(Photo Grete Davey)

Frank Minot died in 1972 and was replaced by Robinson McIlvaine, recently retired from the United States' diplomatic service in which he had been Ambassador to several African states (Zaire, Guinea and Kenya). Like Frank he, too, was a patrician WASP (white Anlgo-Saxon Protestant) from the north-eastern states and saw life through humorous eyes. Rob (or Bob – people seemed to use both and if he had preference for one or t'other he, consummate diplomat, held his peace) kept AWLF more or less on the same course as Frank. Refreshingly, his diplomatic experience kept conservation in perspective with the rest of the world's affairs.

Rwanda had a problem. On one hand two small elephant populations, surrounded and increasingly compressed by dense humans, destroyed crops and killed a rising number of people, making both public and legislators unsympathetic towards conservation. They wanted the elephants removed. At the same time the Government was under international pressure to preserve mountain gorillas, which was difficult in the environment caused by the elephant problem. Obviously the solution was eliminate the elephants. The Government asked several high profile international conservation organisations for help, without success. Because it would be unpopular with donors and lower their ability to raise funds, these bodies turned down the Rwandans' request. Eventually the Rwandans approached AWF and Bob McIlvaine. Appreciating the problem, he took the bull by the horns, arranged the elephants' elimination and set foundations for a successful gorilla conservation programme. My company, Wildlife Services, together with the animal trappers – Seago and Parkinson – was persuaded to get rid of the elephants.

The situation was extreme: two groups each of about seventy elephants had been progressively hemmed in by expanding humanity until, in one group that holed up in a swamp, the elephants were never more than 500 yards from people and with the other, things were only marginally better. We shot all adults and the calves less than a year old, and caught all other calves between the ages of one and ten years. These were transferred to the Akagera National Park, but as I have told this tale elsewhere[144] I will only give it a tail here. Eleven males and fourteen female calves were released into Akagera: as wild a bunch of young elephant as existed, which is understandable given the trauma they had been through.

They tended to keep to thick scrub and away from the open – a sure sign of elephants ill at ease. Yet these twenty-five youngsters not only survived, but by the early 1990s had increased to around sixty. The fourteen females must, on average, have produced at least two calves each to have attained this. The oldest of them will now (2004) be thirty-nine and the youngest caught thirty, and their first female calves will by now have produced their first offspring. We know that their numbers more than doubled. Detail of what has

---

[144] Parker & Amin. *Ibid*

Notwithstanding the order to eliminate Rwanda's elephants, we managed to catch all between one and ten years of age, and move them to the Akagera national park. At the time, catching and translocating elephants of this size was quite a feat. Today, full grown elephants can be moved. The translocation and successfully establishing a breeding population was the first success of its kind.

happened is lacking because Rwanda has been in turmoil through the '90s. Yet, for a while at least, translocating these young elephants was an outstanding success. However, once we completed the elimination and translocation, my colleagues and I wrote up the results as a scientific paper[145] which ended –

> *"The 25 elephants in Akagera will be barely half-way through their potential life-span of 63 years ... when Rwanda will have 8,000,000 people. At that time the 'new' elephant population will be under much the same pressures as those which led to the elimination of its immediate forebears."*

We were prescient: in 1997 the Rwanda Government excised two-thirds of the Akagera National Park for human settlement. We did not have to wait our predicted twenty-five years before the elephants were once again losing land.

* * * *

Bob McIlvaine was as concerned as anyone else in conservation about elephant poaching and its apparent correlation with rising ivory prices. Logically, if trade was causing elephant decline, that trade should be understood. Consequently he was among the first to advocate a trade study. We had talked of this and he knew of my previous work in this field and my interest in expanding knowledge from East Africa to cover the global ivory trade.

At the time IUCN, WWF and the New York Zoological Society (NYZS) were planning a survey of the status of Africa's elephants. Obviously an accurate picture of the

---

[145] Haigh J.C, Parker I.S.C., Parkinson D. A. & Archer A. L., 1979. *An Elephant Extermination.* Environmental Conservation, Vol. 6, # 4.

continent's ivory production would throw light on what was happening to elephants. The outcome was a decision, in which Bob McIlvaine's views were influential, to fund a trade study through the US Fish & Wildlife Service. IUCN put itself forward as the proper body to undertake the study. In turn, IUCN sub-contracted the whole of the undertaking to Iain Douglas-Hamilton, but as neither IUCN nor Douglas-Hamilton knew much about ivory or trade in it, the work was sub-contracted to me. As I had personally drawn up the programme in the first instance, perhaps this was no more than tortuous justice.

The trade study was completed in the stipulated year and while it fleshed out the picture we already had with a lot of detail, little changed in principle. The completed report has been a public document available from the US Fish & Wildlife Service since 1979[146] and I have also written about it in *Ivory Crisis*.

Summarising my findings: the volume of ivory exported from Africa was over 1,000 tons annually. The ivory trade report was completed as the IUCN/WWF/NYZS research on African elephant status indicated that in 1979 Africa had a minimum population of 1.3 million elephants. If this was so, the continent's ivory exports in 1976, 1977 & 1978 represented less than five per cent of total elephants. On theoretical grounds a stable elephant population could have sustained such exports. I made the point to draw attention to the order of offtake, never arguing that exports *were* sustainable. From previous research we knew that Africa's elephants were not stable. For decades their range had been shrinking in the face of human increase. Their conservation involved many issues, paramount among them being recognition that outside of the national parks elephant numbers were mostly unsustainable. Equally important, there was no general crisis. On these grounds I did not support moves for banning ivory imports into the USA.

* * * *

Once the ivory trade study was completed the Merchant Marine & Fisheries Committee of the US Congress under the chairmanship of a Congressman Murphy wished to review the situation and, if warranted, to take up Congressman Bielensen's suggestion that ivory imports into the USA be barred. I was asked to testify before it .

I went to Washington in 1979 with an ambivalent outlook. On the one hand I felt alien solutions for African problems would not work and resented Americans trying to make decisions for Africa. On the other it was a unique opportunity to see the American system at first hand. I met Congressman Bielensen of California who had as much interest in the welfare of elephants as I have of catching rats in space. Though we met briefly he came across as a cold fish whose motive for banning ivory imports was garnering votes from the Californians who thought it a good idea. Testifying before the Merchant Marine & Fisheries Committee was anticlimactic. In the depths of my naiveté I had expected dignity and common sense in some proportion to the United States' position as the world's wealthiest and most powerful nation. It recalled the false assumption years earlier when as a little boy I had thought that the most civilised of the British must be those in metropolitan Britain.

Testimony was given in a room superficially resembling a law court. The committee members sat in a crescent behind desks on a raised dais at one end of the room, facing it.

---

[146] Parker 1979. *Ibid.*

Below them within the bight of the crescent and also facing the room sat a palantypist, recording what she could hear and making up what was mumbled as best she could. Facing the crescent in the well of the court was a table at which the individual giving testimony sat. Behind this, also facing the Committee were rows of chairs for the public.

For the most part testimonies were delivered from written scripts which were then handed in to the 'court'. At one point there were no committee members present, but the testifier was told to continue regardless as they would read what he had said later. The proceedings were frequently interrupted by a shrill electric bell summoning congressmen to vote. Whenever it sounded those committeemen present trooped out.

Douglas-Hamilton had also been asked to testify and we had agreed beforehand that in as far as possible we would not contradict one another. Thus he agreed not to support a ban on ivory trading (I believe that the next day he reversed this position by writing to the Committee). I recollect only one question from the Committee and it was clearly posed for the media. It was whether the President of Kenya's wife smuggled ivory.

Having delivered testimony I was invited to the 'Captain's Cabin' – Chairman Murphy's private quarters. This gave an opportunity to pass on an invitation from the world's ivory traders to members of the Merchant Marine & Fisheries Committee or their aides to visit them and see and speak to people involved in Africa, Asia and Europe. Chairman Murphy seemed to have other things on his mind and, as he was being accused of corrupt involvement in the 'Abscam' affair, he probably did. This was just after Prince Bernhard of the Netherlands, President of WWF, had been embroiled in the Lockheed scandal. Corruption was obviously not an African prerogative and, in the circumstances, American comments about crooked African politicians and Asian traders were ironic. Murphy was in no position to accept my invitation, but several other congressmen in the Captain's Cabin expressed enthusiasm and agreed to come on the tour. In the end, however, only Colonel 'Tiger' Howell, Murphy's aide, recently retired from the US Army and service in Vietnam – a gung-ho southerner whose views were a long way right of centre – managed it. The one aspect of Washington that did not disappoint was the McIlvaine's hospitality.

A week or so later I met Tiger Howell at Heathrow and introduced him to Thom Friedlein, the only traditional ivory merchant in Britain, who then conducted us around Britain, Belgium and Germany. Thereafter we visited Kenya, South Africa, Botswana, Hong Kong and Japan.

Taking men from the US Congress round the ivory markets was to prove that the traders were both approachable and open to dialogue. Ivory trading was not a shady, criminal activity but open, backed by tradition and a deep Oriental sense of honour. If my trade study was a success, it was only because the traders had been so forthcoming.

Normally in national and international trade, traders themselves are involved in designing regulatory processes. There were no grounds for assuming otherwise with ivory. Their absence where this commodity was concerned was not because they were inaccessible, but because conservationists and the media *did not want* them to be involved – a very different matter.

Two ideas were firmly entrenched in my mind by 1979. While I did not favour a ban on the ivory trade on any ground – practical or ethical – I was certain that sentiment against hunting, killing, and using animal parts including those from wild animals would get stronger. One has only to review its growth over the past millennium to appreciate that (i) it is not new and (ii) it has grown exponentially. St Francis of Assisi espoused it nearly 800 years ago and in western culture it seems to have become progressively stronger ever since, with fastest growth over the past 200 years. It does not seem to be a phenomenon on its own: the sense that animals should be treated kindly and most recently that they have rights seems integral to the liberal thinking that, among other things, had earlier brought about the abolition of slavery. Such sentiments lie at the heart of modern ideas about being civilised.

Given rising sentiment against animal use, it seemed likely that wild animal products would become progressively more difficult to sell. The assault on the ivory trade was as much an aspect of this general rebellion as anything specifically to do with elephants. Thus, somewhat paradoxically perhaps, while opposing a ban on trading ivory, I nonetheless advised friends in the ivory trade to abandon it: to get out while ahead; to sell their ivory at a good profit and go into something else less controversial. None took the advice, though one or two subsequently rued not doing so.

* * * *

The trip with Tiger Howell was not without its drama. On one day William van de Velde, our Belgian host, chose a superior Antwerp establishment in which to give us lunch. A neck tie was *de rigueur* so as Tiger, Thom and William disappeared inside for a pre-lunch drink, I repaired to our vehicle to collect a tie. Stepping off the pavement I was knocked into next week by a tram which, by British standards, was travelling on the wrong side of the street. Forty minutes later a puzzled Friedlein emerged, could not see me, but took no notice of the crowd clustered about the tram nearby. Several drinks later William emerged and, being more attuned to the local scene, approached the crowd and was told that an Engelsman had been 'taken out' by a tram.

"Il est mort!" were his first words on rejoining Thom and Tiger. Eventually they located me in the nearest hospital where, so they said, they were relieved to find that things were not quite as dire. Concussed, with a split scalp, broken collarbone and sundry abrasions and contusions, they left me with casualty doctors stitching and patching and made off for the delayed lunch. The idea of a bed in hospital was too awful to contemplate. A night back at the hotel would see things right. Walking battered and bandaged through the hotel foyer I may not have looked my best, but nonetheless felt people rude for staring. On the following morning I saw why. The tram had removed the seat of my trousers completely and clearly it was more what was not rather than what was bandaged, that had drawn eyes the evening before.

The accident did not delay the schedule and in Hong Kong Tiger and I were met in style with a Rolls Royce apiece. Tiger was ushered into his and departed first for the hotel into which we were booked. My departure was less elegant for no sooner had his vehicle pulled away than half a dozen traders piled into my Rolls for a conference *en route* to the hotel.

Tiger and I had parted in Tokyo – he heading east across the Pacific back to the USA, and I westward for Africa. Calling in again on Hong Kong there was a celebratory dinner as the traders had been gratified that someone from the US Congress had come to talk with them. The dinner involved toasts and counter-toasts over tumblers of neat Hennessy brandy. The first tumbler expunged both caution and memory. Reportedly several more followed, greatly adding to the damage earlier inflicted by the tram.

Perhaps this is why no-one took my advice to get out of ivory. Perhaps they felt that being mangled by a tram and the banquet in Hong Kong had upset my reasoning.

CHAPTER TWENTY-FOUR

THE SOUTH AFRICAN CONNECTION

John Ilsley originally left London's Metropolitan Police to join the colonial Bechuanaland Police Special Branch. Africanised, he then went into business in South Africa, dealing in curios and game products – among them ivory. Our outlooks were in synchrony. Not only was he interested in ivory for its own sake, as most who handled it tended to be, but better than anyone else I knew, was sensitive to its political ramifications. Touching some tusks newly arrived in his warehouse, I had been astonished that they were ice-cold. An hour earlier they had arrived under a load of frozen prawns from what was then still LM or Lourenço Marques (today Maputo). The rising trickle of ivory presaged the precipitant independence of both Angola and Mozambique in 1975, after which the flow became a flood.

"Every fleeing Portuguese had at least a tusk or two," was the way John had put it. The same had happened over a decade earlier when the former Belgian Congo became independent. Wherever violence, civil war or economic chaos broke out in Africa, ivory flowed away from it.

By 1970 a broad cordon of ruin was developing across southern Africa between the Atlantic and Indian Oceans. Wars were being fought in Angola, Zimbabwe and Mozambique and as the price of copper fell, Zambia was squandering its inherited colonial bank balance in ill-advised economic and social experiment. It was no surprise, therefore, to learn that ivory was flowing into South Africa at a rising rate. By 1975 initial trickles had become significant flows and those buying it in South Africa had unusual insight into what was happening in the lands north of them. History was repeating itself as something similar had happened almost one hundred years earlier.

* * * *

For two thousand years and more Africa's Bantu people had expanded southward from their equatorial origins. Conquering, drifting, exploring, fleeing and settling, overrunning, displacing but, above all, interbreeding with the Khoi and San people who thinly populated the land ahead of them. There had been relatively little of the continent left to colonise when another group, the Europeans settled in South Africa in 1642, blocking further southerly Bantu expansion. From then on increase in Bantu numbers had to be accommodated in lands already settled, or by competing with the whites expanding from their base at the southernmost Cape.

The outcome was manifest in the very early 19th century with the outbreak of the *mfecane* (from the Zulu for 'crushing') or *difaqane* wars (the Sotho version). As so often in human affairs, events driven by evolutionary forces are nonetheless attributed to a people or personality. In this case a cycle of drought augmented by the same outbreak of Portuguese slaving that rolled up the Zambezi watershed and onward into Tanganyika, disrupted southern Bantu societies in the hinterlands of Delagoa Bay. These forces created the environment in which the warrior leader Shaka welded disparate Nguni clans into the

Zulu nation. In 1818 he turned on neighbouring tribes instigating violent chain reactions throughout southern and central Africa, adding to the chaos already set in motion by slaving.

The Ngoni broke away and rampaged northward, shattering the Shona of Zimbabwe, going on to the shores of Lake Victoria before rebounding and settling around the northern end of Lake Malawi. Soshangane took the Ndwandwe clan into Mozambique to form the Gaza Empire and established the Shangaan people. Mizilikazi took his followers and set up the Matabele nation among the ruins of the Shona left by the Ngoni. A Sotho group under Sebetwane (or Sebetuane) fled via modern Botswana and the Okavango to the mid-Zambezi.

These were the bigger movements. In less than three decades of chaos many smaller groups fought, fled, coalesced, split, recombined, broke up again, repeatedly capturing and losing resources. This was added to by Boers trekking away from British influence, together with other white adventurers, missionaries, prospectors and traders hell-bent on making their fortunes. Whether

Shaka, leader of the South African Zulu people in the early 19th century, popularly credited with setting a wave of mayhem in motion that reached as far north as Lake Victoria. While he may well have played an important rôle in this event, its root cause was more likely to have been the Portuguese decision to start taking slaves, as was the case in the Zambesi watershed.

they sought converted souls or coin was of little consequence: all acquired ivory. The greatest hunters were not the whites with muskets, but the far more numerous black people for whom ivory was wealth which could be converted back into the livestock and resources they had lost and so desperately needed. White men certainly shot many elephants, but they also bought any and all tusks brought to them. That they might later say they had shot the elephants themselves added derring-do to their exploits.

The white traders' ox-wagons carried tusks south to South African ports. The high volume of ivory exported from them in the nineteenth century was the product of the political and demographic mayhem that slaving, famine and Shaka let loose, and to which the competing white men added. This South African trade dried up when the 1884/85 Berlin Conference divided Africa into (relatively) peaceful colonial estates. From then on southern central Africa's tusks went eastward towards the great ivory marts in Zanzibar, Mombasa and Dar-es-Salaam. This was the orderly ivory trade that I first came across as a callow young game ranger in 1957.

By 1885 South Africa had few elephants left. With negligible elephants there was no need for rules to regulate ivory trading. South Africa never ratified the 1900 London Convention for the Preservation of Wild Animals, Birds, and Fish[147], reasoning that, as a developed country whose wild fauna had largely disappeared, there was no need for it to do so. For this reason, in the twentieth century South Africa had flimsy laws to control commerce in tusks.

Colonial Africa fell apart a scant eighty years after the Berlin conference. By the 1970s Angola, Zambia, Zimbabwe and Mozambique were in turmoil and their ivory could no longer go to the eastern seaboard. The outlets of Zanzibar and Dar-es-Salaam were closed by socialist and Marxist experiment and Mombasa fell into disuse as the corrupt Game Department closed down the annual auctions. Again ivory started to flow southward and by the late 1970s the situation mirrored that which immediately preceded the colonial years. For the South African ivory trade, the decades 1875-85 and 1975-85 were ironically similar.

As John Ilsley pointed out, importing and exporting ivory into and from South Africa was easy. The official bodies involved – and it was a different one in each of the country's (then) four provinces – had not kept ivory legislation up-to-date. The small volume of tusks produced by Kruger National Park came principally from officially run culling programmes. Issuing such permits as were necessary was a clerical rather than law enforcement affair.

From the mid-1970s onward I intuitively knew official South African interest in ivory must develop rapidly. Indeed, given its involvement in events to the north, this was a certainty. My experience with ivory elsewhere was not the only root of this thinking. It was common knowledge that South Africa's security and intelligence services were efficient. Perceiving themselves pitted against the world, much of what they did had the urgency of self-survival. Passing through Johannesburg I received a tip that *Ebur*'s existence was known and that the security people would not mind seeing a copy.

This was intriguing. How was it known? The urge to find out was irresistible; so, too, was interest in how the South African system worked. Following up the tip led to a meeting in London and another in the Isle of Man at which I was given a special pen – it looked like any other ballpoint – only it seemed empty. Using this I could write 'invisible' letters that could be disguised with a normal letter written on the reverse side. I was given a London address to which I could send information. Messages to me would be via *Reader's Digest* in the post. A mark on the contents page would indicate where I should search for messages. I should turn to that page and find the tiny square of microfilm that would be concealed somewhere on it. Prised off, it could then be read with a low powered microscope.

My aim as a 'spook' was to get leads into what went on with ivory. After passing a couple of innocuous messages and receiving two or three, I was asked to fly an agent into Uganda, and I developed cold feet. I was not so much frightened of flying to Uganda while

---

[147] Anon, 1900. *Ibid.*

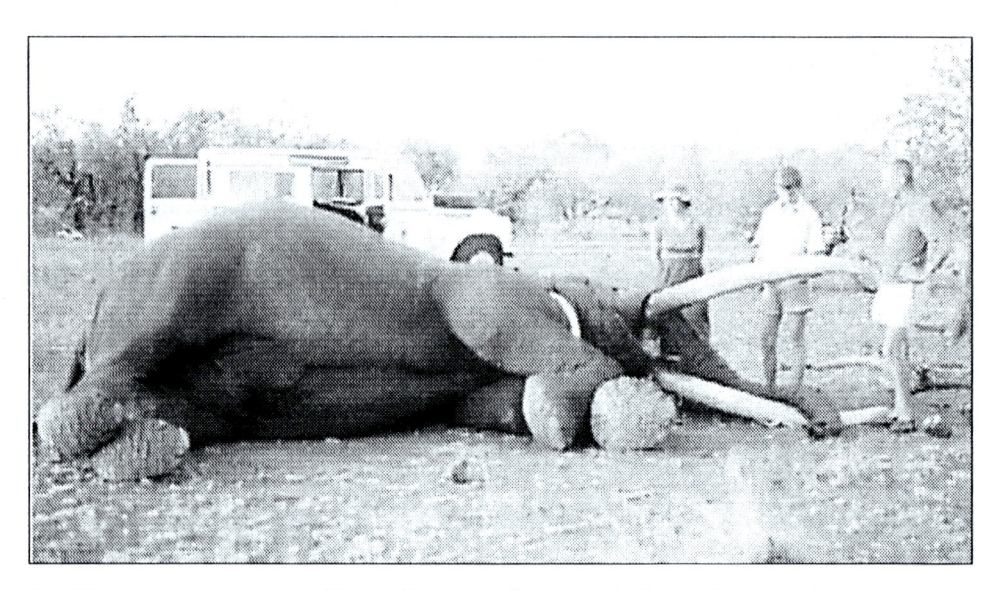

By 1981 the authorities in the Kruger Park were following elephants through radio telemetry and were able to anaesthetise them at will with a skill that had been unimaginable even ten years earlier.
(Photo Anthony Hall-Martin)

Idi Amin was in the saddle, as being involved in I knew not what, for a cause that was politically hopeless. Other than ascertaining South African intelligence interest in ivory generally and in *Ebur* in particular (they never got a copy), I was getting no new information on ivory so I broke off contact. If I wanted to learn about official South African involvement with tusks, I would have to follow other trails.

* * * *

In the 1970s the African states whose economies and administrations had collapsed still had many elephants. Their money may have become worthless beyond their own borders (and sometimes within them), but their ivory was generally available and internationally valuable. South Africa, with its vestigial regulations, was an easier place than most to convert tusks into hard currency or both civilian and military goods. And where the suppliers were political movements whose goodwill and co-operation the South Africans wanted, it made both commercial and psychological good sense to accept the ivory they offered in payment.

The downside of fostering a clandestine ivory traffic was international criticism. To a country that saw itself in a state close to war, this downside was outweighed by the advantages. Putting myself in the South African government's position, the growing ivory flow into the country was not something to be curtailed in the name of conservation. On the contrary, had I been in control I would have encouraged it. To that end the lax rules regarding import and export were decidedly useful. If the conservation authorities slept, let them sleep on. If they woke up, send them back to sleep. In 1974 that was my hypothesis.

Tusks chilled under a load of frozen prawns may have started their journey contravening some Mozambican law, but their arrival in Johannesburg certainly broke no South African conservation law – though it may have been against Customs rules.

A simple example of South Africa's facilitative rôle in regional ivory trading was disposing of tusks from Rhodesia against whom the world in general was observing trade sanctions. In Johannesburg, I saw hard forest ivory that does not occur naturally in Zimbabwe, yet which was covered with Rhodesian permits. It transpired that a Boeing 707 load of frozen Rhodesian beef to Gabon produced a return load of ivory. This was bought by a South African trader, but when sold on to the Far East, it went with South African Permits. The same was done for ivory from the big Rhodesian elephant culling programmes. Even though South Africa was one of the earliest signatories to the Convention on International Trade in Endangered Species of Fauna & Flora (CITES) which demanded country of origin be identified, it did not implement the Convention's requirements until well into the 1980s.

While my first-hand evidence of the ivory flow into South Africa was initially tusks in private firms' hands, I looked for the government involvement logic said had to exist. A strong hint came when John Ilsley was asked to meet the Minister of Finance, Jan Haak, and explain how the ivory trade worked. The meeting, not in official offices, but almost clandestinely in a Pretoria hotel, implied interest in disposing of the commodity. At that time there was a persistent rumour in South Africa that the Government had around forty tons to dispose of and I assumed that John's meeting with Haak related to it.

During the ivory trade study for the US Congress South Africa's Customs statistics and official records were difficult to unravel because they were combined with those of Botswana, Lesotho, Swaziland and South-West Africa (Namibia) as the Southern African Customs Union. Nevertheless local production plus declared imports combined were less than the Union's own claimed exports. Overseas import records showed that exports were substantially greater than the Southern African Customs Union figures. Further, many of them identified the true country of origin and it was clear that South Africa was by far the largest source within the Customs Union. Both import and export figures seemed understated.

While in Hong Kong I saw and measured 727 tusks stamped 'Rundu' in a purple ink and was certain that this related to the town of Runtu or Rundu on the Okavango River where it emerges from Angola into what is now Namibia. All had been imported on South African permits and there was no reference whatsoever on any documents to either Angola or Namibia. The Hong Kong owners had bought these tusks from traders in Johannesburg, among them John Ilsley. On my return to Johannesburg, John confirmed that Runtu was a major source of ivory. There were not enough elephant in Namibia to have produced the number of tusks that were coming forward, for all that they were covered by documents issued by the Namibian authorities. Logically they were coming from Angola.

* * * *

John's partner, Hans Oyrer, wanted to go to Runtu so we chartered a Beechcraft Baron to fly us there. On arrival Hans disappeared to conduct his business while I stayed as the guest of Neville Steyn, the local conservation officer (equivalent of game warden). He had once run a business in Tanzania collecting animals for zoos, but as a South African, had had to leave that country at independence. We had been friends when I was in the Kenya Game Department and, though it had been sixteen years since our last meeting, it was one of those open friendships where we picked up where we had left off, time being of no consequence.

I had been unaware that Angola had any black rhino, and the evidence of 730 horns having left the country in the first three months of 1979 notwithstanding, many 'authorities' denied that it had. Their stone-age logic was what we do not know, cannot be. (Photo Peter Davey)

We had had more time to talk than originally anticipated. Neville's wife was matron at the Runtu military hospital and that evening I could not but overhear her telephone call to authorities in Windhoek trying to get an aircraft to evacuate a casualty. Apparently a nurse returning from leave had been taking a lift in a 'bakkie' (pickup) heading for Runtu when it overturned and she had suffered critical head injuries. As we had a twin-engined aircraft on the Runtu airstrip, it was ridiculous not use it to save the girl. Hans and I offered 'plane and pilot, asking only that the authorities meet the cost. They agreed and we stayed in Runtu while the pilot flew the casualty to Pretoria, an eight hour round trip.

Neville explained the ivory situation succinctly. Most tusks for which he wrote permits came across the Okavango from Angola, as did many rhino horns. I inspected a batch of such horns in a Veterinary Department store where they were to be sprayed with formalin to ensure disease did not enter South Africa whence they were all destined. Clearly with one department issuing permits and another enforcing veterinary hygiene routines, the authorities were not acting covertly. The cross-border trophy trade was open, even if questionable. Further evidence of official sanction was a letter from the local police commander recommending that the trade be fostered.

I examined and copied the duplicate folios of all 114 permits that Neville had issued in the first three months of 1979. They covered 3,923 elephant tusks and 730 black rhino horns. The latter intrigued me as I had been unaware that Angola had any black rhinos. Inevitably I asked Neville how this trade fitted with his rôle of conservation officer.

He answered in two parts. First, regardless of what he thought, he had no choice as issuing permits was part of his duties and he had been ordered to do so. Second, as Angola and Namibia were both going to be run by Africans, the animals were doomed. If the security forces felt the trade was necessary, who was he to argue against the national interest? In his view game in African hands would not survive. African independence destroyed the game, not Neville Steyn issuing permits!

He held this view sincerely. Neville knew what was happening in Kenya, Uganda and Tanzania. If the pattern of conservation generally in Africa had been influenced by developments in Kenya in the first half of the century, they were also equally influenced by what happened in the second. Many white South Africans' convictions may have been rooted in a racialist philosophy, but the Kenya Game Department's track record with ivory would, alone, have produced the same attitude even among the most liberal thinkers. Nothing takes place, least of all with ivory, in isolation.

Some of the permits Neville had issued were made out in civilian names, but most were to men in the army. Somewhat unusually, I thought, they were only valid for three months. Yet more peculiar, there was one permit for 876 tusks, fourteen in multiples of five, ninety in multiples of ten and nine in multiples of 100. I had never seen anything like it elsewhere. The permits for rhino horn included three for 200 horns. Elephant tusks and rhino horns come in pairs or through accident or rare genetic disposition as singles.

It was not until years later when South Africa's President Nelson Mandela appointed a Commission to look into the former apartheid Government's use of ivory, that Judge Mark Kumleben came to what I think was the right conclusion. Dividing tusks and horns into neat fives, tens and hundreds was the product of an orderly but unimaginative military mind. The evidence was far more in keeping, not with individuals crossing the Okavango and acquiring tusks and horns independently and at random, but with a stockpile being broken down and fed into trade in smaller units for concealment. Neville must have known this but if I was too dumb to ask about it, he had no need to explain. At any rate, I got what I had set out to get: proof that the South African authorities were trafficking ivory out of a neighbouring state on a substantial scale. Given the ancient and recent history of ivory in African affairs, I had guessed that they *had* to have been and it was gratifying to prove surmise correct.

* * * *

In Hong Kong I noted that the Rand Company imported exclusively from Ah(Mr) Pong and Sons of Pretoria. In turn most of the Rand Company's applications to import were of ivory originating in South Africa and Botswana. In Johannesburg and Botswana Ah Pong

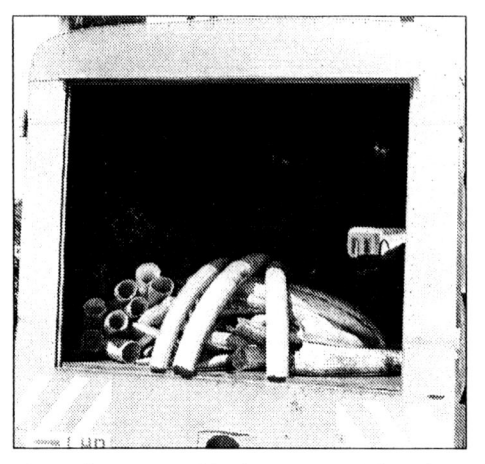

A ton of ivory seized by Customs in Malawi from under a false floor in a truck.

The compartments under the truck floor.

was well known. Specifically, the man best known was Cheong Pong, son of Ah Pong who ran the Rand Company in Hong Kong.

Cheong was spoken of with respect as the biggest ivory buyer in southern Africa with close connections to the South African government. Rumour is pernicious, usually inaccurate, but often holds grains of truth. If a small fraction of what was said about Cheong Pong was true, I had to meet him. John Ilsley arranged it. Urbane and courteous, Cheong appreciated that I was investigating the ivory trade, though I stressed my sympathy for the merchants involved and also that I had both produced ivory in the past from culling work and traded it on BGI's behalf in the Sudan. My notes from the meeting read –

> "Handles Rhodesian ivory; buys Mozambique thro' Swaziland; Zambia through Botswana as machinery parts – Minister; Angola?; Agrees some goes completely hidden – sea freight; Obvious that a considerable quantity goes by sea from SWA; Supplies Red C. through Macao rhino horn; Concedes getting supplies from Zaire thro' Zambia."

Reading these notes subsequently I interpret them to have meant that Cheong Pong admitted handling Rhodesian ivory which he bought openly at the Salisbury auctions. He had brought Mozambique ivory through Swaziland; that he brought Zambian ivory through Botswana disguised as machinery parts and that a Minister was somehow involved. Angola with a question mark indicated that I had obtained no information on it from Cheong. I had thought that he agreed that some of *his* ivory went completely hidden, as would be the case if in a container, and was not declared as ivory. He had said that it was obvious that much ivory is exported from South-West Africa by sea and that he supplied Red China with ivory and rhino horn through Macao. Finally he, Cheong, conceded that he got ivory from Zaire through Zambia.

Seventeen years later I was asked by the Mandela-appointed Ivory Commission if other interpretations could be put to my notes. Specifically, could there have been any possibility that he had been referring to another party who moved ivory as machinery parts? I did indeed know a case in which ivory left Zambia as machinery. A white man had sent it for overhaul to Queensland, Australia. The arrival of tusks in Hong Kong from Australia led me to unravelling that trail. Might this have been what Cheong was referring to? I do not think so. Similarly, was there any possibility that he was referring to other people's ivory being sent hidden by sea from South Africa and/or Namibia?

Knowing my own method and style of taking notes I feel that my interpretations are correct. Yet at a distance of seventeen years, and in all fairness, I could not be certain. Nor was John Ilsley who had been present at the meeting. Both he, Pong and the judicial officials asked why Pong would implicate himself to a stranger? To that eminently sensible question I could only reply that therein lies the skill of an interrogator. Years earlier the same might have been asked of both Woodley and Sheldrick, yet with my own ears I heard them get individuals to admit to crimes without a shred of evidence to start the confession. An element in the art is making a bridge of sympathy and understanding with the interviewee. Even more, is the ability to make him or her think that the inquisitor knows more than he or she actually does.

With the caveat that seventeen years after my notes were written, I could not swear that my interpretations implicated Cheong personally, I nonetheless believe the information obtained if only by the same logical process. Why would he have told me anything about sea shipments, supplies to Red China or getting supplies from Zaire for that matter? Overlooked at the time of the Ivory Commission was a telex lying deep in the copious notes I recorded in Hong Kong. It was from Ah Pong in South Africa to the Rand Company in Hong Kong and read –

"Also please find out whether they [meaning the Hong Kong govt] *accept Zaire permits as we are busy negotiating with a company in Malawi for 1 tons* [sic] *ivory from Zaire."*

Subsequent to getting this and interviewing Cheong, the investigative arm of the Customs & Excise Department in Malawi seized a ton of Zaire ivory being carried by a Zambian transport firm en route to Ah Pong in Johannesburg, back via Zambia and Botswana. Wherever I got corroborating evidence, it all tallied with my interpretation of what Cheong said. Be that as it may, ivory sent by container or in any way hidden at the points of export would, in all probability, have arrived hidden at the destination and not appeared in either export and import Customs records. Then again, there was the issue of how China got its ivory. As that country did not then publish trade statistics, I could not refer to them. Yet I calculated China's large ivory crafting industry consumed minimally 60 tons a year, though intuitively thought consumption far higher. Cheong's evidence took me in the right direction and was later borne out.

With sufficient time I would have pursued the leads Cheong Pong provided. So too would I have followed up the import by Italy in 1973 of fourteen tons of elephant ivory from Uruguay. There were good connections between the respective right-wing governments of South Africa and Uruguay and Argentina. I bet that the fourteen tons had originated in South Africa and gone to Uruguay before surfacing in Italy. Yet while I believed that far more ivory was moving through South Africa than the muddled statistics revealed, in my final report on the ivory trade I did not dwell on the South African aspects.

The facts were there in tabular form for anyone to get the basic picture. Playing down the evidence I concluded –

> *"The South African/Namibian data do not reconcile readily with Botswana's statements and the low volumes I believe to have been produced internally. Suffice it that the information is too imprecise to analyse accurately. This is in part a reflection of the political turmoil racking southern Africa, which creates a climate for concealment of trade statistics – particularly of a commodity such as ivory."*

I reasoned at the time that as the country was an international whipping-boy over apartheid, anything derogatory about it, no matter what, was hysterically splashed across the world's papers. As I saw it, my duty was to produce a balanced picture of the world ivory trade. Substantial as the ivory passing through South Africa may have seemed, it still amounted to between only two and five per cent of the tusks leaving Africa. The overwhelming problem, if it was a problem, lay four-square with the continent's newly independent countries exporting 95% of its ivory production. Had I put a spotlight on South Africa's five per cent, exports from the rest of the continent would have been treated as peccadilloes, or ignored. I was of course wrong. The whole ivory report was ignored and of no consequence!

CHAPTER TWENTY-FIVE

A BROKERING SERVICE

The end of the colonial era saw the end of Kenya's regular public ivory auctions. There was no formal announcement that the auctions were to cease; just progressively greater amounts of ivory disappeared by private treaty until there was embarrassingly little to sell publicly. The last public Mombasa auction was in 1976. An auction catalogue was produced in 1979, but it was 'front' to give an impression that the ivory and rhino horn had been sold by proper process. It was given to eight carefully selected buyers only.

Uganda, which since the nineteenth century had sold all its ivory through the Mombasa auctions, decided to go it alone in 1967. Again, the reasons were the same as in Kenya. Those controlling it could not bear to see such wealth slip by without some (a lot) sticking to their fingers. In due course the amount that stuck grew progressively greater. However there was a problem: Uganda reached the bottom of its elephant barrel in the mid-1970s. Nevertheless its Game Department found an ingenious way to 'stay on the gravy train'. While it had few elephants left, it could still issue ivory export permits that were worth a lot of money. Hence I came across permits in Hong Kong, as did Rodney Elliott in Nairobi, made out in the name of a company registered in Arua, a hamlet in north-west Uganda close to Idi Amin's home.

During Amin's reign there was little likelihood of anyone going to Arua to check whether such a company existed. The brainchild of an Asian lawyer who had fled to Nairobi after Idi Amin's ethnic cleansing, permits were sold as a very profitable little scam by the Uganda Game Department in the name of this fictitious company and used to cover ivory collected in Kenya and elsewhere. The owners then said such tusks were in transit from Uganda to avoid the normal cut taken by Kenyan officials.

Tanzania, theoretically, kept its auctions going until the CITES ban on ivory trading in 1990. Nevertheless a great deal of the ivory which arrived at the Ivory Room in Dar-es-Salaam left it through sales that were neither public nor in the public interest. I knew of only three countries where ivory was openly and properly accounted for: Botswana, Malawi and Zimbabwe. I had little confidence in South Africa after finding that the Kruger National Park directed a portion of the Park's ivory production at a price well below international levels into the hands of one company. The arrangement may have been innocent, yet, with a cynicism born of experience, and noting that the company's principal shareholder was the then Director of National Parks' son-in-law, I would not have bet on it.

Since early in the 1970s, Malawi's Chief Game Warden, David Anstey, sent me a reply-paid telegram asking for the prevailing international ivory prices whenever his department was due to sell ivory. So small a service I provided *gratis*, but it got me to thinking about a wider ivory brokering service. The idea gained strength from a working relationship with the CITES Secretariat and particularly with Chris Huxley whom I had got to know when he was introducing the CITES procedures to Hong Kong as a member of the Colony's Agriculture & Fisheries Department. He brought to CITES a lively intellect, a good sense of fun and a pragmatic and professional rather than emotional approach to conservation. Though it should not have been so, he was unique among the Secretariat staff in combining grasp of the Convention's laws with knowledge of biology.

Djibouti seemed to be the world capital for ospreys.
(Photo Dave Richards)

An ivory brokering service had an outside chance of bringing some order to the mess that African ivory sales were in. The earlier regular auctions let sellers decide when to put stock on the market or withhold it to their best advantage. Similarly buyers also knew what was on offer and when. Everyone benefited from the market's stability and transparency. Once ivory was sold clandestinely, traders were forced to buy opportunistically and speculatively. Inevitably they hedged by lowering prices and increased this loss to Africa further by what was paid into officials' private accounts. A brokering service could restore much of what had vanished when the auctions died out.

I thus offered to broker African government ivory, placing tusks on the market at the most advantageous moments. As important, end-users could be approached directly and middlemen cut out to further raise African returns. I hoped that the higher prices would make it worth abandoning the haphazard selling that then prevailed. For this service, which would be available to anyone and not confined to governments, I would charge the traditional two to three per cent of sales values that had been paid to the Ivory Rooms. Naturally I was attracted to the profits I might earn. Yet, as it would cut out the civil servants currently benefiting from their countries' ivory production, I had no illusions about the proposal's low chances of success. I was not particularly surprised when no government acknowledged my circular. Yet ... nothing ventured, nothing gained.

* * * *

Several times in 1980 Thom Friedlein, the British ivory merchant, had been offered four tons of ivory in Djibouti. Its origins were unknown, but said to be Somalia. The CITES Secretariat knew nothing of it, Djibouti not being a party to the treaty at that time. Thom asked me to go there, examine and value the tusks, establish their origins and see if they had any CITES-acceptable documentation.

Stepping out of the aircraft into a wind so strong and hot that I thought it was effluent from a jet engine blowing across the airport apron, it took a full fifty paces to appreciate that this was just Djibouti. A 46°C temperature was not unusual for that time of the year. Lodged in one of the better hotels, I wondered what the lesser establishments must be like. My room offered the option of a reasonable temperature and an air conditioner that roared like an unsilenced tractor, or quiet and heat of over 40°C. The single bonus was the number of ospreys I could see in the bay in front of the town. In most places single ospreys are worth remark. Djibouti must be the osprey capital of the world as virtually every time that I looked out I saw half a dozen or so.

The ivory lay in a warehouse at the docks. Each tusk needed to be weighed, marked and placed it into different size and sex categories. Then each category had to be broken into qualities: sound, slightly defective, seriously defective and rotten.

As I wrote earlier, let someone pick up a firearm and those brought up with guns can instantly tell if that person is familiar with weapons. So it is with tusks. There is a knack to handling them that revolves about knowing where the centre of balance will be, how to hold the long, curved shape, how to skid the heavier ones across a floor and how to stack them. The way a person handles tusks instantly betrays whether or not he has done it before. It was plain as a pikestaff that the labourers in this warehouse were inexperienced and I was obviously not looking in on a regular ivory route.

As a product of the maritime dhow trade over the past millennium, Kiswahili is spoken widely throughout the Arabian Gulf, the ports of southern Arabia, the southern Red Sea and down the Somali coast. Several of the warehouse labourers spoke it sufficiently for us to communicate, which was useful as I speak neither Somali nor French. The labourers were happy to pass on what they knew about the tusks as we worked. As the owners did not speak Swahili they were unable to follow the dialogue – which was useful too as I had a check of sorts on what these owners told me.

I never knew who actually owned the tusks. Periodically individuals were introduced as part-owners. Not all liked what they saw, apparently not having been told what was happening to their ivory. Being Somalis they assumed that it would be to their disadvantage. Every time one of these strangers turned up the work stopped. They argued amongst themselves and whether it was the heat or regular Somali behaviour, they routinely lost their tempers. Facing one another the vituperation invariably ended with one grabbing the other by beard or chin and holding it firmly to enforce eye-to-eye contact: this local habit is funny to watch, but offensive to the victim.

Merchants and labourers knew little except that the tusks were originally from Kenya and southern Somalia. The ivory had been in Djibouti for a year or so and was part of a larger stock of which forty tons still remained in Mogadishu under President Siad Barré's control. Everyone was vague about how the seven tons in Djibouti had become detached from the main stock, but it was covered by government permits issued in Mogadishu. The CITES Secretariat in Switzerland confirmed that they were internationally acceptable.

In shapes and ages the tusks were similar to what I would have expected from Kenya's Tsavo region and the drylands northeast of the parks across the Kenya border into southern Somalia. Many were stained with the red soils characteristic of that area. From their surfaces and staining, many tusks had been buried for at least one and perhaps two wet periods. All plasticity in the paper-thin ivory at their bases was gone, confirming none were fresh. If they had been in Djibouti a year and buried for one or two seasons before that, these tusks were at least two years out of the elephants and probably more like four. This put the period of killing at around 1976-8.

When an elephant has been killed and the tusks are chopped out of the fresh skull, some marks are *always* left on the tusk surface. Each hunter has his personal style and tusks prepared by individuals can be recognised from this evidence. It is obvious whether the work was by an experienced or inexperienced person. The careless chips, axe cuts and damage to these Djibouti tusks indicated that they had been cut out by beginners. The subsequent scraping and cleaning was likewise done by people with no 'tusk sense'. This was further confirmed by some of the largest and most valuable tusks having been cut in two with an axe to make them easier to carry. No-one with ivory knowledge would have done this as it halved their value. Bullet holes in the ivory were mostly AK47 calibre.

Many tusks had bullet damage in the body of the tusk outside the elephant's head, which was consistent with automatic fire sprayed by amateurs rather then selected shots by experienced hunters. A surprisingly high proportion showed wholly or partially healed bullet damage in the basal section. In other words there had been time for a set of earlier wounds to start healing months or weeks before hunters again came across the animal and killed it.

The picture that emerged was of a band of amateurs knowing little either about ivory or hunting suddenly getting loose among elephants. The whole was in keeping with the wave of Somali poaching that swamped Kenya between 1973 and 1979 when many footloose young Somalis went south to seek their fortunes. Knowing little about ivory they had been unable to dispose of the tusks immediately, which is why they had been buried. Being aliens in Kenya they had chosen to carry them back to Somalia. At every point along this line, debts and obligations had been incurred until the greatest obstacle to selling the ivory had been getting agreement between all who had become involved. A basic knowledge of tusks, how they grew, how to handle them and so on had allowed me to build up a surprisingly good forensic picture of the background to these Djibouti tusks. In due course they were exported.

Several years later I was again asked to go to Djibouti to examine another very similar consignment, but I never got far with it. By then a Sheraton Hotel had been built with improved air-conditioning which made my time there more comfortable than previously, though from several stories up I could no longer see the ospreys. The man who owned the ivory had apparently pledged it as collateral for loans from every bank and money lender between Port Said and Aden. When word went out that a paleface was examining and valuing his tusks, there had been a rash of litigation to prevent them leaving the country. Before I made my examination all interested in the ivory were summoned to the palace of justice.

To impart Napoleonic grandeur to this edifice, French colonial architects had placed it on high ground with a long flight of steps leading up to the entrance. In Djibouti's heat, having to sweat one's way to the top, must have ensured that few frivolous cases reached the courts. At the summit angry litigants had been arguing among themselves as to who had prior claim on the ivory. The tusks' owner who owed money to all was probably meeting his creditors *en masse* for the first time. Tempers flared and one man, angrier than the others, held him briefly by the beard in classic Djibouti grip to ensure he heard the tirade of unkind words. Released and overcome with emotion, the ivory's owner turned his back on the gathering. That had been unwise, for another gentleman was so overwrought that with a short run he launched an almighty kick at the ivory man's backside, propelling him forcefully into space and down the imperial steps in a cartwheeling confusion of arms and legs. Landing at the bottom, amazingly intact, the outraged party came back up the steps five at a time, and fell upon whom he thought had kicked him. That it was the wrong man added to what was now first class opera. The crowd's temper miraculously improved and we all shouted encouragement and advice in the greatest good humour until some gendarmes summoned us before a magistrate. He asked why I was in his court? No wiser than he, I was bidden leave and left. I don't know what happened to the ivory, but I enjoyed the second trip to Djibouti.

Convinced that my brokering scheme had come to nought I had turned to other things when a communication arrived from Somalia asking for my help. As with all things Somali, it warrants a chapter to itself.

CHAPTER TWENTY-SIX

MOGADISHU & ETHIOPIA

I had almost forgotten about ivory brokering when in early 1985 Zimbabwean Rowan Martin, on a consultancy for CITES, suggested writing to the Manager of the National Range Management Agency of Somalia, and offering my services to help dispose of forty tons of ivory still in Mogadishu. In May 1985 I received a telephone call from Mogadishu. Over a very bad line the Manager wanted me to sell fifty and not forty tons of ivory for the Somali Government and I agreed to come and talk with him. Presumably this was the large stock of tusks of which I had first heard in Djibouti in 1980. Since then there had been word of it in many quarters. Chris Huxley knew of it from the traders in the Far East, but Rowan Martin's visit to Mogadishu confirmed it existed.

All through the Kenya Game Department records of the last century, there are references to Somalis taking tusks and, above other things, rhino horn, from Kenya into Somalia. As indicated in Chapter Five, even during the war years, ivory and horn were flowing from Kenya into Somalia. Rodney Elliott's work confirmed that it went on through the 1950s into the 1970s.

In the early 1980s in Djibouti I had seen one lot of four tons of ivory and another of seven tons with Somali permits. Chris Huxley in the CITES Secretariat knew of a consignment of seventeen tons of Somali tusks arriving in Singapore. My own records from customs import statistics were as follows –

Country	Year	Weight kg	Price $/kg
Japan	1981	1,072	83.35
	1982	852	83.80
	1984	7,274	62.75
France	1982	3,000	79.00
	1983	2,000	63.40
Italy	1978	1,760	47.23
	1979	532	27.48
	1980	337	56.07
Total Weight kg		**16,827**	

Yet while these dribs and drabs of rather disparate information confirmed that Somalia trafficked ivory across the twentieth century, relatively little ivory had turned up legally from Somalia in overseas markets and not much was known about the country's trade in it.

Working on the fifty tons in Mogadishu seemed an opportunity I should not miss. In addition to providing new information, it might illustrate to other states the value of the brokering service I could provide them. With these points in mind, together with the prospect of a substantial commission, I accepted the invitation from Mogadishu.

* * * *

The litter of derelict aircraft on both sides of Mogadishu's airport runway was convincing evidence that though heavier-than-air machines might fly, they all eventually succumb to gravity. Four De Havilland Vampires, an assortment of models from behind the Iron Curtain that I had never seen before, a dead Dakota or two together with a scattering of American light aircraft mutely testified to Somalia's regard for the twentieth century. On the apron in front of the airport terminal, a Boeing 707, its fore-end propped up on blocks and minus its nose-wheel, had clearly been there for a long time. Elsewhere, even in disorganised Africa, such evidence would have been discreetly towed out of sight. Here it was left for all to see, not so much as evidence of Somali deficiency, but of the unreliability of machines made in the infidel west! Somali capacity to delude themselves that mishap is the fault of others is infinite.

At the airport Catch 22 manifested itself immediately. All arrivals had to acquire an exchange control form without which one could not approach the Immigration Control desk to formally enter the country. The form was only obtainable from a desk well beyond Immigration Control which one could not get to before checking in with Immigration Control which, as stated, would not entertain anyone without the form. Just to make the point clearer, the form cost 100 Somali shillings. As it was illegal for anyone to take Somali shillings out of Somalia, anyone arriving from abroad with 100 Somali shillings in hand had obviously broken the law and could be prosecuted. As one could not get Somali shillings outside Somalia one had to get this from the bank inside the airport building. As no one was allowed past Immigration without the form, there was small chance of getting the 100 shillings with which to buy the form so essential to entering the country.

Believe me, this is not nearly as ridiculous on paper as it was to those experiencing it. Yet it was not stupidity, for the Somalis are not stupid. To the contrary they are sharp, and intelligent. The seemingly illogical arrival arrangements stem from arrogance. The way out of the conundrum was of course to engage one of the many 'fixers' waiting on the right side of the barrier to go and buy the appropriate form. Thus even before formally entering the country, the traveller was in debt to an individual who would then shepherd him through all other steps to enter the country: naturally for a fee. It is the only place I know where a rip-off was quite so openly built-in to the arrival process. Fortunately the Range Management Agency seemed to want my services urgently enough for the manager himself be waiting with the necessary documents.

Like so many post-colonial African towns, Mogadishu had seen better times. No-one gave a damn, however. The Juba Hotel in which I was lodged at government expense, had once been comfortable, but was now full of surprises. The lift shaft had no doors and I received a powerful electric shock from touching a tap in the bathroom. Both electricity for lights and air conditioners and water in the taps were available only occasionally. No doubt some of this was the outcome of Somalia's flirtation with the USSR and socialism. Yet the west had made its contributions too. One was warned not to swim or paddle in the sea or 'Lido' in front of the town. Indeed I saw several sizeable sharks right in the curl of the waves breaking against the beach. Apparently twenty-three people had been killed by such sharks in the previous eighteen months. Where once safe to swim, it became very dangerous after the United Nation's Food & Agriculture Organisation (FAO) built Mogadishu a modern camel abattoir that discharged blood and offal into the sea nearby.

* * * *

On the evening of my arrival, the manager of the Range Management Agency, the Somali government department responsible, among other things, for conservation, came to my hotel. He was accompanied by his deputy. They wanted me to sell the government's ivory. They knew of my visits to Djibouti, had heard of my brokering service from Mr Rowan Martin who had recently visited them and thought that I was just the person to help them solve their problem. Having explained how I worked they said that it was just what they had wanted and I agreed to sell their ivory. It all seemed very straightforward – until I agreed, that is.

To that point, they had been supplicants: please, *please* would I conduct the sale. On the instant that I agreed to the job this changed. From then on they behaved as though I was a supplicant: if I wanted the work it would all *depend*! By appointing me broker for the Somali Government, they were doing me an immense favour – weren't they? Eventually it got through that they wanted to be in on the action so I asked them what 'commission' they had in mind. It was only fair, they felt, that it should be the same order as mine, say three per cent?

As I wanted to sell the ivory by public auction with payment going directly to the government, such 'commissions' as the two officials wanted would have to be negotiated with their own people and not me. As a matter of principle I wished to avoid handling any transactions and never be in a position to pay commissions. To have made this clear would probably have closed down the operation there and then, so while seeming compliant I avoided any actual commitment. It soon became apparent just how reasonable the head of the Range Management Agency's request for a commission had been. Apparently on presidential instructions he and his deputy had been sidelined and were *not* responsible for selling the fifty tons of ivory. Instead the Shiré Company, owned by Presidential relatives, had been given permission to market it and the Range Management Agency's rôle reduced to writing such permits as would eventually have to be issued. Introducing me into the equation was a private endeavour to regain some real influence on whatever came to pass.

Wrangles in Djibouti should have prepared me for a typically tortuous affair. The 'owners' (they said that they were) of the Shiré Company, Ali Shiré Warsame and Abdi Aden, affirmed that they and they alone had the right to sell the fifty tons of ivory, and showed me a letter from the President to that effect. Neither reading Somali nor understanding the language I assumed it stated what they said. Urbane, cosmopolitan and charming, they looked after me, agreed to my proposals in principle, took me to their banana plantation on the Webe Shebele River and to the ancient ports of Merca and Brava, once held by the Portuguese. But they signed nothing and we saw no ivory.

As with the Range Agency's officers, their attitude was that I was not so much an agent working for them, as a buyer from whom they would have to extract every penny. They wanted the highest possible price, to get which, I explained, there was only one route: a public auction. The only way to get top prices was to divide the tusks into trade categories so that the different grades could go directly to specific users and get overseas buyers to bid against one another. The idea of an open, public auction held either in Somalia or overseas seemed strangely repugnant to them.

The ivory was in four six-metre sea containers held at a police station and could only be viewed in the Station Commander's presence. A busy man, he was routinely not available. His predecessor, a colonel, had apparently swapped small tusks acquired elsewhere for larger specimens in the containers. Caught and prosecuted, he had been sentenced to death and was in custody awaiting execution. If his family could pay back the value of what he had stolen, his life would not be forfeit. This explained the present Commandant's reluctance to give anyone access to the containers, but if I could not see the tusks, I could not sell them. Eventually, on my third trip to Mogadishu I got to see them. From what little I saw of the tusks stacked in the four containers, they were similar to the material that I had seen in Djibouti. None seemed fresh. However, somehow the total weight had gained a ton: now everyone spoke of fifty-one tons. The containers had been standing in the open for at least five years. Each would have been a veritable oven once the sun was high, which could have caused many tusks to crack and lower their value. This made me particularly keen to examine them.

Interestingly, each container had five locks of which the Shiré Company had keys for only one. Those for the other four were held respectively by the Range Agency, the police station Commandant, the Central Bank, and the Ministry of Commerce. Representatives of each party had to be present when the containers were opened. Clearly the Shiré Company's control of the ivory was less than they had let on and the Range Agency's influence was equally diluted. Reasons for the difficulties in seeing the ivory and getting anyone to agree on how it should be disposed of were now becoming apparent. Every party with a key to a lock on a container had the power of veto over any decision.

Ali Shiré Warsame and Abdi Aden, eventually admitted that they owned thirteen of the fifty-one tons on offer. Other parties owned the rest and they had by common agreement been appointed negotiating agents for disposing of the total. Govern-ment's control over the whole arose from the exchange control regu-lations under which the price accepted had to be approved by both Central Bank and Ministry of Commerce.

Initially these two bodies set $63 kg as an acceptable price and subsequently raised it to $64, but it had to be paid in Somalia. The owners would be allowed to hold 65% of the receipts in dollars. They were officially allowed to convert this back into Somali shillings at the black market rate of 85 shillings to the dollar. The balance of 35% had to be converted into shillings at the official rate of 40 to the dollar. Any difference between the $64 kg set price and what they could get would obviously be held outside Somalia. The reason for everyone's reluctance to sell by open auction was now obvious. It would remove latitude to take part payment overseas, and all monies would be paid under the gaze of both the Central Bank, Ministry of Commerce and any others who might have an interest in any of the parties for whom the Shiré Company was acting.

While at the police station to inspect the ivory in the containers, I was asked if I would like to see another lot of tusks that had nothing to do with the now fifty-one tons. Naturally I did and was taken to a room in a police station building. There, in the middle of the floor, was a stack of about four tons of rather fresh tusks. On shelves around the walls were some 300 or so leopard and cheetah skins. In stacks on the floor were 400 sun-dried zebra skins. The tusks' owners were not in town and no one could speak for them. The real object of getting me into the room was because Abdi Aden wanted me to find him a market for the spotted cat skins and the zebra hides. As the former were banned from international trade under CITES I told him not to even think of selling them. The deputy head of the Range Agency interjected I was wrong; CITES was no problem and he would issue export permits.

To show goodwill in the face of the otherwise negative reactions they were getting from me I gave Abdi Peter Becker's address and sent a telex to him about the zebra hides. By letting him know that they were sun-dried, I knew that he was unlikely to touch them – which proved the case. Indeed both zebra hides and the spotted cat skins were so badly prepared that they were worthless. It was the four tons of fresh ivory that interested me. Yet I never saw them again and they never showed up in any trade, Customs or CITES statistics.

I was shown a mixed bag of around 300 leopard and cheetah skins. (Photo Peter Davey)

In early 1985 Rowan Martin, as a CITES consultant, had been told that forty tons was the country's stock of ivory. When I was first contacted in the middle of the year, it was fifty tons. Several weeks later it had grown to fifty-one tons. It was not the only ivory in the country. I had seen another four tons with my own eyes, but had heard of much more, perhaps as much as 120 tons. Of this total there were fifty-one tons under government control in Mogadishu and thirty-one tons in Djibouti owned by 'Omar'. In addition I believed that there were definitely other stocks in Djibouti, Kismayu and Brava. All of the ivory had originated in Somalia itself or Kenya, with the bulk coming from the latter.

I also got to see the counterfoils of exports permits issued over the preceding three years by the Range Management Agency. While they did not reveal the seventeen tons exported to Singapore which Huxley in CITES was so interested in, they did show many permits had been issued for smaller amounts.

Early in my trips to Mogadishu I had notified some seventy-four ivory traders in the Far East and Europe of my intention to auction the Somali ivory. Their response indicated that this particular ivory was well known in the trade. It had been offered to several different buyers over the previous six years, the most recent being the Takaichi Company of Japan which had opened a letter of credit for over three million dollars. The Hong Kong traders' experiences were so bad and their hostility towards dealing with the Somalis such that they would only bid *en bloc* and offer the lowest possible prices. Traders advised me to be wary of the ivory in Mogadishu. Clearly, inability to sell the fifty-one tons arose neither in Somali ignorance of the market nor the market's ignorance of the tusks. Problems over its disposal lay in the generic Somali condition of the parties involved being unable to agree details.

There was no point in spending further time on the project. After five visits to Mogadishu I had no written agreement with my 'principals' and they had not let me inspect the ivory properly. Further agreement was necessary with the Central Bank and the Ministry of Commerce – in addition to the Shiré Company and the Manager of the Range Agency. Two officials wanted a piece of the action and no one wanted an open public auction where, not only would prices be at least fifteen per cent higher than by bulk tender sale, but the whole business would be transparent. While the Range Management Agency said that Somalia wished to conform to CITES procedures, it was clear that what was agreed on paper with distant officials in Switzerland was different to what one did in Mogadishu. The original forty tons had within months become fifty tons and then fifty-one tons. Greater quantities had been moving through the country and some were still in it. Trying to act as a broker for these muddlers, I would go broke myself if I was not careful. So I cut my losses, informed the Head of the Range Management Agency that I was quitting and the Shiré Company that I no longer wished to represent them.

Shortly afterwards the ivory was bought by a Hong Kong firm. What brought this about was intelligence from the CITES Secretariat that if the ivory was not properly registered by 6th December 1986, no CITES party would be willing to permit its import. That risk at last overcame the inability of the various parties to agree details. In the end, every tusk was registered under Secretariat supervision.

There was one more thing for me to do: verify whether the President really controlled the fifty tons. If he had control, surely it would have over-ridden the disagreements that blocked its export? I made a direct approach to him, apologising for pulling out and giving my reasons for doing so. Shortly afterwards a Somali lady contacted me in Nairobi as the President's personal envoy, saying that he wished me alone to handle the fifty-one tons. All others involved had been ordered not to interfere.

This message was delivered two days after the ship carrying the ivory had sailed for the Far East. When I gave the ship's name and all details of the sale, she and her male companion seemed not at all put out and sounded so convincing that, had my previous trips to Mogadishu not been so unprofitable, I might well have been taken in. I never did find out whether the President was involved.

The Head of the Range Management Agency was not very happy about my contact with the CITES Secretariat, and displeased by my withdrawal from the Somali project. When he learned of my endeavour to contact the President, he was moved to write a most uncomplimentary letter about me to CITES.

* * * *

In the middle of the Somali saga Thom Friedlein asked me to fly to Addis Ababa in Ethiopia and examine several tons of tusks that had been offered to him. According to the IUCN's African Elephant & Rhino Specialist Group, there were few elephants in the country and, if this was anything to go by, the ivory must have come from elsewhere. Ethiopia was always worth a visit. In many respects it was not an African but a Mediterranean country caught in a medieval time warp. Vignettes from my first visit there in the mid-1960s remained indelibly: the local squire, white-robed and bestride his nag trotting to town, while several serfs ran ahead of the horse and several more behind. One of my partners, Tony Archer, and I had been in Addis Ababa at the invitation of John Blower, ex-Uganda and advisor to the Emperor's government on wildlife matters. Would we reconnoitre western Ethiopia in our own aircraft and then quote for a series of aerial surveys? Our first reconnaissance was nearly our last.

The then head of the Wildlife Department had been a major in the Ethiopian Air Force – an organisation with a reputation for airmanship and panache. Thus when he asked if he could come along we had been delighted. The maps were only a trifle better than blank paper and it would be an asset to have someone who knew the country. Heading west from the capital and with long-range fuel tanks, we had somewhat over eight hours endurance. Not long airborne, the Major asked if he might take the controls. I was happy to let him as it freed me to stare at the ground and not bother about navigation. We saw much interesting wild country in the Blue Nile Valley, but when we ran out over flat land I took stock and found that we were on a westerly heading and without doubt a long way into the Sudan. The major hadn't a clue where we were. Heading south-easterly back towards the distant highlands we arrived over a large river. Flying up it we came overhead a town with a large airstrip and the word 'Gambela' written in white stone. The river was the Bara.

The Major now knew where we were and tension in the cabin eased. Gambela was a port where in colonial times Nile steamers from Kosti, near Khartoum, came to collect Ethiopian coffee and produce for export via the Nile, Khartoum and Port Sudan on the Red Sea. Although technically in Ethiopia, when the British governed the Sudan, they had treated Gambela almost as a British out-station until the Italian invasion of Ethiopia in 1936. Deep in the continental interior and before the days of commercial aircraft, this little-known river port anachronistically had better contact with Europe via the Nile than most of the rest of Ethiopia had with its own capital.

Reassured, I let the Major again take control and head for Addis Ababa. Tony and I were fascinated by a sweep of forest that ran unbroken from an altitude of 1,000 feet to over 10,000 feet above sea level. It stretched away south as far as we could see: a dark green skirt covering the flanks of the Ethiopian massif from as high as trees could grow, down to flare out over the lowlands between the Bara and Gila rivers. Much of it clothed steep, rocky slopes, on which we saw no sign of people. It seemed truly unexplored. *This,* we told one another, was where there might be chimps, gorillas even or bongo: animals of the West African rain forests and others new to science. There and then we decided to arrange an exploratory expedition into this wilderness. We also felt that there just had to be elephants, many elephants, in such a habitat. The terrain had taken my mind off navigating and flying and I only emerged from another world when I noticed the time. We had ninety minutes' endurance left. It was then the Major again said that he was lost. Completely lost.

I took over and headed towards Addis Ababa on the ADF (Automatic Direction Finder: a device that indicates the location of radio navigation beacons), wondering if we had the fuel to make it. The needles in both fuel gauges had ceased to move and pessimistically registered against E (unlike a very logical Kenya Air Force trainee, I could not persuade myself that the E was for 'enough' and the F for 'finished'). Every open patch along our route was being sized up for landing possibilities when we came upon a hamlet with a long airstrip on which we landed. Once down we learned from a notice board that it was Buna. On our map we could only find Buna Bedele. If this was Buna Bedele, then we were within twenty minutes' flying of Jimma – a larger airfield where we could get fuel. But was this Buna Bedele? The Major talked to some locals standing nearby, while I climbed onto the

Long, straight, downward-pointing tusks typical of the forest form of elephant.

Curved and forward-pointing tusks characteristic of savanna elephants.

wings and, removing the fuel caps, tried to determine exactly what fuel we had. With a stretch of the imagination I thought we might get twenty-five minutes in the air.

The Major came back and said, "Buna Bedele;" so we headed for Jimma with the fuel mixture as lean as I could make it and still fly. Airborne for somewhat over ten minutes I had asked the Major why the sign said Buna and not Buna Bedele. He didn't know.

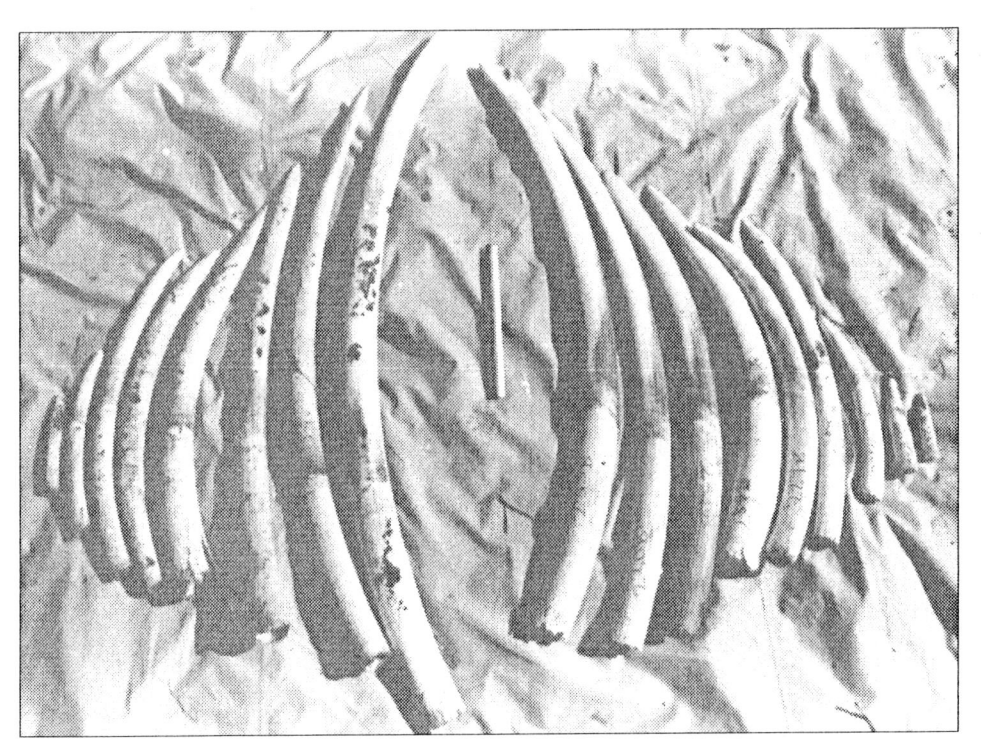

Elephants associate in family herds which, as in humans, display common characteristics. These two photographs each show the tusks of an entire family to make the point.

Assailed by a very uncomfortable feeling I sought confirmation that we had indeed just left Buna Bedele and was horrified when he said that he didn't know. Hadn't he asked the people on the ground? He had not, as a man in his position should not betray ignorance of where he was. Tony's face had reflected my thoughts.

Further talk was pointless. We were committed. Acrid armpit vapours betrayed states of mind. We would find out soon enough if the strip we had left was not Buna Bedele. When Jimma's long runway appeared ahead the sense of relief was beyond ready description. Several salutary lessons were learned that day. If we had come down it would have been my fault. If in command never assume another's competence. And *never* underestimate the ego's power to set aside common sense. Rather than admit ignorance, the Major would have risked his life and ours on the gamble that we had been at Buna Bedele. Only later did John Blower tell us that the man had left the Air Force because he lacked flying aptitude.

Thom Friedlein's ivory was some of the most interesting I ever came across. In shape it was clearly from 'bush' elephant – that is of the form that occurs in most African savannahs as well as the forests outside the West African and Zaire basin rain forests. It was the whitest ivory I had ever seen, with little trace of the creamy colour normally associated with tusks. The points were smoothly worn and sharp, with little uneven breakage, a characteristic of lush vegetation, as was the black vegetable staining around the tusks where they emerged from the elephants' lips. There were relatively few 'shakes' – the trade term for minute, longitudinal, hairline cracks on the tusks' upper surface that are far more dense on tusks exposed to strong sunlight. This was again in keeping with an environment of heavy shade. The undersides of all the bigger tusks, male and female, were heavily scarred with deep grooves and scratches, features which only occur in rocky habitats, particularly on steep slopes. The assortment of tusks from small to very big indicated that the hunters were not selective: they had taken whatever elephant they could. Many of the tusks were very big. There were several from males that must have weighed upwards of 120 lbs (55 kg) and there were even more female tusks over 25 lbs (11kg) indicating that in both sexes there was an unusual proportion of animals in their last decade of life. This suggested a population that had not been hunted for either a long time or too lightly hunted to reduce life expectancy much. As with the Somalis' ivory in Djibouti, the hunters had axed the big tusks in half to make them easier to carry which, together with the deep cuts and slashes in the butt sections, indicated that they knew little about either the value of ivory or extracting tusks.

The combined evidence implied a forest origin, most likely in the eastern half of Africa, whose ivory type I had not seen before, which had not been hunted for a long time and was now being hunted by amateurs who knew little about tusks or their value. Further the forests were likely to be located on steep rocky slopes. The obvious prospect were the forests that Tony Archer and I had flown over twenty years earlier. I put all this to the officer supervising the inspection and drew childish delight from his amazement. The ivory had been seized from a donkey train carrying it out of the forested region in the south-west of the country. And, yes, the Department had only just become aware that there were a lot of elephants in those forests.

Thom Friedlein never got this ivory and had never been meant to. The Head of the Wildlife Department sold it to a French firm with whom he had had a long-standing arrangement. Apparently this firm had become a little too confident of its standing and he had offered it to another buyer merely to bring the Frenchmen to heel and negotiate terms that better suited the seller. I was sorry that Thom could not really share the pleasure that I experienced at confirming our earlier certainty that Ethiopia's south western forests did hold elephants, regardless of the Elephant Specialists' equal certainty that they were barren.

* * * *

While ivory was the only commodity I offered to broker, I occasionally looked at other trophies for Peter Becker. Thus in the mid-1980s he asked me to pass judgement on some colobus monkey skins held by the WCMD in Nairobi. It was a stock of around 20,000 that originated in Ethiopia, but had been confiscated in the days before dealing in game trophies was banned. As expected, the skins were on shelves where they had been put years earlier. Through insect and bacterial activity, they had parted company with the long black and white hair that made them so striking. They were useless so I had casually looked over other items in the store. The scouts supposedly supervising me, stood just outside the door paying little attention.

On the floor beneath rows of shelves that ran down the centre of the room, there were many rhino horns. Some stood on the floor and others were in tin trunks – some with lids open and several with their lids shut. When you've seen one rhino horn you've seen them all, so the only thing that initially snagged my mind was the number, which I guessed was certainly over 100 and possibly as many as 400. One stood out from the rest: it had angular planes and odd pale patches, so I picked it up. The lack of weight instantly indicated it wasn't a rhino horn, but a fake made of pine wood. Badly made and stained black, it might have fooled the casual eye at a distance, but would not pass detailed scrutiny. Curious, I squatted down beside the rows of horns and examined them. Every single one was wooden and most were extremely well made. A heavy wood had been used to give weight, the staining was nigh perfect and some even had the wood grain roughed up around the base to give a very passable imitation of the bristly hair that occurs round every black rhino horn base. The horns in the tin trunks were made of wood too, and it was the sound made as I prised open one of the trunk lids that had made a scout look in to see what I was up to. I was asked to leave.

The horns would certainly have fooled any accountant taking stock. Indeed many people who knew something about rhino horn would

Imitation rhino horns are easy to carve out of wood, as I found out.
(Photo Peter Davey)

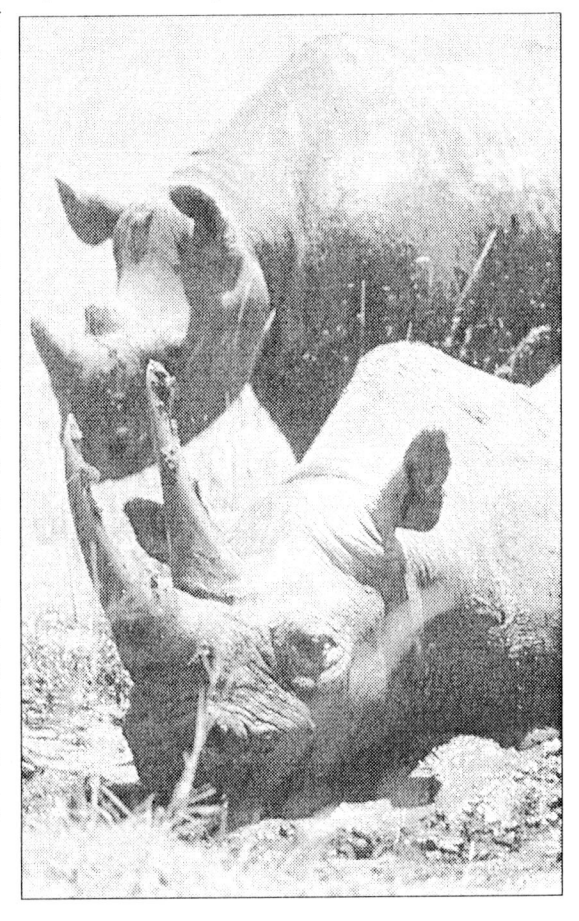

have been fooled too. CITES did not permit trade in rhino horn and anyone checking Kenya's stock of horns would have been convinced that it was being withheld from trade, not realising that all had indeed been sold. When, in a publicity gesture, they were later burned, they would have burned rather better than regular horn.

I mentioned my find to Rodney Elliott who, when through laughing, confessed to being the possible origin of the scam. Years earlier when investigating the Depart-ment's illegal activities he had wanted to trap certain parties in the act of buying rhino horn. At the same time he had not wanted risk losing valuable horn and had six substitutes made of wood – and very realistic they had been, too. They were left in the Department when he abandoned his investigations and were possibly the origin of the ambitious project I had stumbled across.

CHAPTER TWENTY-SEVEN

BURUNDI

It must have been some time in June 1986 that the telephone rang and Chris Huxley was on the line from CITES headquarters in Lausanne. He asked if I knew a Mr Rahemtullah? I did not. He had gone on to say that I probably would meet him fairly soon as he had given a Mr Zulfikar Rahemtullah my name and that he would no doubt call me when he returned to Nairobi. He thought that I might help Rahemtullah and at the same time 'sort out' Burundi.

Burundi, landlocked in the centre of Africa, county-sized at 27,834 square kilometres (10,747 square miles) is, along with its neighbour Rwanda, a country that should never have been. It is an accident of colonial history compounded by Africa's subsequent commitment not to change inherited colonial boundaries. In the lands west of Lake Victoria encompassed today by southern Uganda, north-western Tanzania, contiguous eastern Zaire, lie Rwanda and Burundi, the original inhabitants of whom were small, pygmoid hunters and gatherers – the Twa. At least five or more centuries ago the area was settled by Bantu farmers. Displacing and intermarrying with the Twa, they acquired shorter statures and founded what are now the Hutu.

Subsequently a wave of tall Nilo-Cushitic cattle people entered the region from the north and north-east. Uniformly they intermarried with the Bantu agriculturalists already in residence and adopted Bantu languages. They evolved into a cluster of polities or kingdoms in which pastoralism was the key to the ruling classes and farming was the business of the lower orders. With the prefix ba denoting the people and bu their land, in example, there were the Baganda of Buganda, the Banyoro of Bunyoro and the Barundi of Burundi. The distinction between the two founding classes is still most apparent among the Banyarwanda of what is now Rwanda, and the Barundi of Burundi where, in both, the upper castes still identify themselves as the Tutsi, and the lower castes as Hutu.

In the sense of modern nation-states, this cluster of kingdoms spanning south-western Uganda, north-western Tanzania, eastern Zaire as well as Rwanda and Burundi might have formed a single viable nation. Yet this was forestalled by the division of Africa into colonial spheres at the 1884/85 Berlin Conference. It was the cluster's misfortune that British, Belgian and German interests met in their centre. The northernmost kingdoms went into Uganda, the westerly fringes were incorporated into the Belgian Congo, and the lands of the Banyarwanda and Barundi into German East Africa.

In 1914 the Europeans fell out and fought. Germany lost and was deprived of its African estates whose fate was then determined by the League of Nations. All of German East Africa was ceded to the British as Tanganyika. While Belgium itself was occupied by the Germans in the First World War, its colony – the Belgian Congo – remained in the opposite camp and Belgians there helped the British chase the German forces out of German East Africa. Certain circles felt that they should be rewarded territorially for this service and when Britain took over Tanganyika, what the Germans had called Ruanda and Urundi were detached and given into Belgian administration. As League of Nations mandates, they could not be incorporated into the Belgian Congo, because it was a colony. In due course the United Nations Organisation assumed the defunct League of Nation's responsibilities, and sustained opposition to incorporating the two kingdoms into any of the three surrounding states. When independence fever swept across Africa, they were perforce given sovereign nation status as Rwanda and Burundi by Belgium and have been festering sores on Africa's back ever since.

With no resources other than good soil and climate, Rwanda and Burundi are at the same time the two most densely populated nations in Africa. Their 1997 population densities were 316 per square kilometre (809 per square mile) and 228 per square kilometre (584 per square mile) respectively. Unlike most other African countries, they have little space into which people can expand, and both suffer from severe population pressure. It is this, above all else, that exacerbates the ethnic conflict both have experienced more or less continuously since independence. In Rwanda the departing Belgians arranged for political power to be in the hands of the majority Hutus, though most recently, the Tutsi have wrested it back from the Hutu. In Burundi they left the minority (±13% of the population) Tutsi in power.

Burundi's predicament as a non-viable, overcrowded mini-state was alleviated to a considerable degree by post-colonial mismanagement in the surrounding bigger states – Tanzania, Zambia and Zaire. By mismanaging their economies these countries guaranteed widespread smuggling because it was the only way their citizens could convert the commodities available to them (gold, gems, ivory, rhino horn, etc.) into the manufactured goods they wanted.

Burundi took advantage of this situation by fostering (relatively) free trade and not asking questions about commodities' African origins. Thus it regularly exported quantities of *robusta* coffee that grows in hot, wet, lowlands, whereas all its home-grown coffee was the *arabica* variety suited to its cooler uplands. The *robusta* came from neighbouring Zaire. Belgian diamond buyers had offices in Bujumbura, the capital, although Burundi had no known diamond deposits. The country produced a little alluvial gold, but exported far more than could ever have come from these limited sources. Unsurprisingly, it also exported ivory although it had no elephants.

Burundi's ivory exports started to rise steeply when Belgium acceded to CITES and closed Antwerp as the entrepôt through which most of Francophone Africa's ivory passed to the Far East. Such information as came my way suggested that by 1983-4 the ivory going through Burundi might amount to 300 tons a year, a figure questioned by Chris Huxley in CITES. Yet though he felt it less, Burundi's traffic undermined CITES' efforts to control the ivory trade.

By 1985 all ivory producing nations and all major ivory consuming nations were Parties to CITES. There were, however, three entrepôts that bucked the system. Singapore was not a Party and placed no restrictions on importing and exporting ivory. As an exceptionally law-abiding country it was felt, rightly as it turned out, that international criticism would redress this situation. Dubai was the second non-conforming nation, although it was nominally a Party to the Convention. Here again it was felt that diplomatic pressure would, eventually, get it to stop unrestricted imports. Burundi, however, was altogether different.

The ivory traffic through Singapore and Dubai constituted very small proportions of their economies whose loss would make little difference to them. Making Burundi stop handling ivory, on the other hand, challenged the basis of its whole entrepôt traffic with neighbouring countries. Tanzania, Zambia and Zaire all took a gloomy view, not only of Burundi's ivory exports, but even more its gold, gems, *robusta* coffee, hides and skins and other goods that, in their eyes, had left their territories illegally. Conceding that one aspect of its entrepôt traffic was in any way 'wrong' would set precedent for stopping others on the same grounds. Where Singapore and Dubai were wealthy, Burundi was one of the world's poorest states, and surrender over ivory could have serious general economic implications. Consequently Burundi resisted pressure to stop trading ivory and was averse to joining CITES.

Huxley wanted Burundi to join CITES, but as important, that it did so without what is known under the treaty as pre-convention stocks. Here it is necessary to digress briefly. When a country joins CITES, any wildlife items listed on the Convention's appendices but legally owned by its citizens are classified as 'pre-convention stock'. These can be traded with the words 'pre-convention stock' written where country of origin was normally entered on permits. Individual governments had to satisfy themselves that stocks really were pre-convention. As joining parties did not have to register their pre-convention stocks with the Secretariat, dishonest governments could use 'pre-convention stock' as a laundry for new acquisitions. It was a loophole in the Articles.

To block pre-convention status being used to get new ivory onto the market, at their fifth biennial meeting in Buenos Aires in 1985, the Parties asked all governments to register pre-convention ivory stocks with the Secretariat by 1st December 1986. The resolution was neither binding nor mandatory: to have made it so called for amending the Articles – a lengthy process. They also agreed a formula wherein at the beginning of every year Africa's countries would notify the Secretariat of how much ivory they expected to produce: such amounts were referred to as each country's quota.

Hence Huxley's interest that Burundi not only join CITES, but did so without pre-convention stocks of ivory and, as it had no elephants, a nil quota. He felt that through Rahemtullah there was an outside chance I might get the country to join the Convention, but at least I might get valuable information on Burundi's ivory business.

Apparently Rahemtullah had first approached Chris Huxley and Joe Yovino, the CITES Ivory Officer, with his 'problem' while they were in Mogadishu supervising the registration of the Mogadishu fifty-one tons after I had thrown in the towel. Subsequently he had again talked with them in their Lausanne headquarters. His story was that he had forty tons of ivory in Burundi. Originally it had come from Uganda lawfully and with permits. These had been mislaid or destroyed, and he now wanted to know whether if he got duplicates and a letter of confirmation from the Uganda authorities, his forty tons would become 'CITES acceptable'. The answer had been an unequivocal 'no!' However, Chris said that if anyone could help him it would be me, and that Rahemtullah should see me. The compliment was not quite as backhanded as it sounds!

* * * *

Zully Rahemtullah invited me to lunch. An Ismaili, short, balding, heavily moustached, volatile, dynamic: Zully was a 'character'. Not only was he a man with fingers in numerous commercial pies, but he had also been a prominent motor sportsman and had competed several times in the East African Safari Rally. Direct and to the point, it was immediately clear that, where he was concerned, red tape was unnecessary. In the local idiom Zully was a bit of a *mkora* – the word meaning a 'wild one' – but as used in colloquial Kenya English translating more closely in the sense of being a bit of a lad. Naturally I had done a little homework on who he was, and a white rally driver who knew him summed up Zully – a *mkora* maybe, but a man who would stick by his friends and whom friends could trust. I have never had reason to doubt this.

There was no haggling. My fee, I said would be three per cent of the gross value of his ivory when it was sold. He agreed to it without a pause. Mark you, he never paid the full fee so perhaps he never intended to. His need was simple. He had forty tons of ivory in Burundi which he could sell in Dubai or Singapore, but although it was perfectly 'legal', being covered by a Burundi export permit, he could at most get only half the going rate for ivory with CITES-acceptable permits. He naturally wanted to sell the ivory at the going rate which, at the time was around $120 per kilo. Huxley had told him that I would know how this could be done.

I did. If Burundi joined CITES all the ivory legally in the country at that time would be 'pre-convention stock'. As such it would automatically become CITES-acceptable. While some countries might prevent its import under their national legislation, technically it should be internationally tradable. So I told Zully he would get what he wanted if Burundi became a member of CITES.

He could hardly believe it could be so straightforward. I don't doubt that he must have felt three per cent of the value of forty tons of ivory a steep price for such simple information. If so, he didn't let on, but asked how I intended getting Burundi to join CITES. As the country was run by a dictator, I would have to meet him and persuade him to bring his country into the convention. Zully had thought on this for a while before saying bluntly that he didn't think that I would have much success. Did I mind if he handled the matter? And that is what happened. Ten days later Zully called me and said that Burundi would join CITES. I was not indelicate enough to ask what suasion he used as it did not really matter. Shortly thereafter a somewhat surprised Secretariat confirmed what Zully said. Huxley's strategy was working!

Zully Rahemtullah had pulled off what had defeated CITES and the conservation lobbies over several years. The coup was incomplete, however. To be worthwhile, Burundi's pre-convention loophole had to be firmly plugged. So when Zully asked if anything else remained to be done, I was economical with the truth and forgot to tell Burundi that it had the option *not* to register its pre-convention ivory. If I could get it to declare its stocks and then get those stocks exported quickly to ensure the country had no pre-convention stock, its use as an entrepôt would be over.

Believing that the ivory *had* to be registered, Burundi appealed to the Secretariat for an 'expert' to help do this. Huxley and Yovino visited Bujumbura and, playing along, suggested several names, mine among them. I was selected, opening phase two of the Burundi project. The objective: register all tusks in Burundi and get them out of the country as quickly as possible, thereby ensuring it had no pre-convention stocks. In addition I would try to learn where tusks had come from and how.

Summoned to Burundi, I went as quickly as a visa could be arranged (which in Zully's hands was a fraction of the normal time required). I wanted some idea of how much ivory was involved and to formalise my status with the Burundi Government.

* * * *

Like all of Africa's great lakes, Lake Tanganyika is literally an inland sea, complete with a pounding surf and white beaches. With swaying coconut palms, Bujumbura could well have been a small port at the edge of any tropical ocean: the whole marine ambience added to by the lake steamers and fishing boats coming and going.

Rather than stay in town, I opted for a room at Le Club du Lac Tanganyika, that fronted onto a white sandy beach and the sound of breaking waves. At night the horizon twinkled with hundreds of small lights as though there was a veritable city

Perhaps because we can see so much of ourselves reflected, one of the saddest of sights is a chimpanzee chained to a pole.
(Photo Peter Davey)

out on the water. It was the fishing fleet out for *dagaa* (*Stolothrissa tanganyikae* or more recently *Limnothryssa miodon*), a small freshwater sardine endemic to Lake Tanganyika and which was introduced to Lake Kariba where it is known as *kapenta*. They come to lights hung over the water and shoal so densely that they can literally be scooped out of the water by the ton and sustain lucrative fisheries on both Tanganyika and Kariba. The Club's attractive and exotic atmosphere was only marred by the chimpanzee tethered in the gardens, neurotic and bored – literally out of its mind.

Though I slept at Le Club, I was looked after by Zully's brother, Amir, and his family. Zully and Amir had been brought up in Mpanda in western Tanganyika where their father had been a trader. The Rahemtullah enterprises now stretched across eastern Africa, into Britain and Canada – the country of which most of the family now seemed to be citizens. These encompassed money-lending, supermarkets, transport, car sales, trading in general merchandise, buying and selling gold and gems, as well as running restaurants and hotels. The commerce was managed by the family, but Zully was clearly its entrepreneurial spirit and constantly on the move between countries and projects. The involvement in ivory was a sideline. Indeed, Zully said that they had got into it more by chance than design. His interest had been caught both by the volume of money ivory traders borrowed and their reliability in repaying debts. As ivory was so obviously good business, he took a closer look at it. However, it was only when a borrower defaulted and Zully had to take a hand in managing the borrower's business that he became involved on any scale himself.

Problems arose when Belgium joined CITES and stopped accepting ivory from Burundi. That was the seminal point at which a two-tiered ivory pricing structure emerged. Countries which produced CITES-acceptable tusks and permits obtained full prices for their ivory, while those whose output was not CITES-acceptable were forced to sell their tusks at around half the going market rate. Burundi, Dubai and Singapore replaced Antwerp as the entrepôts for ivory that was not CITES-acceptable.

By 1986 two Hong Kong companies dominated the Burundi-Dubai-Singapore trade: that owned by K. T. Wang and the other by the Poon family. Zully had sold to the latter and believed that he had not received a square deal. At least in part, his desire to get ivory out of Burundi in CITES-acceptable form was knowing that it would spoil the Poons' business. The prospect that once Burundi had joined CITES it would no longer be an entrepôt did not worry the Rahemtullahs. They wanted to be shot of the forty tons which, in any case, was clearly not theirs alone, but owned by a syndicate. After that, the Rahemtullahs said, that they did not care if they never traded a tusk again.

*   *   *   *

In less than a day I was appointed consultant to register Burundi's ivory. My terms were agreed – a fee of $9,000 (which was never paid) and that the Government would do my bidding. I had feared that if ivory trading in Burundi was free there would be many people to deal with. It transpired that trading was not free. To import and export ivory from Burundi one needed a special licence of which only four were current in mid-1986. The Government's attitude towards importing contraband from other countries was interesting. It was not, it said, responsible for enforcing any other countries' laws. If they were unable to enforce their own laws and stop the export of commodities – be they ivory, gems, gold or coffee – that was their problem. The officials likened Burundi to Switzerland: Switzerland banked anyone's money so long as it entered Switzerland legally as determined by Swiss law. Burundi traded any goods for as long as they were imported according to Burundi law.

The four parties licensed to import and export ivory into Burundi by the Minister of Finance were Tariq Bashir, Lucien Basabose, Caspar Egebu Ndikumasabo and Jean-Baptiste Matangana. To get a licence the holder had to be a Burundi citizen, agree that $25 per kilo of ivory exported would be repatriated to Burundi in hard currency and to pay a government levy of $15 per kilo to the Central Bank. This information was provided by Audace Kabayanda. Subsequently I heard so many different versions of the licence conditions that I assumed they were 'negotiable'.

Tariq Bashir was a young Asian who had been born in Burundi, was a citizen of the country, was fluent in French and Kiswahili, but spoke no English. He owned the licence under which Zully's syndicate operated, but was clearly more than just a front in the business. Lucien Basabose was the Burundi front for two Lebanese – Jamal Nasser and his brother Ali Suleiman. Caspar Egebu fronted for two Malian traders called Tanaba and Mamu. Jean-Baptiste was very much his own man. All four groups were general merchants and ivory was not their sole – or even their main – business. Of them, only Jean-Baptiste did not buy gold.

While ivory had received much publicity, its value must have been a small fraction of that earned for Burundi by the gold trade, most of which came as alluvial dust from all over eastern and central Africa. Some was brought by individuals, some by small groups. Some came on foot, some on bicycles or in new Mercedes. All seemed to know what they were doing and carried transistor radios to keep abreast of gold prices via BBC and South African broadcasts. Sales were made almost by ritual. First the seller would arrive at the buyer's premises and sit at a bare wooden table opposite the buyer or his agent. In tones so quiet that I could not hear what was said from five paces away, they talked about gold prices.

Usually no deal was concluded during the seller's first visit, for he would visit the other buyers. He might visit all several times before a price was agreed with one. Until that point no gold was produced or seen, though the seller had given some idea of the quantity he had for sale.

Once the price was agreed, the seller was escorted to a shed and given a container, a demijohn of nitric acid and a pair of tongs. Everyone except the seller then stood well back for there could be no risk of someone else accidentally knocking or spilling anything that held gold. With everyone well clear, the seller produced his gold. It might come out of a briefcase, or a variety of receptacles: paper envelopes, plastic bags, paper twists, camera film cassettes, and the traditional African gold dust container – hollow stems of vulture wing quills. The owner emptied his gold into the container, taking his time. When satisfied that no particle remained, the receptacle was cast aside.

The seller then took the demijohn of acid and poured a liberal quantity into the container with his gold. Taking the container with the gold and acid, he placed it on a charcoal furnace. Cranking a hand blower, the smouldering charcoal quickly became white-hot and the acid boiled viciously, giving off fumes more acrid than tear gas. After five very unpleasant minutes, the seller then took the tongs and grasping the container with its precious load, carried it to a large porcelain sink and gently turned on a tap. The thin stream of water first diluted and then washed the acid away. This had to be done very carefully as the water could all too easily wash out some of the gold dust. The water was poured off and the container returned to the furnace where all remaining dampness was driven off to leave dry gold dust and the silica from sand that would not dissolve in acid. The process had, however, removed all other metals and impurities.

The container with the dry gold was now taken to another bare wooden table at which the seller seated himself opposite the buyer. He tipped the gold into a brass mortar and with a brass pestle pounded hard for ten minutes or so. The buyer then took over and he, too, pounded and ground away at it. This operation reduced gold particles to flakes fractions of a millimetre thick, and crushed all sand grains held by the gold into a very fine dust. The contents of the mortar were then tipped onto a shallow dish which both seller and buyer took turns blowing upon very gently to get rid of the fine silica dust. Finally, only gold remained and it was weighed.

Payment was in whatever currency the seller requested. At this point I had been astonished to observe that contrary to expectations, a large proportion of payments were asked for in African currencies and not hard international coin. The money involved in this gold trading was far more than could be readily supplied by the Central Bank of Burundi and the dealers were authorised to bring in hard currency from abroad. I had been surprised by the lack of security and witnessed $250,000 in a brown paper parcel arrive from Zurich on a Sabena flight. This gold dealing revealed aspects of Africa that are obscured to most western eyes.

* * * *

From what I saw in my reconnaissance it was obvious that more ivory had been moving through Burundi than the CITES Secretariat wanted to believe. If anything, I had erred on the low side in my estimate of 300 tons annually. The amount of ivory to be registered could be anywhere between sixty and ninety tons.

Obviously the registration exercise had to be carried out as quickly as possible as the longer it took, the greater the chance of things going wrong. I had no illusions that there would be strong opposition to closing down the trade.

There is no practical way of permanently marking a tusk. The system that I had produced for CITES and which had been adopted internationally, was to punch the serial code into the surface of the tusk just below where it had emerged from the elephant's head. Numbers punched into the tusk matrix were more difficult to remove than any written on with marker pen or paint, but they took time to apply. Consider what was required. Burundi tusks would have the two-letter country code followed by a two-digit year code, followed by a five-digit serial number (I was certain that there would be at least 10,000 tusks) followed by weight. An example would thus be BI/86/10,000/15.5, which with its spacers contains seventeen separate elements. For each the appropriate die had to be selected, picked up, placed in position, struck with a hammer and returned from whence it had been picked up. When used routinely to register tusks coming into a store a few at a time, the work would take a few minutes. However, when thousands of tusks were to be registered at once, stamping with punch dies would involve many thousands of minutes and the time needed to stamp the numbers would compromise a speedy resolution of the Burundi problem.

Before leaving Burundi I timed stamping the serial numbers into a set of tusks and it took, on average, between one and two minutes per tusk. To cut down the time I welded the standard elements of every number BI/86/ – that is six dies – together so that all six could be picked up and applied simultaneously with one rather than six hammer blows. The advantages of one pick-up and a single stroke were outweighed by the time it took to align such an awkward punch on a tusk's curved cylindrical surface. At between one and two minutes per tusk, the registration rate would be thirty to sixty tusks an hour or 167 to 333 hours work per 10,000 tusks. If there were between 10,000 and 20,000 tusks to register, the job might take as many as 600 hours or seventy-five working days. For this reason I decided that, in the circumstances, it would be more practical to number the tusks with ordinary indelible marker pens as, with them, it took only ten seconds per tusk. There were other advantages too: the most obvious of which was that if I wrote each number, every tusk would have my personal handwriting on it.

The longer the registration process took, the greater the opportunity for registered tusks having their numbers removed and being presented a second or more times. This would create a false stock on paper that could be filled later with new tusks from outside. The case for both speedy registration and getting the tusks out of the country as quickly as possible was all the more obvious. All of this I kept from both the government and the traders – other than Zully on whose help I was relying heavily.

Before taking any steps on the ground, however, I wanted to ensure that the plans were understood and approved by the CITES Secretariat and that there would be no hitches once registration started. I flew to Switzerland and got not only the Secretariat's approval, but also its assistance in ensuring that the governments of Switzerland, Belgium, Britain and Hong Kong (the four most likely countries for the ivory to go to) would accept tusks coming out of Burundi in the registration exercise. If there were any queries, let them be made once the tusks had left Burundi.

I confess that I had great plans for a follow-up analysis, and taking advantage of the vast amount of information that the ivory would reveal. I needed time to examine and measure them at leisure. Access to them would be difficult if the tusks went to out of the way places in the Arabian Gulf. I would have preferred the tusks to have gone to Britain and even went as far as trying to steer it into Thom Friedlein's hands, which would have given me greatest access in congenial surroundings.

Having secured all the agreements I needed, including approval to use marker pens rather than punch dies, I returned to Burundi in September 1986 and laid down the rules. The exercise was to be completed within ten days. Within this period I would register any and every tusk brought before me without questions. Once the ten days were up, any unregistered tusk in Burundi would be forfeit to the Government. Once registered, all ivory had to be out of the country before the 1st of December or it, too, would be forfeit.

While I was arranging matters in Switzerland, Zully had not been idle in Burundi. As his competitors were unaware of what was about to befall them, he had bought up a further twenty tons of their stock. I suppose this was insider dealing at its most blatant for they sold unaware that their stock was about to become CITES-acceptable overnight and worth twice what they had been anticipating. While these business tactics may have been 'not cricket', the more of the ivory that was in Zully's hands, the more confident I was of it being registered properly.

On commencement day I was taken to each of the four parties licensed to handle ivory and introduced to them formally by the Director General of the Institut National pour la Conservation de la Nature (INCN), Audace Kabayanda, who was responsible for wildlife matters and ivory. He told them the 'rules' and that they were to do everything I requested as I was working at the President's personal direction. My requests would be the President's wishes. Democracies have no analogue to a dictator's order for motivating civil servants to dissolve red tape (which is a pity). Without this very bald reference to the President, I do not think that I would have achieved much. Zully's party excepted, the other three dealers were shocked by the registration command. While their stocks would fetch double what they had expected for what they had in hand, it would be no recompense for losing their ivory businesses.

* * * *

As on my first visit, I stayed at the Club du Lac Tanganyika and motored from there daily into Bujumbura, returning at night. Naturally I started with Zully's ivory as it seemed to be the greater amount. I had not been marking tusks for long before I came across one with Tanzanian registration numbers: TZ/MG/788/85/14·7. This indicated that it originated in Morogoro District in 1985, was the 788th tusk to be registered there that year and had weighed 14·7 kilos. In all I came across a further fourteen Tanzanian tusks that had been registered in five districts – Arusha, Tabora, Morogoro, Iringa and Lindi. These districts were scattered the entire length and breadth of the Tanzania, which made it more likely that the tusks had come from the Dar-es-Salaam Ivory Room where ivory from all over the country was collected, than to have been acquired independently in the districts of origin. There were eight Uganda tusks, all from the Kidepo Valley National Park and dated 1984. It was first hand evidence of what I had heard by word of mouth: that official ivory leaked out of the Uganda and Tanzanian government systems as it had in Kenya.

The tusks registered in Tanzania and Uganda would have been CITES-acceptable had they been lawfully exported from their countries of origin. In being moved to Burundi they had become CITES-unacceptable and lost half their value. This was understandable if they were stolen property, which is commonly sold for less than true value because it is illegal. No-one remembered how the Uganda tusks had been acquired and they probably had been stolen by a Uganda National Parks employee and then sold on to a trader. Those from Tanzania came from the Head of the Game Division as repayment of a cash loan. Without reference to their weights, he had simply said, "Take those tusks there," giving insight into the prevailing lack of accountability.

I have been asked why I did not insist that these tusks be returned to their lawful 'owners'. As this would merely have given those who had already sold them once an opportunity to do it again, there was no point. So I removed the Tanzanian and Ugandan registration numbers and replaced them with Burundi marks.

The work went quickly and smoothly as Zully provided a team of helpers. And from the way they handled tusks they were clearly experienced! With so many willing hands assisting, I had the opportunity to look over the ivory carefully and make forensic notes about their origin. This I later analysed as a check on what I learned through conversation. The two sources of information broadly tallied, giving me confidence in what I was told but could not otherwise corroborate. Having registered Zully's sixty or so tons, I ran into my first obstacle. I wanted it out of Burundi immediately to reduce the possibility of tusks being presented again for re-registration. I telephoned the Secretariat to notify it that I wanted the ivory to move immediately, but was told it would not approve any ivory leaving Burundi until the whole stock had been successfully registered.

The Secretariat had the whip hand because if it withheld approval, no CITES state would accept the ivory. The bureaucrats feared falling between two stools in which some of the ivory got out of Burundi as CITES-acceptable and some was never registered. It was reasonable apprehension, given the Secretariat's clerical outlook and where fear was confined to verbal brickbats in the media. It was not reasonable from where I was in Bujumbura as it undermined the agreed plan and my position. The time to have laid down such conditions was when I had been in Lausanne when all had unconditionally agreed that the ivory should be got out of Burundi as quickly as possible. Now the programme was under way, the Secretariat was qualifying its position. Fortunately no harm was done, but it illustrated how unqualified the Secretariat was to administer investigational work. As anyone with military or police grounding would appreciate, 180-degree changes of direction once an operation is running create huge difficulties for operatives in the field.

On the third day of the registration exercise I was waiting for my transport in the lobby of the hotel when a curious incident took place. As I entered the lobby a large, dark-haired, white man of perhaps forty got up and went to the telephone. I had the impression that he had been waiting for me. He then spoke in English so loudly into the 'phone that he must have wanted to be overheard. Identifying himself as either Mike or Marc, he had talked to 'George', whom he said should come and transfer one million dollars, that he had the ivory ready and that no, he was not worried at all about 'what was going on'.

Presumably this had been laid on for me and I was to assume that there was a million dollars' worth of ivory about which I knew nothing about to leave the country. Staff at the reception desk said the man was a Belgian doctor.

That night dining in a Greek restaurant, a large and unusually burly Tutsi came to the table and in a loud voice said, "Do you know who I am? I am the Chief of Customs. Get out of Burundi. Get out of Burundi quickly. Get out and stop your meddling."

The restaurant's patrons all craned to see who was addressed thus and I was the centre of unwanted attention. I had replied, hopefully in a firm voice, that I would pass his comments to His Excellency on the morrow, whereon the Director of Customs turned on his heel and disappeared. My 'mission' was clearly not approved in all quarters.

Next on my list was Jamal Nasser's ivory. Audace Kabayanda had introduced me to Lucien Basabose, Jamal's Burundian front man, and had seemed really frightened of him. He was, so Audace said, "A criminal . . a real criminal . . a big criminal." In marked difference to everyone else, Lucien was openly hostile. Speaking reasonable English intermixed with Kiswahili and French, he had been contemptuous of the President, making Kabayanda cringe at the disrespect. Tall and with the Cushitic features characteristic of many Tutsi, Basabose had unkind thoughts on my ancestry. He did not believe the registration scheme came from President Bagaya, but from me and 'those Indians.' He had ranted on about yet another white man's plot to screw Africa.

My riposte that the Presidents of Zaire, Zambia and Tanzania wanted to stop Burundi trading their ivory triggered an outburst. "How can that be? Do you know how much of this ivory actually comes from those Presidents? If they really felt it should be stopped, do you think that they could not do so? How many roads are there into Burundi from Tanzania? From Zaire? One! One from each! How many trucks with tusks come here each year? Can you hide a truck of ivory? How many are caught at their border posts? I tell you none. If those Presidents did not want the trade they would stop the trucks. They don't. And you lie!"

He had, of course, hit the nail on the head. For the Burundi trade to function, the Tanzanian and Zairian authorities had to be involved to a substantial degree. Some of Nasser's tusks were in Lucien's house and he was reluctant to let me register them. He was going to Nairobi and he would see about it in two weeks' time when he came back. Nasser was clearly worried by Basabose's attitude. Like Kabayanda, he was not keen to be associated with disrespect for the President and it was his intervention that eventually got the ivory out of Basabose's house for registration. Basabose had left us in a shower of gravel and dust as he gunned his Mercedes away and I did not see him again until the evening of the tenth day.

Jamal Nasser was a big man and one of those unfortunates who was also grossly overweight. I guessed that he must have weighed well over 400 pounds (180 kg). He lived a bachelor existence in a house lacking the slightest feminine evidence. As an air gun buff, it seemed that an air rifle was never far from his hand. At the back of his house a wide, spacious veranda faced a large walled in yard. Scattered about the yard were several packing cases, to each of which a dog was tethered. They were so scrawny and close to death from starvation that they were horrible to behold.

Three rooms full of ivory opened onto the large back veranda and I had asked if I could weigh and mark the tusks there, using Nasser's own scales. He had agreed readily enough, but warned me with a sly grin that the scales under-weighed by 400 grams. He would thus have been gaining eighty kilos per ton of ivory worth some $4,000 at the prices Burundi ivory was going for. Obviously he felt that if weights had been checked, his scam would have been exposed and resulted in later problems, hence taking the trouble to inform me of it.

Nasser, himself, took no part in the weighing though he provided labourers to give me a hand. I recognised them as the same men who had also worked as Zully's team in the preceding days. It transpired that they were ivory handling 'specialists' who had formed themselves into a guild of around 100 strong and who handled all ivory movements in Bujumbura. Most of them came from Mali. I paraphrase from memory how one of them explained this –

> *"Most of us have been here for some years. In Mali there is no work for young people. That is why so many from there and from Senegal trade across Africa. But some, like us, have no capital. We came here working for traders and took any work when we got here. Some of us started to carry tusks eight or nine years ago. So we learned about tusks and we learned Kiswahili. As the trade grows, everyone with ivory wants our help. Soon we make our own guild. We know how to handle tusks without breaking them."*

This was a stroke of good fortune, for these youngsters were a decent bunch and when the ivory's owner was not present, talked freely. We talked as specialist to specialists, for they were truly interested in ivory. I told them much about tusks, and they told me who brought the ivory, from where and how. As they unloaded all substantial consignments arriving in Burundi, and loaded all those leaving, this had been a real windfall.

Occasionally Nasser would emerge from inside the house and watch. He always brought an air-gun and his eyes never stopped searching for small birds. If one landed in the yard, or in the scraggly trees that grew in it, he shot it. If it fell in the yard a servant was instructed to throw it to one of the dogs. He was clearly driven by an unusual obsession. Even if he could not recover the bird, he would still shoot it. On one occasion his glance took in the neighbouring house where a sparrow hopped sideways along the roof guttering. The air rifle came up, there was a 'phut' and the sparrow fell backwards out of sight into the gutter. Nasser had then turned and walked back into the house without a word.

On another occasion Audace Kabayanda had been present and had remonstrated, pointing out that it was against the law to shoot small birds, which had surprised Nasser. The ivory handlers saw the compulsion as a kind of madness. No doubt they saw my bringing the dogs loaves of bread as madness too, but it briefly assuaged my guilt over not being able to better their condition.

During the midday break Nasser invited me into the house and offered cool drinks. He was obviously concerned about what would now happen to his ivory business. More urgently, with only five days to go before registration was closed, what was going to happen to ivory *en route* to him, but still in other countries? He had $600,000 outlayed, so he claimed, though he later changed this to 'over one million'. I did not really believe him, as in the picture of the business that was building up, there had been no need to induce ivory to come to Burundi; it came of its own accord.

Two Omani Arabs who lived in Kenya, but who traded salt from Tanzania to Zaire, arrived at Jamal's house to get a cash advance against four tons of ivory that they had assembled near Dar-es-Salaam, but for which the transport had yet to be arranged. He refused the advance and told them to speak to me. Neither spoke English so we conversed in Kiswahili. They wanted me to keep registration going until they managed to get their ivory into Burundi. They were told that if they could get it there within the remaining four days, I would have no option but to register it. If it wasn't here by then, there was no way that I could oblige them.

"We beg you to be reasonable. Most of our money is in this ivory which we have coming. It is impossible to get it here within four days, even if we could arrange a 'plane to fly it here. What you are doing will ruin us. Please make some more time ... just another week; just one week."

I had reasoned that they must have known that what they were doing was illegal in Tanzania. That brought forth an outburst. "Illegal? If you have paid the officials it is legal and we have paid. And we will pay you too ... if you help us."

The two Omanis stressed that getting the material to Burundi in four days was impossible. Apparently the ivory had to move slowly. Each route sector had to be prepared in advance with police and other officials approached and paid off. It was a laborious, step-by-step process.

"If only you had given us some warning, then we could have made sure that all ivory was here and ready," Jamal Nasser complained.

In the evening I placed the unique 'Parker Lock' on the ivory that had been registered so that I could see if it was disturbed during the night. Every time that tusks are stacked, their shapes ensure that the stack is unique. Because of their shapes they slip and slide against one another. Move one and all the others move relative to one another and it is impossible to get them back into their original positions. The Parker Lock was a couple of lines drawn across the outer surface of a tusk pile. On the pile the lines appear as just that, two lines: move a tusk out of the pile, all other tusks shift and the lines break, some parts of it moving left, others right. And there is a terrible tendency for the breaks to become bigger rather than smaller when one tries to push them back into position to reconstitute the line. All present, as well as Jamal, appreciated the purpose of the lines. The handlers had laughed at the lock's simplicity.

When next I arrived at Jamal's house it was to find a gentleman from Hong Kong ensconced. His first name was George and I had met him in Hong Kong: perhaps he had been the 'George' that the Belgian Doctor had asked to come immediately. His family had been buying a lot of Jamal's ivory and George was the family member most directly involved with Burundi. He had a reputation for being abrupt, but on this occasion, he was courteous and pleasant. We went through the pleasantries of greeting and enquiring after one another's health before he broached what was on his mind. What I was doing was not in Burundi's interests. I was depriving it of revenue and employment. It would benefit no one. And it was spoiling his business too.

He felt that I should stop. There was no way in which I could possibly succeed in closing Burundi as a source of ivory. If he publicised evidence of a single export of ivory from

Burundi after the registration exercise, western governments would criticise Burundi, which was all that would be needed to make it go back on its word. And he was sure I appreciated that such evidence of a shipment need not actually involve any ivory. Paper would be quite enough: a false airwaybill would be sufficient. He went further and said that he was prepared to pay the Burundi government a million dollars to reverse its decision to join CITES and was certain that it would. He also realised that if I did not get the job done within ten days, the conservation community would automatically suspect that I was allowing more ivory to come in, the Secretariat would panic and importing countries would react by refusing imports of Burundi registered ivory even if Burundi had joined the Convention. He had put a finger on one of CITES' great weaknesses: the Secretariat's and western governments' unreasonable fear of public criticism. If any Burundi registered ivory was refused entry into a CITES state, the country would automatically go back to trading.

George was right. If the registration was prolonged, many countries would suspect the worst and ban imports of the ivory even if Burundi had joined CITES. The solution to the problem was that I should disappear. A doctor's certificate stating that I had malaria and was unable to continue the registration could be provided – presumably from Mike or Marc. And of course there would be reasonable compensation for my loss of earnings. Indeed when the figure was later softly mentioned by Jamal, faking a bout of malaria would have made me $250,000 better off. By then I knew that the Burundi ivory trade was worth $35 million annually to the raw ivory buyers (though obviously not to Burundi). This was so blatantly worth defending, that I might easily be made unfit! Suffice it that I wedged a chair under the handle of my locked hotel room door before going to sleep, just in case. That night a ruckus had me thinking, "Here they come," before my mind cleared sufficiently to appreciate that the 'rough stuff' next door was voluntary and pleasurable and not my concern. True to his word, George tried to stop other traders from co-operating, but the presidential edict was more persuasive.

At one-thirty on the morning of the sixth day the room telephone rang. It was Jean-Baptiste Matangana calling from Dubai, though for the first minute or so as I clambered back to consciousness the name meant nothing. He was of course Burundi's fourth ivory trading licensee. To put it mildly I was irritated to be called at such an hour, but he begged me not to ring off. He had, he said, been trying to contact me for the past two days. "The matter is very urgent. Please can you come to Dubai to register my ivory."

The humour notwithstanding, I declined: my job was registering tusks in Burundi and nowhere else.

"But this is Burundi ivory, genuine Burundi ivory, I brought it here from Burundi. When you are finished in Burundi you come here and register it as Burundi ivory. Please. I beg you. I pay your ticket. . . a first class ticket."

Again I had declined and rung off and left the 'phone off the hook. At breakfast next morning the man I knew as Matangana's secretary came to my table, asking if we might talk. He sat down and said that his boss was most upset with me. I had to go to Dubai to register his tusks. He could be very reasonable about this. I had been most unfair in not giving any warning of the impending registration. The ivory was Burundi ivory: real Burundi ivory and it was my duty to register it as such. If I did not he regretted that they would have no option but to take me to court. In colloquial terms I gave him short shrift. Not mistaking the tone he was immediately contrite.

"M'sieur, I mean no offence. But my boss is desperate. His fortune, his future is in that ivory. . . it is only four tons. If he flies it back to Burundi, will you register it?"

If he could have got it back and registered, its price would have doubled and the extra expense justified. All tusks brought before me before the end of the tenth day would be registered. Thereafter the operation was over. If the tusks arrived back in Burundi after the close of the tenth day, they would become government property. In his opinion I was a hard, unjust man. Why did I want to crush them? I felt sorry for Matangana. Neither he nor his men understood what was happening or believed that their ivory business was in any way reprehensible. By their own lights and within their own community it was admirable. The chasm of misunderstanding between us was vast.

Before completing Jamal's ivory he had said somewhat cryptically, "We Lebanese ... we Lebanese know how to do things," then a short while later, "You live on Malima Road in Langata with your wife, don't you? And your telephone number is ..." giving our home number. I had no doubt that the two separated comments were an oblique threat. That if I did not drop out, they knew where I lived and could take retribution on me or possibly my wife. I may have been wrong, but did not react.

Doing the final lot of ivory owned by Tanaba and Mamu the routine went smoothly, but again I was requested to go away, get malaria for which a doctor's certificate would be provided, and take $250,000 in lieu of lost earnings. Obviously they had talked with the others and I was hearing the same proposal. I wasn't clear whether it was the same $250,000 that each party referred to or whether each was offering me this sum that combined would have come to half a million dollars. And they too tried to frighten me, only this time there was no ambiguity about it. They asked me into Tanaba's minute office, where there were a dozen or so Malians and Senegalese. The heat and stench of sweat were stifling. In reasonable tones Tanaba said that if I would not accept their advice and depart, they would take other steps. At this point one of the men rather melodramatically placed his hand on what looked like a Somali dagger. With what I hoped passed for nonchalance I replied that if that was the case, tough; it is written that every man has his day. At that someone laughed and the tension was gone. The whole business was so corny that, with hindsight, it was as though someone had said, "Let's give him a fright." Indeed, of all the bunch, there had been only one man who I judged really dangerous and that was Jamal Nasser's front, Lucien Basabose. The rest were not violent people; they were merchants. The likelihood of someone sticking me with a dagger in Tanaba's office cubicle was so improbable that the motions gone through were quite funny.

On the evening of the ninth day I thought that I had finished the tusks and was horrified to see a man followed by more than a dozen or so others all carrying sacks enter Tanaba's warehouse. The leader came up to me and anxiously asked if he was too late. Despondently I had said that as he was still within the ten-day limit his tusks could be registered. He had turned to the men behind him and signalled them to undo their sacks. They did so and out tumbled not ivory, but rhino horn. The owner – whose name was Dabanani – said that there were some 700 horns for me to register. He was most put out when I declined. It was yet another example of how immensely ignorant many of these people were about conservation and that in as far as CITES rules were concerned, one might have been talking about the traffic laws on Mars. In this one consignment I saw with my own eyes almost as much rhino horn as the WWF and IUCN experts believed was being traded annually world-wide.

The last day of the exercise was devoted to writing up registers of all the tusks that had been numbered, there being just under 18,000. These registers, each sheet of which I signed, constituted the entire ninety tons of ivory which comprised Burundi's pre-convention stock. Late in the afternoon Audace Kabayanda asked me to go with him to Jamal's place as some more ivory had arrived. Cursing, I went to find Lucien Basabose with Jamal. Where were the tusks? They were on a truck which had broken down half way between Bujumbura and the Rwandan border. How many tusks were there? About 300 and Lucien had produced several sheets of paper on which they were all neatly listed together with their weights. Coming from the north through Rwanda, it was likely that the tusks were from Uganda or from Kenya. As Basabose had been away in Kenya, the latter seemed more likely, so I asked if this was where they had come from? He had answered, "Of course."

I suggested that they take another truck, transfer the tusks to it and bring them to Bujumbura and I would register them that night. This was impossible Basabose said. The truck was not only broken down, it was stuck in deep mud. Other vehicles could not get there. I offered to go in a four-wheeled drive vehicle and register the tusks on the spot. This too was impossible. Kabayanda was unable to provide me with transport, nor could Basabose or Jamal. They clearly did not want me to go to where the ivory was stuck for the very good reason that it wasn't there. I was intuitively sure that the ivory was on its way, but not yet in Burundi. The obvious response was, "Sorry chaps, too late," as had been the case with the Arabs and Jean-Baptiste Matangana.

Kabayanda was terrified; he was so frightened that he looked as though he had been dusted with ash and his condition verged on severe shock. Lucien Basabose was the cause of this terror though I never found out why. What did they want me to do? And it had been Kabayanda who begged me to take the list of tusks and fill in the registers as though I had seen and numbered them, and he would write the numbers on them himself as soon as the tusks reached Bujumbura.

What would be the outcome if I refused? My reasoning went thus: once I had left Burundi I would be accused of having behaved unreasonably and refused to register tusks that were already in the country, but which could not be got to me in time through *force majeur*. This was the thin end of a wedge that would soon exceed 300 tusks. Intuitively I sensed Basabose was counting on me to refuse. If I extended the registration period for this one small consignment there would be a cascade of subsequent lots that I had 'missed' including Matangana's tusks from Dubai and the Arabs' from Dar-es-Salaam. It would be my word not against Basabose's but Kabayanda's that the ivory had not been in Burundi. The complications were endless and could lead to the whole operation being discredited which was, of course, what the traders wanted.

On the other hand registering the 300 tusks on the list before me – less than two per cent of the stock now registered – would not change the overall situation greatly. Was it worth wrecking the whole 18,000 tusk project for 300 tusks? The goal of Burundi having no pre-convention stock was almost achieved and 300 tusks (less than half of one per cent of the number annually moving through Burundi) seemed a small price to pay. Rightly or wrongly, I registered the tusks on the list Basabose presented. The look on his face had confirmed that this was the right decision.

The 300 tusks I never saw were registered as though I had seen and personally numbered them. Only when the registers were drawn up and typed in quadruplicate – each sheet signed by me – did I realise that I had made a mistake. On the sheets I had transposed the national identification number and dates. Instead of reading BI/86/number/weight, on the registers they appeared BI/number/86/weight. The idea of retyping the whole lot again was more than I could face. As all the essential information was there on the registration sheets, I decided to leave things as they were. It also passed through my mind that the mistake might serve a purpose. If people later tried reproducing Burundi permits or duplicating numbers onto other tusks, they would have to decide which was the right sequence ... that on the registration sheets or that on the tusks? If they copied the sequence from the registers onto a tusk, or *vice versa,* used the sequence off tusks as the model in a false register, both would reveal the falsification. Only if someone had both original tusk and the register relating to it would they appreciate that the sequences differed. This would actually help CITES authorities in importing countries pick up any scam. And in due course we did pick up an extraordinary forgery – but of that, more later.

My final task was to help Audace Kabayanda make out export permits against which import permits would be issued. And here as yet another aid to getting the ivory out in several large shipments and mindful of what I had been told about the more permits issued the easier it was to run a scam, I kept the export permits as few as possible.

* * * *

From Burundi I went via Nairobi to Lausanne, bringing duplicates of all the registers and export permits. My strategy over the last 300 tusks that I had not seen was explained and approved. The mood in Lausanne had been congratulatory. Burundi was joining CITES, its ivory had been registered, there would be no pre-convention stock left. All that remained was to encourage the ivory owners to export it as quickly as possible. Yet while that is how I had seen the project when I had embarked on it, as it developed I could see, as is so often the case, that solving one problem created another.

As indicated, the market value of Burundi's raw ivory trade was of the order of $35 million annually. The exercise had caught some ninety tons moving through the country and gratuitously doubled its value. Yet to those involved, this did not redress the loss of their businesses. I believed George all too well when he said that they would reverse the Burundi President's decision to join CITES and stop the ivory trade.

The euphoria in Lausanne evaporated when I said, "Well, now my job is done, yours begins." What had I meant? I told those around the table that now it was very urgent to get behind the Burundi President, congratulate him, laud him in the world's media, get all the bigwigs in conservation to contact and praise him; if possible have them visit Burundi and make him out to be God's gift to conservation – in short, make a hell of a fuss over him. If this were done quickly, he would be pasted into a position that he would find difficult to back out of. It should have been no problem; after all, top conservationists love publicity and travelling to out-of-the-way places. With the Secretariat's weight behind it, the request should be heeded. The essential thing was that the Burundi President should be smothered with praise *now*.

The man to have made this his top priority was Joe Yovino, the CITES Ivory Officer. When he, looking somewhat nonplussed, responded that he would take it up as soon as he got back from three weeks' leave that he was about to start, I knew the Burundi project was on the rocks. Wrapping up the Burundi President was so urgent that Joe's leave should have been postponed. The Secretary General, Eugene Lapointe (ironically for a man in his position, *la pointe* is French for 'the tusk', hence his nickname 'Gene the Tusk') did not seem to absorb what I was saying either. Only Chris Huxley understood the issue, but he was having his own problems with the organisation and being sidelined away from matters to do with ivory. Clerical civil servants simply did not know how to cope with the real world out there and I left Lausanne feeling rather sick.

Zully Rahemtullah's ivory was despatched to Antwerp where I had my last contact with it. The Secretariat was keen that Joe Yovino had some lessons on reading the forensic evidence that tusks provide. I agreed to provide a day's consultancy and flew over from Britain early on the appointed morning to put in an eight-hour day, returning to the UK that night. Seven of the eight hours were spent alone carefully going over six tons of tusks that Rahemtullah had put at our disposal. Joe arrived at four o'clock in the afternoon so did not benefit from a full day's instruction. He hoped that I could change my plans and put in an extra day which, unfortunately, I could not. We marched to the beat of very different drums! Ironically I had accepted three consecutive consultancies on the same ivory – first for Zully, then for the Burundi Government and finally for CITES itself – and only for the last was I fully paid.

Within three weeks of completing the registration operation, I had heard a rumour that 30 unregistered tons of ivory had left Burundi for Singapore. I assumed that George was doing what he said would be done, but never pursued the matter.

* * * *

On 2nd December 1986, the day after registration of all pre-convention ivory world-wide was supposed to have been completed, I had received a call from none other than Lucien Basabose: would I come and see him at Nairobi's Intercontinental Hotel? Rowan Martin from Zimbabwe, who was undertaking a pan-African consultancy on leopards for CITES, was my house guest on that day. Feeling that it would be educational for him, I took him along. Lucien was the epitome of friendship and his greeting was as though we were exceptionally close. No-one hearing it or seeing the expansive smiles would have believed the unkind things he had said about my ancestry in September.

Lucien invited us to his room so that we could talk in private. Once there he produced a bottle of Johnny Walker Black Label, sloshed it into hotel tooth mugs and bade us drink and drink deep. He topped his up with Pepsi Cola and was soon very loquacious: in the immortal idiom of Asterix the Gaul, he loquaced.

He was sorry that he had not been able to make a 'better offer for my friendship while I was in Burundi. . . it was all Jamal's fault. . . these Arabs were very inbred, they married their sisters and their cousins. . . you could tell this by their eyes.' These observations suggested that he had perhaps fallen out with Jamal. *In vino veritas*, and all that: if Rowan and I kept the dialogue flowing, friendly and entertaining, we might have a very profitable meeting.

My 'very very dear friend' Lucien's problem and the reason for calling me was that he had more ivory in Burundi that he would like me to 'legalise'. It had of course been there all along, but his wife ("Stupid woman") had failed to deliver it to me at Jamal's. Would I oblige him by interceding on his behalf with Mr Yovino and Mr Huxley and get this tiresome matter sorted out? I said that I would convey his predicament to the Secretariat, and did so the following day – though not in terms he would have approved.

As the meeting proceeded he said he had exported fourteen tons to Singapore at the end of October. It went with eleven tons of Tanaba's and was flown by an old Boeing 707 freighter registered in and operating out of Ghana. I was not sure if this was true. If it was going out in such quantities and with such ease, why would he want me to intercede with CITES over the tusks that his wife had not handed over to me? Of particular interest was unsolicited confirmation of what I had suspected all along – that India equalled all other parties as a consumer of Burundi ivory. The Indians paid very badly but bought any quantity. It had to be thus: India still had more ivory craftsmen than any other country and there had to be a substantial supply to keep them in business. Since 1979 I had been insisting on this, but was slapped down every time that the subject came up.

Seeking information on leopard skins, Rowan asked Lucien if he knew much about the trade. I thoroughly enjoyed watching Rowan get the facts from the horse's mouth. Leopard skins? Yes he bought them occasionally but mainly because they were brought by the people selling ivory and rhino horn and one bought the leopard skins simply to keep them happy. But why fool about with leopard skins? It was messy and demand for them was not good. Of course friends in Europe still liked leopard skins, so he took skins as presents. Rowan, amazed, had asked how he hid them.

"Hide them? I don't hide them. I just put them in my suitcase. Each time I go I take one." Infused by Pepsi Cola and Johnny Walker, he had somehow got the idea that Rowan was trying to sell leopard skins, so offered to help. And when asked about CITES he replied in genuine bewilderment, "What about CITES?"

Rowan and I agreed afterwards that this was no act. CITES was only a barrier to those trying to comply with the rules, but no impediment to those prepared to disregard them.

I gather that Lucien Basabose did get his come-uppance a year or so later. He had gone to Belgium where he sought out Ali Suleiman, Jamal Nasser's brother and shot him dead over a long-standing dispute. For this he was jailed. As Audace-anything-but-audacious Kabayanda had said, he was a criminal: a real criminal.

Perhaps the world's most widely distributed large predator, leopard are still abundant.
(Photo Peter Davey)

CHAPTER TWENTY-EIGHT

THE EMERGING OVERVIEW

The Burundi registration in 1986 marked the end of three decades' involvement with ivory. The African elephant's range still ran to millions of square kilometres encompassing every country south of the Sahara except the four smallest: Burundi, Djibouti, Swaziland and Lesotho. Sub-Saharan Africa also had ± 490 million people, most of whom were rural and of whom many lived close to elephants. Wherever the two species lived side by side, local people still hunted them.

Regardless of whether motives were killing elephants for meat with ivory as a useful by-product, or primarily for their tusks, there was a strong pattern across the continent. Hunters tended to stick to their own areas. In Kenya the traditional elephant hunting Wata and Kamba hunted in their own tribal lowlands while the Kikuyu who hunted did so in the highland forests that their farmlands surrounded. In Zambia the Luangwa Valley Bisa people hunted in the Luangwa Valley and so on. Situations where different tribes either farmed or hunted alongside one another were exceptions rather than the rule. Only in towns do tribes mix on any scale.

The way elephants were hunted tended to be uniform and reflected the animals' biology more than anything else. Men went by themselves, or as pairs or small groups, stalking their victims and shooting them from close range with a variety of firearms or with bows and poisoned arrows. There were of course variations on the theme. Nilotic people, for example, still used spears and sometimes ring-fires in which the vegetation around a herd was first set alight. In others heavy wire snares had been adopted, though only rarely as the technique of choice for elephants. Snares were usually used for smaller species. Some people still used traditional drop-spears hung over paths. Regardless of technique, the salient point was that ivory came from virtually everywhere elephants existed and was produced by literally thousands of peasants the length and breadth of sub-Saharan Africa. The pattern of killing elephant and collecting ivory was ancient. If it differed from earlier times it was simply on a greater scale because there were more people.

Here and there across the continent, other patterns were superimposed upon this traditional picture. Arabs from the northern Sudan took heavily armed caravans into the southern Sudan and on into northern Zaire, and Somali gangs operated across eastern Kenya and into northern Tanzania. Some of the latter even went into southern Tanzania and Mozambique. Neither Arabs nor Somalis had much of an elephant resource in their homelands. To hunt for ivory they had to go into other territories and be prepared to fight opposition from both local tribes and governments. As occurred in Tsavo during the mid-1970s, Somalis driving competing Kamba out of the field exposed this intertribal aspect.

Where civil war prevailed – as in southern Sudan, Mozambique and Angola – military units behaved like the Arabs and Somalis. By force of arms they went where they wished taking ivory as revenue for their respective causes. Yet war itself restricted freedom to hunt. Africa's bush wars have had broad zones of confrontation rather than clearly defined front lines. Such zones were roamed by enemy patrols and hunting by one side might alert and attract the other. Places where one might hunt freely were actually far smaller and usually well back from the areas of interaction.

Large scale elephant hunting needed peace, not war. Game departments went after elephants with the authority of government, which overrode tribal inhibitions against outsiders. They did so in peacetime without having to be circumspect about attracting an enemy and were thus all the more pernicious for elephants. Here and there military or game department units had massacred elephant on a spectacular scale. The collector's permits in Kenya came into this category. Overwhelmingly, however, elephant hunting was done by local hunters who lived alongside their quarry and was not by organised criminal syndicates.

Reasons for hunting varied. In Zaire and in West Africa demand for meat was often the primary reason with tusks a by-product. Elsewhere desire for tusks was the sole reason and carcasses were left to rot, particularly where hunters were Muslims, as not having cloven hooves, elephants were unclean and forbidden as food. Situations where hunters were induced by traders to go after elephants did happen. The ecologist Richard Bell recorded such a case. Arriving in Mpika (in Zambia) a group of Senegalese traders quietly asked the local people who among them were the better hunters. Having identified a well-thought-of hunter who, until that time was taking buffalo and antelope for meat, they approached him and as an earnest of their goodwill gave him a transistor radio. After a cautious courtship, he agreed to take elephants and from then until arrested by Bell he hunted elephants for these Senegalese merchants.

Bell also found that the decline of copper prices in 1975 and the collapse of the Zambian economy had thrown many people upon their own resources. Some became petty traders and one business that flourished was buying smoked and dried fish on the shores of Lake Tanganyika then carrying it to the mining towns. Occasionally a carrier would be asked to take a tusk or two concealed with his fish and deliver it to a town merchant – usually from Senegal or Mali. The profits were attractive and some fish carriers became regular ivory carriers, with several actually hunting elephants. It did not surprise Richard that these carriers-turned-links in the ivory chain also collected and peddled gems, which were bought by the same men who bought ivory. When elephant stocks declined they shifted away from ivory and concentrated on precious stones. It seemed that the incentive to hunt for ivory was the same as that to acquire and sell gems or dried fish: the resource was available so it was used.

In Zambia some Senegalese and Malian merchants did put hunters under obligation by providing firearms and ammunition: exact parallels occurred among Somalis, where traders also provided services as well as outlets for tusks for some gangs working in eastern and northern Kenya. However, over far wider areas even this relatively low level of commercial sophistication did not exist. For the most part, as we knew from the Kenya and Sudan evidence, tusks moved from hunter to a petty trader, and from petty trader to a larger scale buyer in a system that was remarkably 'vascular'. The hunters were individual cells, the petty traders provided the capillaries and the bigger buyers the veins along which tusks moved.

I noted the degree to which tusks had been smoked, particularly those from Zaire. This smoking arose fortuitously from the hunters' habit of stashing tusks in the thatch of their smoke-filled huts. The longer they were there, the heavier the smoking (smoke only

My impression was of veritable river of ivory that became ever larger as it slowly moved down the trade channels.

stained the tusk's outer surface) the darker they became. The presence of such smoked tusks, and they were common among Zaire ivory, indicated the hunters were in no particular haste to get rid of them.

The speed at which ivory entered the trade pipeline varied enormously. That from the aggressive, fast-moving Somali gangs operating well away from Somalia may have been in Burundi within weeks of elephants being shot. Yet other tusks such as some of those in Mogadishu could be held up for years. Once the initial movement between hunter and first buyer had happened, movement along the pipeline accelerated. Yet even then it could be quite slow and involve months in a cache underground, for example. However the pattern was of sources gradually combining like the tributaries of a river.

Just as gravity and gradient controls the direction and pace at which water flows, so the local person's need rather the prevailing international price of ivory dictated the manner in which tusks entered trade. From what I saw, I estimated that not more than 15% and probably less than 10% of the tusks registered in Burundi came from elephants killed within the preceding three months. The rest showed a variety of ages from within the preceding year to several years previously. From the quality of the ivory I wrote in my report to CITES that –

> *"I have the impression that one looks in on a great, slow-flowing river in which individual tusks move from place to place at a rather slow and erratic rate."*

In sum, the picture of traditional Africa with its people constrained to their own tribal territories and trading as they had for centuries still dominated the ivory scene in 1986. Traditionally the successful hunter had high prestige in local society – which was incentive enough for many to hunt. Both sustenance and commerce played their rôles, but motives to hunt were as much locally generated as 'imported', coming from the hunters themselves, their communities and cultures. Superimposed on this broad and somewhat bucolic background, some places had been ravaged by men prosecuting civil wars, or by armed marauders, or by civil servants corruptly taking advantage of the powers vested in them. However, striking as these events may have been, they influenced relatively small parts of the elephants' vast overall range. This is a picture at odds with that reported by the western, conservation-influenced media.

* * * *

In Burundi I found that tusks from natural mortality as opposed to those from elephants clearly killed by man formed around 5% of the total. I had expected it to be higher. In 1979, in Hong Kong, the corresponding figure was nearer to 23% suggesting that in the 1980s fewer tusks were being recovered from elephants dying of natural causes. The largest source of all ivory was Zaire which produced 33% of Burundi imports, followed by Zambia with 30%, Tanzania with 23% and all others combined at 14%. Among the other countries from whence ivory had come were Southern Sudan, Central African Republic, Cameroon, Mozambique, Malawi, Kenya, Botswana and Zimbabwe. The latter two are interesting because at that time (1986) there was thought to be little ivory poaching in Botswana and virtually none in Zimbabwe. It was the first hard evidence I had that the Zimbabwe Department of National Parks and Wildlife Management's poaching control was not as effective as made out.

There appeared to be four tributaries joining the ivory river in Burundi. The first two, from Zaire and from Tanzania, were minor and comprised small scale merchants living close by smuggling a few tusks at a time over the border past the Burundi Customs authorities, selling them to one of the four licensed buyers in Bujumbura. Technically such ivory entered Burundi illegally by the country's own laws. However, once in the hands of a licensed dealer, they submerged out of sight in the bigger volume of 'legal tusks' (by Burundi's measure). And there were two major tributaries – one from the west and Francophone Africa managed by the Senegalese and Malians and one from the east and Anglophone Africa managed mainly by Somalis and a few Arabs. Of the two, the Somalis' was larger through the Somali strength in long-distance road haulage.

From the soil particles in their hollows, many Tanzanian and Zambian type tusks had been buried in riverine alluvium. I had commented on this and one of the ivory handlers who had travelled into Tanzania with the Somalis confirmed it. He said that the tusks were buried in a seasonally dry watercourse near Iringa township in the middle of the country. According to the tusk handlers, Iringa was *the* centre through which much ivory from Zambia, southern Tanzania and Mozambique passed *en route* to Burundi. And all sources agreed many members of the Tanzanian Game Division and National Park authorities both parasitised the general ivory traffic through 'protection fees' and were themselves major sources of ivory and rhino horn. Seemingly Tanzanian authorities were just as corrupt as their Kenya counterparts. This was coincidentally confirmed by safari operators. Each hunting safari had one or more game scouts attached to it to ensure no one broke the law. These men openly admitted shooting elephants on their own accounts and that the tusks were taken by Somalis to Burundi: and, as stated in the previous chapter, I found Tanzanian registered tusks in Burundi.

Perhaps the most telling confirmation of official collusion between officials and the traders was that made by Lucien Basabose: it was the lack of contraband ivory seized at Customs posts on Burundi's borders. There were four such entry points into Burundi: one from Rwanda, one from Zaire, one from Tanzania and Bujumbura port on Lake Tanganyika which took traffic directly from Tanzania, Zambia and Zaire. The northern route handled a relatively small trade from Kenya. Much of this Kenyan ivory was carried in the tankers that brought petroleum products to Burundi. Tusks absorb liquids in which they might be submerged. Later when taken out and stacked, they gradually release them. Diesel stains on the floor or smell alone betrayed such ivory from Kenya. The big imports of consequence came into Burundi through its other three entry points – with that from Tanzania handling most.

Burundi had imported around 300 tons a year for the five years 1981-6, totalling at least 1,500 tons. Assuming average weights of five kilos per tusk and average loads of around two tons, both figures ivory handlers said were not unreasonable, a load came through one of the three posts every two to three days without hindrance by the authorities of the countries from which they were departing. Such volumes of a commodity that is difficult to conceal indicates regular, organised collusion. As the two Arabs who had wanted me to extend the registration period indicated, this was indeed the case and it was further confirmed by the ivory handlers.

Perhaps the most striking aspect of the Burundi ivory trade came from prices. When Antwerp closed as an ivory entrepôt and Burundi took over its rôle, it was not long before nearly half Africa's known ivory exports were going through the country. Because the Burundi buyers had to sell their tusks outside the CITES-approved system, they could only get around half the prevailing world price for them. Consequently they, in turn, were buying it at around one quarter of the prevailing international price. Having to use the Burundi route inflicted a huge price reduction upon both the hunters and traders inside Africa who were producing this ivory. It was of the same order as the drop in price that occurred during the Great Depression of the early 1930s.

If the theory that price drove demand was valid, halving prices should have depressed production; yet it did not – either in the 1930s or the 1980s – thereby destroying the hypothesis that price drove ivory production. Perceiving this helped explain its rôle in African economies.

With a very few exceptions in Mombasa and in Zanzibar in my very early days, none of the merchants I met or heard of in the Burundi ivory trade or elsewhere in Africa for that matter depended on ivory: it was always one among several or many items they traded. The Somalis, for example, traded all manner of commodities in Tanzania. As profits accumulated so would incentive to convert them into a negotiable currency. At this point gold, ivory, rhino horn and gems were sought because they were freely convertible. Their convertibility equalled their intrinsic value in making them attractive.

Ivory prices were understood in US dollars per kilo. The commodity was bought in Tanzania shillings at around 15% of the prevailing world price, the conversion from United States dollars made at the official conversion rate. This ivory would then be sold in Burundi at 25% of world price – $25 per kilo – giving the seller a gross return of 67% on his original outlay. Some of this profit would be used in Burundi to buy goods like torch batteries, cooking oil, motor spares, soap and so on that were hard to get in Tanzania. The balance of the dollars would be converted into Tanzania shillings at the black market conversion rate, which in shillings represented a profit of several hundred per cent. Taking the local currency back into Tanzania, traders would use it in general commerce – buying and selling hides and skins, for example, to restart the cycle of accumulating profits. In due course, these would be used to buy something internationally convertible like ivory, which in turn was traded into manufactured goods or local currency at black market rates.

Ivory played two rôles. Through it Tanzanian profits could be converted into other currencies or goods: it was, *de facto*, a home-produced international currency in its own right. And through the differences in official and black market exchange rates, by buying at the former, but selling in the latter, it multiplied the local currency's power. It and commodities like it became essential once the Tanzanian currency was debased and the formal economy mismanaged. So long as it served as a 'converter commodity', it did not really matter what the actual price of ivory was. Contrary to the common view, the ivory trade was not a business managed by slick Mafiosi independently of the prevailing economies. To the contrary, and Wilbur Smith notwithstanding, it was an integral, inseparable part of general African trade. Ivory profits were not squirrelled away in Swiss bank vaults, but recycled vigorously in commerce within the continent.

Burundi illustrated that in *economic* as opposed to conservation terms, ivory was used beneficially in sustaining interstate trade, as a source of capital, and as a means of getting manufactured goods that had become otherwise unobtainable. By fostering the business Burundi acquired foreign exchange and trade goods that were unavailable in the centralised and depressed economies around it. If people wished to stop Burundi trading ivory (or any other of its neighbours' commodities for that matter) the solution lay with the root of the problem and had nothing to do with CITES. If its neighbours' economies flourished, the whole reason for using Burundi as an illicit entrepôt would vanish. Its ivory trade was a product of misguided social and economic experiments by its neighbours in the wake of their independence. Why sell Tanzanian ivory, gold, coffee, or gems in Burundi if you could get a fair price in a viable currency in Tanzania?

Since the late 1980s, World Bank and International Monetary Fund pressure has made Tanzania, Zambia and Zaire decentralise and privatize their economies, making their currencies freely convertible. This dissolved Burundi's value as an entrepôt overnight. Because so much of this economy was unrecorded, the full extent of this decline will never be measured. Yet I am sure the collapse of its entrepreneurial position added economic pressures that had no small influence on reviving the endemic Tutsi-Hutu conflict and the civil disorder that has paralysed the country ever since.

* * * *

White folk look at Africa through a template of European expectations. Still lacking self-confidence from the colonial experience, Africa pays lip service to these expectations. Governance across the continent is still outwardly European in shape and content. The detail may be slightly different, but bureaucracies contain the same old components of ministers, ministries, directors and secretaries-general, permanent and principal secretaries, under-secretaries and so on down to the lowest clerical ranks. We judge efficiency according to how these systems function. A government can only be well run if it conforms to white concepts of what a government should be. Nyerere's and Kaunda's experiments with Tanzania's and Zambia's economies, for all that they ruined both, were accepted because they retained 'appropriate' civil service structures. Our attitudes towards alternative modes, or even the possibility that there may be alternative modes, have changed little since David Livingstone's day.

In dysfunctional Burundi the template was thinner than in most African countries. It was easier to see past it, and I had glimpsed an older, more confident Africa functioning free of imitated European red tape. The Senegalese, Malians and Mauritanians who traded across the continent, and to whom modern boundaries meant little, conveyed this more strongly than others. Their trading traditions stretch back to and beyond the great Malian Empire of the thirteenth and fourteenth centuries. What white men believed concerned them little. That they were of Islam no doubt contributed to this self-confidence.

The two merchants, Tanaba and Mamu, were not so much independent operators as informal agents of a web of Senegalese, Malians and Mauritanians that spread from their homelands in westernmost Africa across the French speaking states and south into Zambia. I had met some in 1972 exporting tusks to Belgium via the Sudan (see Chapter 19). In a network of formal and informal relationships these people traded anything that was tradable. Just as they had little respect for the boundaries that are whitey's most obvious legacy to Africa, they paid scant regard for his divisions between what is or is not contraband. If a commodity could be bought with potential for resale at a profit, they bought it and sold it on. It might be hides and skins and coffee, but their interest naturally included diamonds, other gems, gold, ivory, rhino horn and items like crocodile skins. They had African agents resident in Europe, particularly Belgium. Money-lending and money-changing were very much part of the business. And most of this extensive commercial activity went unrecorded, beyond taxmen's reach in any country's official statistics. The Somali network was less sophisticated than the West Africans', yet it comprised a net cast right across East Africa, took in at least part of Mozambique, as well as Zambia and into Botswana. Therein lies a tale.

A Somali transporter based in Gaborone approached a young white District Officer in Botswana and asked where he could get ivory. In jest the DO had said, "Break into the Wildlife Department's store: there is plenty in it." A few days later someone made a hole through the wall of that store and removed all the tusks. They never found out who did it, but by then the DO was too embarrassed to tell this story, though he had his suspicions.

Asian traders were more sophisticated than either the Senegalese or the Somalis. Yet they, too, conducted a great deal of their business beyond the law. As they pointed out with impeccable logic, look what had happened when misguided governments impose currency controls and tried to dictate how economies should run. They, too, had an ancient history of dodging white legislation. Yet it is the growth of 'black' (undocumented) economies in Italy, and now to a surprising degree in Britain, that suggests that there may be something wrong with the systems we still try to enforce.

In European terms the traders were smugglers dealing in contraband and, because they were outside the law, the European solution was bring them to justice. It recalled Willie Hale, Kenya's erstwhile Chief Game Warden, stating in his 1953 annual report that every African was a poacher and wrapping up the observation by stating that all must be prosecuted. Burundi revealed an Africa going about its ancient ways efficiently and confidently, regardless of the template left behind by us whites. And perhaps it was a glimpse of the future. After all, aliens have come and gone for the past three millennia and after each departure Africa reverted to ways more in keeping with its own desires.

* * * *

By the end of 1986, the overview of Africa and ivory trading that I had put together was radically different from that generally in vogue. It had not arrived as a flash of light, but by gradual evolution down the years. Nevertheless Burundi fleshed out the picture as no other experience had done.

CHAPTER TWENTY-NINE

CITES – THE UNWORKABLE TREATY

It will by now be apparent that from the late 1970s and into the 1980s I had considerable contact with the CITES Secretariat. I had undertaken consultancies for it and was in frequent touch with its officers. Yet while we worked closely and got on well at personal levels, I never concealed my belief that CITES was unworkable. As it is still a major international conservation tool, I set out the reasons for this opinion.

The idea that trade in wildlife products causes wildlife to decline has been in vogue throughout the twentieth century. Sections 7 and 10 of Article II in the Convention for the Preservation of Wild Animals, Birds and Fish in Africa signed in London on 19th May 1900 demanded its signatories only allow wild animals to be hunted on licence. Further, they were to impose export duties on game trophies to make trading them unattractive. As a result most African states had rules to control commerce in trophies.

The 1933 Brussels Convention on African Conservation, brushing up the 1900 London Convention, introduced hunting/trade categories: Class A being species that could not be hunted or traded, Class B those species that could only be hunted or traded on permits. Yet despite the Convention, animal declines continued. Their causes were never analysed but it was nonetheless assumed that trade must be the reason.

That commerce has caused some wildlife declines is indisputable. Prime examples were seal populations followed by whales. Both were reduced by industrial hunting until they were so few as to be no longer worth exploiting commercially. Towards the end of the nineteenth century the vogue for feathers led to some birds being much reduced – egrets among them. Today oceanic fisheries are examples in which, conservation sentiment notwithstanding, and despite unequivocal trends of decline, the developed nations drive wild stocks down for commercial reasons. North Atlantic cod, herring and the world's tuna fisheries are shrinking because they are over-fished.

Yet decline is *not* the outcome of trade *per se*. There are many cases where commerce has not reduced populations and there are other reasons for decline – commonest among them being changed habitat. Around the world people have harvested wild animals and traded their products for centuries without destroying the resource – the sub-arctic fur trade being a case in point. Thus if trade *per se* does not necessarily diminish wildlife, in what circumstances does it do this?

The 19th century vogue for feathers led to some bird being much reduced in numbers, egrets among them.
(Photo Peter Davey)

The crucial factor seems to be the absence of a sense of ownership. Unsustainable harvesting happens most commonly with wildlife (and resources) over which there is no sense of ownership. By definition whales and fishes in international waters belong to nobody. A sense of ownership seems to have been established where whales are concerned, but it is tenuous. The same nations who promote it most rigorously, however, are among those who are the world's worst despoilers of fisheries. Because wild resources are there for the taking, an attitude prevails that 'if I don't take them someone else will. Why should I husband them when others will merely take advantage of my diligence and remove what I have husbanded?' While this is starkest on the oceans, it also applies on land.

Ancient juridical opinion holds that, like the air, what is wild and possessed of its own free, independent will, cannot be owned. Yet at the same time as acknowledging what is wild is unownable, societies around the world have also held wild animals to be resources that should not be used incontinently. Therein lies conservation's paradox: looking after something that nobody owns. Its rider is that societies have to decide who may use the resource. If what is wild cannot be owned, that is not true of the space in which it lives. Few senses of ownership are more fiercely expressed than that relating to land and space. Because ownership determines who may or may not enter upon land, it also determines who may have access to wild animals that live on it. It is logical therefore, to let land ownership determine who may use those animals. Across the ages the formula which appears to have served society best is to allow those who own the land to have the right to use the wild animals on it. If a community owns the land, then that community has the user rights. Where private title exists, then the titleholder has that right. Consequently, while what is wild cannot in the strict sense be owned, giving the right to use it to those who hold the land where it occurs creates a sense of ownership in the wider sense. Where the connection between land tenure and user rights has this result, trade is least likely to produce declines. The owners of the land feel that they own its wildlife and it is no longer free for anyone to take.

Where user rights have not been married to land tenure, conservation has seldom worked. Ravaging the world's oceanic fisheries, and over-harvesting seals or whales arose because there were no senses of ownership, and users take as much as possible before it is taken by others. Spain pillages fish stocks in the north-west Atlantic outside Canada's 200 mile nautical limit because they are owned by no nation. Canada sends a gunboat to stop the Spanish within the 200 mile limit because Canada 'owns' the waters. Large scale poaching is not a problem to National Parks in the United States because there is a real national sentiment that they are 'owned' by and for the people. It is a huge problem in Africa because among many reasons, this sense of national ownership does not exist. Too often, land was taken away from the people who owned it to create National Parks. Hence the Kiswahili saying '*ni mali ya Serikali tu* (it is only government's wealth)' is synonymous to 'nobody owns it.' At the village level in Europe, the landlord had a strong sense of owning the game on his estate and preserved it, but peasants poached it because they did not believe anyone could own what is wild.

Understanding why trade produced declines in some cases and not in others and, in particular, the central rôle played by senses of ownership, should have influenced conservationist thinking. It did not and an unremittingly simplistic outlook prevailed: trade is the cause of wildlife decline, *ipso facto* trade is bad and must be eliminated or, if that is

Mute mummies protesting misfortune, these giraffe had the bad luck to be found in an open place by a Land Rover full of policemen. Driving in circles round them, the policemen shot and shot and shot and left the carcasses to dry in the sun. Animals do not have to bear valuable trophies to warrant human attention.

not possible, tightly controlled. When laws enacted as a result of the 1900 and 1933 Conventions did not slow African wildlife decline, no-one noted that decline in species *not* traded at least matched and sometimes exceeded those which were.

* * * *

In 1973, fifty-one nations signed the Convention on International Trade in Endangered Species of Wild Fauna and Flora (CITES) in Washington. Because of the venue it was also sometimes referred to as the 'Washington Convention' (it still is by the Japanese, for example). Yet while Americans and IUCN were happy to claim founders' status and draw attention to the former's Endangered Species Act and the latter's Red Data Book listings, their thinking was not new. Those involved were unaware that sub-Saharan Africa already had tighter and more uniform conservation regulations than any other continent. Instead of Class A in the Brussels Convention, the new treaty listed animals that could not be traded as 'endangered' and put on CITES' Appendix I. Those formerly qualifying as Class B became 'threatened' and put on Appendix II. And a third category, Appendix III, was created for species that individual states simply wished to make subject to international trade controls regardless of status.

CITES' planners disregarded certain facts. Conservation ineffectiveness had nothing to do with lack of law in Africa, but everything to do with lack of government probity, or will, or inadequate funding, or because the rules were simply inappropriate. If authorities did not enforce existing law for such reasons, why would they enforce CITES rules any better? If laws failed because authorities were too under-staffed or under-funded to apply them, the solution was adequate staffing and funding: not new law. To my knowledge no-one has estimated the manpower CITES needed for it to work.

Conservationists rushed to list species on the Convention Appendices. Seemingly, the longer the appendix schedules, the better they felt they were conserving. In 1979 when there were 527 species listed, I wrote in my ivory trade study –

> *"As more countries accede to CITES and as animals and plants are added to its lists, it will become yet more unwieldy. At best it is only partially enforceable and is in danger of becoming totally unenforceable; a state which is worse than no law at all. While the motivation behind it is straightforward enough, it is too ponderous and impractical to succeed as it is."*

I recalled the Kenya Game Department when the manpower needed to issue permits required by law ended in the bizarre farce of the very people the permits were supposed to regulate – the traders – issuing these documents on Government's behalf. Clearly those creating CITES had no thought for the manpower needed to operate the laws. Rowan Martin was independently making the same points repeatedly after 1985.

Such feelings, drawn not from theory but observation, were already strong when a mere 527 species were listed. Now, after an orgy of silliness, over 50,000 plant and animal species are on the CITES lists, each having to be recognised by its scientific name in archaic Latin or Greek. A glimpse of reality came at Hong Kong's Kai Tak airport whence I went with Chris Huxley (when he was Hong Kong's first CITES enforcement officer) to speak with Customs officials about ivory, but had been mesmerised by the import of live frogs and snakes. They came in wickerwork baskets, literally by the ton. Take the lid off one of these containers and the inmates would have been off leaping and slithering through the warehouses as had happened accidentally in the past. As the Customs Officer said, there were 'flogs ev'wheah, flogs an' more flogs.' No-one was going to go through consignments flog by flog or snake by snake, looking for listed species.

* * * *

When we undertook elephant reduction in Uganda, Dick Laws insisted the cull be random to yield a cross-section of the population. This provided both an accurate picture of population structure at that time as a base-line from which subsequent changes could be measured. The first two histograms show the age structure of two culled populations from which we collected basic data.

The columns represent successive ten-year age classes, those on the left being 0-9, the next 10-19 etc. The histogram on the left is Murchison South (MFNP) representing an old population with high average age, a contracting range, nutritionally stressed, low fertility, falling recruitment, no hunting, and an average tusk weight of 6.5 kg. On the right is Mkomasi East (MKE) a young population with a low average age, abundant forage, consequently higher fertility and recruitment, some hunting and therefore a lowered life expectancy and an average tusk weight of 3.81 kg.

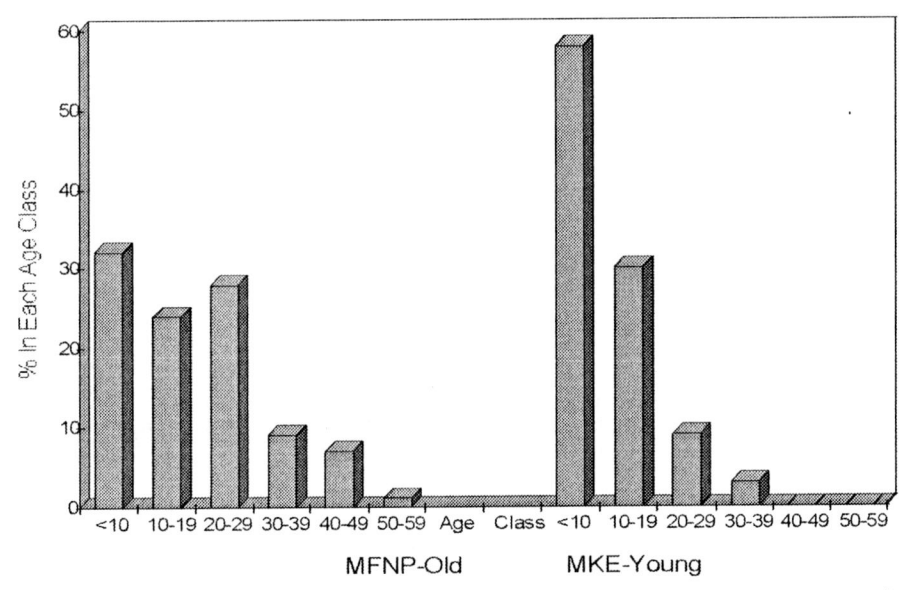

During the trade study I obtained ivory data of elephants culled in the Kruger National Park between 1972 and 1978. From this I was able to construct the two histograms below to compare with the two above.

The Kruger authorities said they culled randomly to produce a cross-section of the population for the same reasons we had followed in East Africa. As it was a young population the pictures should have been close to the new Mkomasi East population on the right above, the only difference arising from the fact that there was negligible poaching in the park and thus life expectancy should have been higher than in Mkomasi. Converting tusk weights back into ages, then reconstructing the populations in ten-year age classes as in the histograms above, this expectation is apparent in the 1972 histogram on the left.

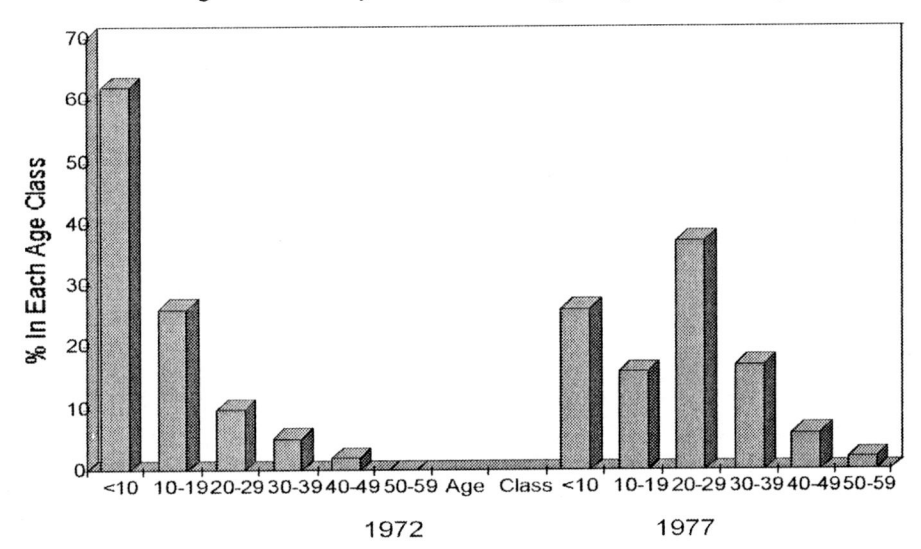

The 1977 histogram on the right presents a very different tale. Either the population structure had changed dramatically in the intervening five years or culling was no longer taking a cross-section of the population. More elephants of between twenty and thirty years old had been taken than those of less than ten, which, theoretically, should have been the population's largest class. The average tusk weight had jumped to from 2.8 kg to 6.7 kg. This was greater than the average weight of 6.5 kg for the old Murchison Falls South population and alone suggested that mature males had been selected while the take of younger age classes had declined.

Returning to Kruger I presented my conclusion that, contrary to what I been told, the 1977 cull had not been random, but deliberately biased towards bulls. This was immediately acknowledged and satisfactorily explained (there had been no intention to deceive anyone). Nevertheless it was a lesson on what ivory data could expose. Given the vast quantity of forensic evidence that tusks can provide, if CITES had wished to monitor the ivory trade, there could have been no better source than 'reading' the tusks as they arrived in Hong Kong. This would have provided a mass of biological and conservation information more accurately and at less cost than any other technique. Yet while this was appreciated (recall, I was employed to show Yovino how to read Burundi tusks in Antwerp), the huge volume of information was left untapped. As pointed out by the eminent New Zealander, Graeme Caughley, the whole business of elephant conservation had quite simply been hijacked. International meetings, 'workshops', seminars and the production of weighty reports absorbed such funds as became available at considerable profit to those involved, but with no influence upon Africa's realities.

* * * *

From the beginning the Secretariat faced an impossible task. The most able of its functionaries was Peter Sand, a German lawyer. He knew little about wildlife, but his incisive mind grasped principles. I felt that if any intellect could inject pragmatism into the organisation, it was his. Yet he moved on to a post in the United Nations Environment Programme before any injection was completed. Though I never asked him (and it would have been indiscreet for him to have answered in any case), I suspect he knew CITES was a basket case.

While CITES was ineffective, it nevertheless made political sense for legitimate international ivory merchants to try to help it. I argued for them to be involved in drawing up CITES procedures and urged them to offer their services. There were two responses. On the one hand conservationists detested it, saying traders 'were too untrustworthy' to have anything to do with regulating (notwithstanding that, world-wide, it is normal to involve businessmen in the regulation of their commerce). On the other, traders agreed with the logic of regulating themselves, but without enthusiasm.

Deeply conservative, the merchants observed that trade had flourished for centuries without regulation, why should it need regulation now? Proud of their traditions, they had been offended by conservationist and press attacks on them and their businesses and had no stomach for closer engagement with such people. In addition they were profoundly ignorant about elephants and Africa. As a result traders inclined to a passive rather than active rôle in influencing official policies. Three who were more positive and tried to help CITES were Thom Friedlein, in England, William van de Velde in Belgium and Dominic Ng in Hong Kong. The outcome of pressure on the traders to become more active was that they appointed me their official delegate to the 1981 Third Conference of Parties to CITES held in Delhi. It seemed an opportunity to see CITES functioning and possibly make a little progress.

If traders were to police their commerce they had to be able to boycott ivory from dubious sources. An effective boycott needed buyers to act in unison. To do that they had to be assured that their refusal to buy from a targeted source did not merely open the door for someone else to do so cheaply. In 1980 anyone could import ivory into Japan, Hong Kong, North America and the European Union and if established dealers refused to buy from Zaire, for example, it would merely have offered speculators from outside the circle a wonderful opportunity to buy ivory cheaply. The trade had to be closed to such opportunism. To create a base from which traders could act in concert, free trade in raw ivory had to be stopped and restricted to people licensed to do so.

Since the 1900 London Conference, Africa's dealers in game trophies including ivory had had to be licensed, so principle and precedent in which the wish to conserve overrode the principles of free trade were long established by the former free trade colonial powers – Britain, France and Germany.

On the ivory traders' behalf, therefore, I asked the Parties to the Conference to make importing and exporting raw ivory subject to licensing. Any dealer breaking the laws could be deprived of his licence. In addition governments should restrict the number of licences issued to limit competition for raw material. The industrial powers were being requested to do no more than implement the system that they had earlier imposed on Africa and which was still in use.

No government was more committed to free trade than Hong Kong's, yet it agreed to implement a licensing system if it was asked to do so by Hong Kong's established ivory traders. They made the request, so my proposal was supported by the most important ivory market in the world. Despite this, the colonial precedents and the legacy of those precedents in Africa, the request was turned down.

Britain and Germany opposed it because it compromised principles of free trade. Yet there are circumstances where even the staunchest advocates of free trade have to concede controls are necessary. Take airlines for example. In most countries, running an airline is subject to licensing. Established operators can oppose issuance of new licenses on the grounds that cut-throat competition might compromise passenger safety. Restrictive licensing is also a feature of national fishing quotas. There were plenty of precedents in free enterprise economies for the sort of licensing requested and both British and German delegates talked twaddle. Rejection confirmed the traders in their pessimism: "They will block anything we propose."

Yet more ironic was the debate on rhino horn that followed. Rhinos were on CITES Appendix I and their horn could not be traded. Zambia, with a substantial stock of horn, was angling for a way to dispose of it profitably and test the Parties' resolve to prohibit all trade. I opined that they should burn their stock of horns, which the Zambians rejected on the grounds that the horn was a 'cultural' heritage. No doubt rhino and rhino horn figure in the grab bag of some Zambian tribal traditions, but in the twentieth century the overriding 'cultural use' of rhino horn had been to sell it overseas. Again both the German and British delegates intervened and 'opposed burning a cultural heritage'. The Zambians, aware that I had been twitting them had, equally tongue-in-cheek, said that the horn was culturally valuable – a point that most from Africa who were present appreciated. The serious and unexpected German and British support for the Zambian position added to the humour no end.

No party at the Third Conference of the Parties was more deceitful than the host nation, India. The Indians talked at length about the need to control African ivory entering trade because it gave cover for illicitly taken Asian elephant ivory. This was rich as Indian elephant ivory was openly sold by the Government itself. At the time India had more ivory carvers than any other country in the world (7,000) and was a major consumer of African ivory: a fact widely known in both India and Africa. Further, unlike all other big ivory consumers, by far the greater bulk of India's ivory imports were illicit. When touring Delhi with several conference delegates we found the shops full of ivory carvings and bric-a-brac and the shop-owners never tried to hide the fact that the bulk of their ivory had come from Africa. Realising that they were talking to someone who knew a little about it, several brought out raw tusks. That I could tell them where in Tanzania and Kenya they had come from (station markings made this simple) the dealers willingly talked about their sources and one asked if I had any contacts who might supply him.

Seizing the opportunity, I asked if the CITES rules made importing ivory difficult. The trader had looked blank, but assured me that there were no rules that could not be taken care of. Indian official hypocrisy was thus obvious. Back at the conference venue, three officials threatened unspecified trouble if I went on spoiling India's name.

This casts a light on how the African elephant got onto the CITES lists. When the idea was first proposed it was turned down flat as there were far too many elephants in Africa for the species to be 'endangered'. Instead, India argued that it be listed under the Convention's 'look-alike' rules because its tusks were so similar to the Asian elephant's. It only did this to appear conservation-conscious before the world. The real lack of scruple was apparent both in having more ivory carvers than any other country and the fact that Indian authorities openly auctioned their own 'endangered' Asian elephant ivory to these carvers. Yet delegates and conservationists did not want to know such facts.

"India," they said, "is serious about conservation and no good will come of embarrassing the country."

* * * *

In 1983 I was again asked by the ivory traders' associations of Europe, Hong Kong and Japan to represent them and their interests at the Fourth CITES Conference of Parties in Botswana, with full plenipotentiary powers. My goals were three – (i) to get acceptance that the ivory trade should contribute financially towards African conservation; (ii) to get the parties to coerce African countries into a uniform marketing system – preferably public auction – for their ivory and, (iii) to gain acceptance that regulating international ivory trading should focus on raw ivory – that is whole, unworked tusks – because they were both easy to identify and difficult to conceal.

Trade regulation and control is commonly funded by levies on the products concerned. By the late 1980s I believed, taking all aspects of the ivory trade world-wide, that its annual turnover was worth $634 million[148]. A five per cent cess on this turnover would

---

[148] From data in I. Parker, 1979. *Ibid.*

have been nearly $32 million and a significant contribution to African conservation. Talk of taxes made the traders gloomy, but they accepted the idea in principle. They would oppose it, however, while Africa's disorganised, erratic, ivory production persisted. Some transparent marketing system had to be adopted. Once again we were back to what had worked so well through the colonial era, and which I had so totally failed to reintroduce privately: open public auction.

I had no great hopes of success this time round. While technically a trade treaty, the Washington Convention was used by the conservation lobbies to stop trade. Any steps to streamline and improve ivory commerce would be opposed no matter how well founded the arguments. Talk of introducing a uniform marketing system was squashed, developed countries saying they could not force Africa to do what they did not want to do.

Talk of taxing the ivory trade led one ardent conserver to bray, "Blood Money ... It will be Blood Money!" and I felt, truly, it was a concourse of asses. Though supposedly gathered to confer, most Parties came with positions and ideas formed in advance which they either had no latitude to change, nor intention of doing so. Thus most of the 'work' could have been done better by post and a large international jamboree was unnecessary. Listening to the deliberations in Gaborone I knew I was wasting my principals' money.

At two conferences the traders had made proposals that went to the heart of the matter. They wanted ivory trading to be internationally licensed and freedom to do so restricted. They had suggested that their business be taxed to redress a major African conservation problem: lack of money. They asked that Africa market its ivory in a uniform and transparent manner. On spurious grounds all proposals were rejected out of hand and I was shamed.

The only thing of use to come out of the conference was the green plastic briefcase with the CITES logo (the copyright for which was paid for by the ivory traders – blood money or not) issued to all official delegates. Twenty years on I still have it, though for the life of me I cannot see what conservation value it had.

* * * *

While CITES' conferences may be where silliness is most openly displayed, it is inherent throughout its work. Every CITES country is supposed to submit an annual report detailing any exports and imports of the more than 50,000 species of plants and animals or parts thereof listed in the treaty appendices. Combined, the reports should document international trade in listed species. In theory, exports from one country should have been traceable in the import statistics of those to whom they were sent. If exports and imports matched it would have proved the verity of one against the other and, *vice versa*, if they did not match, one or the other or both was in some way incorrect. Conservationists call this 'monitoring' which, to them, is an end in itself. Everything has to be 'monitored'.

To ordinary people, monitoring is a tool to reveal where and what action to take. Thus one supposed that if in the CITES annual reports an export did not tally with a corresponding import, someone would do something about it. The Secretariat was only a clerical centre with no powers to order, reprimand, demand or otherwise force parties to redress defaults. All the Secretariat could do was to report its findings to the Convention's

Parties. At best, months after a default was detected and long after the items involved had vanished down untraceable market channels, it might be publicised.

In practice, annual reports, even from the 'developed' nations, were seldom submitted promptly. Delays of two years were common and some countries did not comply at all. At both conferences I attended, the Secretariat gave the outcome of the monitoring effected through annual reports over the previous two years. Less than 5% of animal exports reported tallied with any imports reported. In other words over 95% of stated animal exports did not correspond with any import statistics. With plants, exports matched imports in less than 2% of cases: that is 98% of exports did not correspond with any imports. When the assembled delegates were presented with these extraordinary data, I expected some reaction. After all, they were statements of total monitoring failure. Yet there was no protest. If ever there was proof of monitoring being an end in itself, this was it. That the outcome proved the whole annual report exercise a waste of time mattered not at all. True, the delegates present were probably the self-same civil servants responsible for either failing to submit the reports or presenting the unmatchable trash if they had. Their silence could have been self-interest. Yet the media never picked it up. Lack of match between exports and imports was a non-issue.

In the chapter on Burundi I explained that joining Parties did not *have* to register their pre-convention stocks of trophies with the Secretariat and dishonest governments could use 'pre-convention stock' as a device for laundering new acquisitions. To block this, the Parties at the Fifth Conference of the Parties in Buenos Aires asked all governments to register pre-convention ivory stocks in their countries with the Secretariat by the 1st December 1986. The resolution was *non-binding* and no country could be forced to register pre-convention stocks. Whether it achieved anything therefore depended on the Parties' goodwill. That the resolution was non-binding was what I concealed from Burundi (and Huxley and Yovino concealed from Somalia!) Yet few nations registered their raw ivory with the Secretariat.

It was not as if raw ivory stocks did not exist outside Africa. Throughout this century – other than for the ten years of the two world wars – a stream of North American sport hunters have shot elephants in Africa and taken their tusks as trophies (even after the international ban of the ivory trade in 1990). If this ran at only 100 elephants a year (probably an underestimate) x 90 years = 9,000 elephants = 18,000 tusks. Similarly there was a flow of trophy tusks out of Africa into Britain, France, Portugal, Germany and the rest of Europe that could easily have exceeded the flow into North America. Thus 36,000 tusks may have been taken as sporting trophies through the first 90 years of the twentieth century. As trophies the average tusk weights would have been high. Again, plucking a figure from my mind, say it was 20 kg (44 lbs – in Kenya 1925 to 1972 it averaged 33 kg or 72 lbs), the gross weight would have been 720 tons. While I may be out in detail, it is likely that Europe and North America's trophy tusks totalled hundreds of tons of pre-convention raw ivory stock.

When originally imported in to North America and Europe, trophy tusks were not for sale. During the life of the hunter that is how they stayed. Yet trophies seldom have the same sentimental appeal to the hunters' heirs who tend to sell them. Indeed they do so with such regularity and predictability that most trophy ivory in due course does enter trade – albeit decades after the sportsman killed his elephant.

Britain, France, Germany and the United States should have set an example by registering their large sportsmen-generated pre-convention raw ivory stock. They never did and if tweaked on the matter referred to the wording of the resolution[149] –

> "*1. The Parties are reminded that all **current stocks of raw ivory which might enter international trade** must be registered with the Secretariat no later than 1 December 1986....*"

The words 'which might enter international trade' formed their disingenuous escape clause as they said they did not think their stocks 'might enter' trade. In truth these developed nations never intended registering their pre-convention ivory stocks with the Secretariat. The magnitude of the bureaucratic task was too great and, in any case, the resolution was really aimed at the African ivory-producing countries.

Perhaps the most cogent comment on CITES has come from actions taken by the United States of America and the European Union. Both enacted national laws that override CITES and render the treaty obsolete. If CITES was workable, there would have been no need for such additional national rules. Conversely, if one now has to comply with different rules for different nations, why have CITES?

In the Third World CITES most profound effect has been to divert attention and limited manpower from the field where basic conservation has to take place.

* * * *

As indicated at the outset of this chapter, I had a working relationship with officials in the CITES Secretariat despite my criticisms. Such tolerance was not quite so apparent in some of the non-government conservation bodies associated with it. They wanted the relationship terminated and this was nowhere more obvious than with those 'monitoring' the ivory trade.

Both the Secretariat and the Wildlife Trade & Monitoring Unit (WTMU) – a body set up to monitor trade statistics based in Cambridge, U.K. that analysed CITES data under contract – said that the volume of ivory moving internationally in the 1980s was declining because of the steps CITES had taken to control the trade. Any evidence that declines in ivory exported was less than they claimed was very unwelcome. It was particularly disliked by WTMU as it threatened their CITES contracts.

Theoretically by 1989 all Africa's ivory exports were covered by the 'quota' system so that any amounts in excess could be challenged by importing nations. Yet the concept was flawed. If a producing country suddenly got more ivory than anticipated (as when it seized a big illegal consignment) all it need do was notify the Secretariat that its quota for the year had been raised in order to export it. Using the word 'quota' loosely was unfortunate as it was interpreted in a strict sense by a public and media that did not know that its original use was loose. As will be recounted in a following chapter, Mozambique's quota was raised from under 1,000 tusks annually to over 19,000 tusks. South Africa's went from under 2,000 to over 8,000 in one jump. Both were to accommodate illegally acquired ivory,

---

[149] Secretariat's *Ivory Notification* to all Parties dated 27th May 1986.

but far from raising public outcry, both notifications bequeathed the increases with an aura of spurious legitimacy that took in both CITES Secretariat and their WTMU monitors.

Ivory importing countries were asked to have all ivory consignments verified by the Secretariat Ivory Officer. A copy of every African export permit was supposed to be sent to him so that when an import query was made, he had the relevant data at his fingertips. Hopefully, evidence of the entire legal trade out of Africa would pass across his desk. Theoretically slick, this was unrealistic.

I had seen ivory move across the Zambia/Botswana border. Customs declarations were properly made, the necessary export and import permits issued, but no notification of any sort was made to the CITES Ivory Officer. It was not done because the officials concerned saw no need for the red tape. In addition, important though undeclared, the verification was correctly seen as a statement of no confidence by developed nations in the exporting African states.

The Ivory Officer was not informed of all ivory movements for other reasons too. One was cost. A week's delay in verifying a trade transaction worth millions of dollars incurred high interest. The Ivory Officer went on leave or visited other nations on duty – or attended CITES conferences – making delays in verification inevitable. On their own such interest costs were strong incentives to by-pass the system.

As pointed out elsewhere, the traditional way to get illicit ivory out of Africa was by getting the right permits from the appropriate conservation department. If an African department corruptly issued the right paper work, no amount of verification from behind a Swiss desk was going to rectify it.

As was apparent before CITES came into being, Customs and Excise reports worldwide provided detailed records of the international ivory trade. Extending back for more than a century, they stated origins, weights, values and destinations. Produced by professional law enforcers and trade analysts, they were uniform and reliable. Where anomalies did arise, they were usually relatively easy to interpret. Yet CITES and WTMU did not start off using Customs reports, available month by month, for the cost of photocopying and postage. I know because I used them extensively myself. Instead of collecting the Customs statistics, CITES contracted WTMU to create a new system based on CITES data. If ever there was classic feather-bedding, this was it.

WTMU's interpretation of CITES data showed that through the 1980s the volume of ivory exported from Africa progressively declined. Both it and the Secretariat took this as evidence that the measures introduced under CITES were bringing the international ivory trade under control. My doubts persuaded the Secretariat to commission an independent study of the trade – in essence extending the work I completed in 1979 up to 1987. I wanted to compare Customs statistics with CITES data and asked WTMU if it had the Customs material to hand. It didn't as, to quote the reason given, the Secretariat had never contracted for its collection. WTMU said that for $5,000, its staff would collect the international ivory Customs figures for me. After accepting this offer and agreeing to deliver the information within two months, WTMU changed its mind, saying that I need not pay for the Customs data: it would be provided *gratis*. Upon the Secretariat's instruction, WTMU also handed over its analysis of CITES data.

In the two sets of evidence, Customs records of minimum ivory production were greater than CITES' in every year from 1979 until 1987. As more and more countries were joining CITES through these years, the differences should have grown less both from the greater number of countries involved and as the CITES staff learned how to handle the information. Yet the reverse happened. CITES estimates of trade were only 67% of Customs' estimates in 1979 and had fallen to only 30% in 1987. Far from getting better at monitoring the ivory trade, CITES/WTMU had been becoming progressively worse. To claim that the results showed CITES had gained progressively greater control of the ivory trade was scandalous.

My draft report detailing the evidence was circulated in confidence for comment and to give both Secretariat and WTMU the opportunity to correct facts I had got wrong. Throwing ethics to the winds, the recipients never commented or criticised the report to me, but tried to stall publication as long as possible before the 1989 CITES biennial conference. Always sensitive to public opinion, the Secretariat sat on its hands until, many months later, I went ahead and published the report. As Jacques Berney, CITES Deputy Secretary General, admitted two years later: the Secretariat behaved very badly. With hindsight, the reaction is less surprising than it was at the time. After all, the evidence exposed CITES' and WTMU's ineffectiveness.

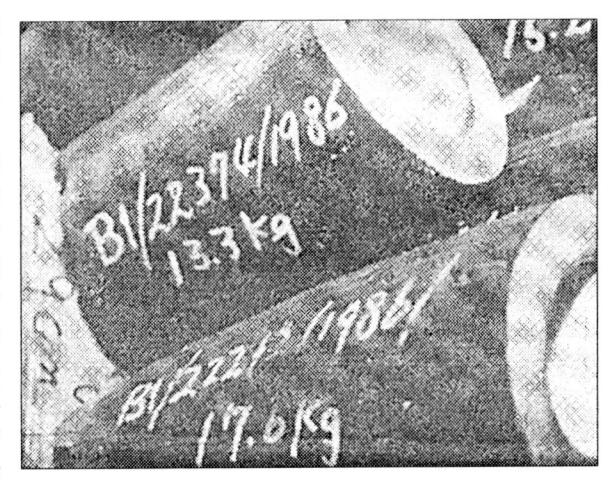

As part of a plan to 'stitch me up,' these tusk butt-ends were marked to make it appear that I had registered them. However, knowing neither ivory nor the numbering system those concerned made several elementary mistakes that revealed the attempt for what it was.

It was at this point in September 1989 that a journalist – Jo Revill – asked if she could have an interview to talk about CITES. I consented if the interview was off the record, to which she agreed. The outcome was a startling article in a double-page spread in London's *Mail on Sunday* of 1st October 1989, the gist of which was that I was a crook who had set up the Burundi registration exercise as a scam in which I had conned the CITES Secretariat. It was followed a week later by an article by Richard Ellis and Brian Jackman in London's *The Sunday Times* in which the charges were repeated. A third article by the Environmental Investigation Agency (EIA) repeated the message in yet greater detail. In 1992 Jackman edited a book by Iain & Oria Douglas-Hamilton (*Among the Elephants*) repeating the allegations yet again. When I wrote to *The Sunday Times*, rebutting the accusations the editor responsible for the letters column – Dorothy Ravenscroft – refused me the right of reply.

Someone in the Secretariat gave the journalists involved selected extracts from correspondence in CITES files that, presented out of context, supported the contention that I was corrupt. However, as an 'axe job', Revill's article was professional.

Nowhere was this more apparent than in two photographs accompanying the article. One shows the foot of a registration sheet from Burundi purporting to be a 'vital' certificate – that is an export certificate. In fact it was the bottom of a registration sheet listing tusks by number and weight against which CITES could later check queries about Burundi tusks. It must have been provided by someone in the Secretariat who will have been well aware of its misuse in Revill's article.

The second picture purported to be examples of 'raw tusks coded and weighed' by me. Here the newspaper fell into a trap. As detailed earlier, in Burundi I had incorrectly put the sequence country code/serial number/year/weight on registration sheets whereas the correct sequence of country/year/serial number/weight was on the tusks. It was left this way as I simply did not have the time to run off a new set of registration sheets. Only CITES officers and I knew this. Anyone illegally numbering tusks faced the quandary of which sequence to follow. The correct sequence on a permit would identify it as false, as would the wrong sequence on a tusk. Instead of trapping an illegal dealer, it caught the *Mail on Sunday*.

There were other flaws. The picture showed not whole tusks, but butt-ends: that is the sawn-off hollow first quarter of a tusk's length. Registration marks are not placed on butt-ends but in a tusk's centre section. At most, the butt-ends shown would have weighed between two and three kilos each. The false numbers on them indicated 17.6, 13.3 and 15.2 kilos respectively, all far greater than the cut bits of tusk pictured could have weighed. Then the numbers themselves were in excess of 22,000 whereas the highest number I recorded was just over 18,000. The year showed 4 digits whereas the ivory numbering code used two only. On top of this, the marking was not in my handwriting, which was the second trap. Finally, the bits of ivory were marked with white numbers and lettering. One does not number tusks white on white. One uses a colour that shows up on white and even where a tusk is stained dark, the staining is scraped away to provide a white background before numbers are applied. There never had been any numbers painted on the three bits of tusk shown: the letters and numbers in the photograph had been applied with white ink to another photograph which was then re-photographed. Whoever did it had to do this because, in a photograph, the dark butt-ends originally pictured could not be scraped clean.

While the general public is not aware of these technicalities, it was very obvious to those in the know – among whom there must have been at least some of the CITES Secretariat. As will become apparent, there were other reasons for setting me up in the Press, but the most curious aspects of the whole business was the Secretariat's failure to defend itself, for the articles critical of me were equally critical of the Secretariat.

CHAPTER THIRTY

A PRINCE AND A PRESIDENT

In Switzerland in 1987 I met Professor John Hanks, the WWF Africa Projects Manager, at his request. He knew that I did not think highly of WWF, but he was a new broom and he would like to start afresh. He was complimentary about the Burundi registration exercise and wanted advice on stopping the rhino horn trade. My parrot-like response had been that if conservation was the goal, adequate funds and effective manpower in the field made trade studies superfluous. But John persisted with the need to stop trade and, my foolish ego flattered, I tried to oblige him.

The first obvious step was to learn something about the business. WWF knew that rhino were being killed at unsustainable rates and that population after population was disappearing. Beyond that it knew little. As the trade was entirely illegal, getting knowledge about it was a job for professional detectives doing routine, boring, time-consuming sleuthing. Engage good men and turn them loose, but WWF should not expect quick results. John said that they did not have the time so wasn't there anything I could do through my contacts to describe the trade?

I knew a little, but my knowledge was dated. As with elephants, in the main, rhinos were hunted by the people who lived nearest to them. In Ethiopia it had been Ethiopians, in the Sudan the Sudanese, in Kenya the Kenyans and so forth across the continent. Within each nation's tribal divisions it had been tribesmen of those areas who poached within them. As with ivory, there were exceptions like the Somalis who poached across tribal and national boundaries. Within each community there will have been specialists - like Maitha and Wambua in Chapter Ten. The rhino horn trade mimicked the ivory trade: indeed most ivory buyers bought rhino horn. Some traders may have specialised somewhat - Dabanani in Burundi being a case in point - but as with ivory, it was basically generalised and opportunistic. I have never heard of a rhino horn buyer in Africa who did nothing else.

As indicated earlier, for at least the past thirty years and probably longer, the international postal service was the prime rhino horn transporter. There were of course exceptions. An outraged Indian told me of shipping a consignment to the Yemen in second-hand tea chests marked textiles. On arrival his agent had opened them and found the markings correct. Someone with a sense of fun had replaced the rhino horns with cotton T-shirts. In the past fifty years or so I recall perhaps four consignments numbering tens of horns and one of several hundred being seized by law enforcers. I had seen seven hundred or so of Dabanani's in Burundi. Normally, illicit rhino horn was moved in far smaller numbers: ones and twos, rather than tens and twenties. Somali gangs and game departments themselves came closest to the concept of 'Mafia' rhino horn hunting and trading syndicates. The concentration points were obviously the processors and manufacturers who used horn in Yemen for dagger handles and the Far East for medicine. Until it arrived at these points, it had been a generalised rather than specialised business. Knocking out one or several traders in Africa would not really change this as others would immediately take their places. Consequently, there *was* no substitute for effective protection on the ground.

Contrary to media claims, in Africa, few who bought rhino horn specialized in the product. The bulk of what was traded was moved by traders who had many lines of business.

As pointed out, this overview was dated. As rhinos became progressively fewer, so they were no longer available for the ordinary hunter to take. As supplies fell away while demand remained high, so the incentive for buyers to take the initiative and stimulate production would rise. Such circumstances might well produce specialists who disregarded tribal and national boundaries. To a degree the Somalis were already performing this rôle. Yet, treating them as exceptions that proved the rule, I had no evidence of such a trend.

The assertion that rhino horn was used as an aphrodisiac reappears again and again in the media because *anything* to do with sex attracts attention. It is wrong. Its primary use in traditional Chinese medicine is as a febrifuge, fulfilling the role that aspirin plays in western medicine. Some idea of how difficult it will be to get two billion people in the Far East to abandon something so widely and effectively used (even if the effect was psychosomatic) must be apparent from the high annual advertising cost of selling aspirin in the Far East (probably many millions of dollars). I suspected this would be far greater than the conservation purse. *Possibly,* a scare could be thrown into rhino horn users: release news that some irradiated horn had been released onto the market and anyone using it risked infertility. Manic fringe conservationists are not beyond trying such actions. Given the knowledge that smoking causes cancer, it seems such scares do not readily get people to kick habits. I was certain that spoiling demand would be more expensive than effective protection where the rhinos lived.

Against my better judgement I agreed to help Hanks get information on the rhino horn trade. The men who bought and sold horn did so for money. And for money, they would sell information on how the business functioned. I could approach some whom I had met (I had Dabanani in Burundi in mind, for example) and offer them cash for information on how they and others both acquired and sold rhino horns.

It would be expensive and so would my time. Having just completed one venture that had not been without risk, I did not want to start another without adequate reward. In personal terms the inducement to help WWF describe the rhino horn trade would be money. John said he understood my sentiments.

I wanted complete confidentiality. Hanks assured me that this was no problem and made two points that later became important. The first was that only he and his superior, Charles de Haes, would know of the project and of my rôle. As John was aware that I knew Charles, and might bump into him subsequently, he would surely not have mentioned him if he was not to be privy to what was proposed?

The second point was that he had access to 'unaccountable' funds and no payment record would lead back to me. Here I imply no criticism of such 'unaccountability' *per se*. Many intelligence objectives can only be attained secretly and secrecy is inconsistent with transparency. 'Right' or 'wrong' doesn't enter the equation: reality supersedes right to know in our flawed human affairs.

* * * *

Back in Nairobi I put my proposal to get information on the rhino horn trade in writing and waited. Weeks later I was asked to Switzerland to brief Prince Sadruddin Aga Khan - a WWF vice-president about to visit African Heads of State - on stopping the rhino horn trade. Presuming the real reason was to talk about 'our' project I went, but I was wrong in my assumption. They wanted me to talk to Sadruddin.

On the way to meet the Prince John reported that the rhino horn information project was on track, but the finance was taking time to arrange. Only himself and de Haes knew about it as agreed. This created a quandary. I was to meet a WWF vice-president who wanted my advice on stopping trade in rhino horn, yet I must keep quiet about what was being planned with his subordinates. Seemingly, though Prince Sadruddin was senior to both Hanks and Charles de Haes, they saw themselves as professional executives and the prince as a titular figurehead not entitled to know all the facts. WWF was clearly as *ad hoc* as ever.

At the meeting with the prince, whom I liked, no mention was made of the plan to covertly buy information. Sadruddin wanted the names of contacts in Burundi. As an Ismaili leader, it was inconceivable that he would go there and not meet the Rahemtullahs who were prominent Ismailis. As men who had dealt in rhino horn and ivory, it was equally inconceivable that he would not bring up this business with them. Indeed the possibility of getting his brother, the Aga Khan, to issue a *fatwa* against trade in either product was mentioned. Zully and Amir Rahemtullah had been unique sources of information on the trade in both horns and tusks, they were friends and I wanted to keep on good terms with them. Least of all did I want them to get the impression that I had tattled about them to their leaders. Thus I asked Prince Sadruddin to tread lightly with the Rahemtullahs.

Two months or so later I was surprised to be asked to go again to Geneva - this time to 'debrief Prince Sadruddin on the outcome of his visit to Africa'. As I was in any case going to Europe, visiting Geneva did not take me out of my way and it would give me the

In the late 1980s, WWF's attention was very much focused on the black rhino
(Photo Peter Davey)

opportunity to press Hanks for a decision on the covert information buying proposal. On our way to meeting Sadruddin, John said that he had had to tell several other people about the project as a lot of money would be involved. He never listed those who now knew of it, but said Prince Sadruddin was one of them.

Several other WWF functionaries in addition to Hanks attended the meeting. While we were told of the African Presidents visited and how they had agreed to take rhino conservation seriously, the meeting did not so much discuss Prince Sadruddin's safari as what WWF's rhino conservation strategy should be.

WWF was divided into two schools. One, led by Hanks, favoured funding field anti-poaching operations directly, the other led by the public relations officers was chary of direct involvement. WWF had helped Zimbabwe's attempts to stop rhino poaching in the Zambesi Valley by paying for a helicopter's hire. Poachers had been killed and those opposing direct involvement were worried that if these deaths were linked to abuse of human rights, it could be poor publicity for WWF.

Hanks reported that Zimbabwe's Department of National Parks & Wildlife Management claimed that the rhino poachers all came from Zambia across the Zambesi. A senior warden, Glen Tatham was in charge of operations to stop them and was Hanks' principal informant. According to the Zimbabweans, the Zambians came in small groups of a hunter or two - armed with AK47s - and several porters or helpers. They would slip in, take one or several rhino, then head back to Zambia. Apparently they were aggressive and, if followed, one or two members of a gang would hang back and ambush their trackers to allow the rest to escape.

This suggested high motivation and good training. No-one in the region could have been unaware that as a result of their recent bush war, the Rhodesian-now-Zimbabwean wardens and scouts were exceptionally well trained, aggressive, supremely confident and probably the best guerrilla infantry in the world. Any brace of Zambians caring to ambush a 'stick' of Zimbabweans knew the odds in a fight were against them. Casualties the Zimbabweans quoted bore this out: Zambian dead out-numbered those wounded and captured, reversing the normal war pattern where woundings substantially outnumber kills. Those laying ambushes have the advantages of choosing location and opening fire first: that initial volley inflicts most damage. The lack of Zimbabwean casualties versus high poacher mortality was not in keeping with claims that Zambians ambushed their pursuers. A contrary interpretation almost certainly nearer the truth was that militarily inoffensive men were being shot as they ran away and then finished off.

If the Zambian poachers were just peasants, they would lack both motive and training to routinely ambush their pursuers. Simple countrymen would head for home if followed. The routine ambush story was the classic 'they fired at us' justification for shooting them. I doubted Zambians behaved like Somalis in 'shoot or be shot' situations.

The Zimbabweans said that the Zambian gangs tended to come across the Zambesi whenever the helicopter on anti-poaching duties was withdrawn for servicing. If true, it meant they had excellent intelligence from someone highly placed in the anti-poaching forces, and again, was not in keeping with the poachers being simple Zambian peasants. All their claimed attributes - aggression, military competence, and access to Zimbabwean intelligence - were far more in keeping with the rhino hunters being Zimbabweans[150] and if Zambians were involved, it was as junior partners. One way or another, the stories coming out of Zimbabwe that Hanks supported (and which I had already heard from Zimbabweans) did not ring true.

Be that as it may and regardless of who the rhino hunters were, the death penalty was apparently being summarily inflicted for poaching. I pointed out that, historically, when the death penalty had been used against poachers, it had invariably produced social ructions. WWF involvement would be criticised by civil liberty lobbies and the publicity staff were right to be concerned.

Discussion drifted to who or what could be behind Zimbabwe's Zambesi Valley rhino hunting and I had suggested South Africa. At the time South African involvement in the horn and ivory trade was not widely known. My first-hand evidence from Rundu in 1979 had not been publicised, but I did describe it at this meeting to substantiate my suggestion. Hanks was instantly dismissive, implying I was fabricating a story and subsequently was to refer to it as the 'conspiracy theory'.

Eventually Prince Sadruddin's real interest in my presence emerged: Burundi. By March 1987 it was clear Burundi had gone back on its word and was once again allowing unlimited trade in ivory. The Secretariat had never contacted the President saying, weakly, it had been unable to get him on the telephone. It did not enlist WWF and IUCN to praise him for joining CITES. It had done nothing to block the country from backsliding out of its agreement which I had warned might happen.

Even if the Secretariat had acted as advised, Burundi may not have stood by its original commitment to stop trading. With hindsight the chances of failure had always been high. Worth $35 million annually to those whom it benefited, reinstating the ivory trade was worth fighting for. When they took action to get the President to reverse his decision they had no opposition. The nabobs of international conservation just let them have their way.

I had learned of the resurgent Burundi trade from Zully Rahemtulla. Having agreed with Prince Sadruddin to keep clear of it, his family were sticking to their word. Nevertheless his erstwhile associates like Tariq Bashir (who was not an Ismaili) were not of the same mind.

It was against this background that Prince Sadruddin reported 'fruitful' discussions with the President of Burundi who had once again agreed to put an end to the resurgent rhino horn and ivory traffic. However he needed an expert to advise him on how to do this and Sadruddin had promised that WWF would supply such an expert. Would I go? Once again, the grand, empty gesture by both parties. I was unimpressed. If an African dictator wanted

---

[150] Later this proved to be the case. In 1989 in Sengwa National Park rhino were being shot and the culprit turned out to be Gawa, the senior game scout from Mana Pools National Park (in the Zambesi Valley), operating 300 km from his duty station, using his National Parks issued weapon, monitoring pursuit with his National Parks VHF radio, and travelling with his National Parks uniform in his kitbag. This man would certainly have ambushed anyone pressing him in pursuit. Arrested and handed over to Glen Tatham, the anti-poaching warden responsible for the reports Hanks quoted, Gawa committed suicide before he could implicate accomplices.

to kill the ivory and rhino horn import/export business he, better than anyone, knew what to do. Personally controlling customs, police and the military, if he could not do it no-one could. What was I supposed to do?

His Highness had been placatory saying that he understood all this, but he had given his word that someone would be sent and I would be doing him a personal favour if I went as WWF's representative. Reluctantly I agreed to go as a gesture and providing that I could leave the country on the next plane out. It was also agreed that Jacques Berney, the CITES Deputy Secretary-General, would accompany me.

Because other people were present, I could not talk openly about the proposal to buy data on the rhino horn trade. During a lunch break, however, I mentioned it quietly to Prince Sadruddin. Naturally, from what Hanks had said on the way to the meeting, I assumed that he knew about it. It didn't take long to realise that he knew nothing of it whatever. After the meeting I confronted John and asked what the hell was going on. Flustered, he replied that he must have been mistaken, but could have sworn that he had told Prince Sadruddin. WWF had not changed. On my return to Nairobi I submitted a memorandum to Sadruddin placing my advice on record.

By January 1988 I had heard nothing of the information buying project and formally withdrew my proposal. John's response was interesting. It was that my original advice had been taken and that WWF *had* engaged professional investigators to collect information on the rhino horn trade. At the same time he sent me an internal WWF document[151] which confirmed this decision. As he said, it was my original advice that WWF engage professional investigators. The lack of courtesy in not telling me and keeping me dangling for a year was of no great import.

Certain facts emerged from these contacts with Hanks and WWF in Switzerland that later events proved of consequence. They were that (i) by his own word Hanks had access to unaccountable funds, (ii) this was kept hidden from the WWF vice-president Prince Sadruddin Aga Khan, (iii) WWF had engaged unnamed covert investigators to look into the rhino horn trade, (iv) Hanks advocated funding field operations in the Zambesi Valley and, (v) he was hostile towards and smothered suggestions that South Africa might be involved with what was happening in the Zambesi Valley, despite my 1979 evidence of its official involvement with traffic in rhino horns and ivory.

* * * *

In September 1987 Burundi's President Jean-Baptiste Bagaya was deposed in a coup which I assumed released me from my promise to Prince Sadruddin. The new president immediately imposed a ban on both ivory imports and exports, freezing what was *en route* through the country. Among his motives was desire for international approval. So, too, was hope that there would be another registration exercise. WWF insisted that the 'mission' to Burundi should go ahead despite the change of president. Jacques Berney and I went to Bujumbura at the end of March 1988.

---

[151] Hanks J., December 1987. Discussion Paper Prepared by John Hanks (Projects Manager – WWF International) on the Need for a Continental Strategy for the Conservation of Rhino in Africa. Revised edition for consideration by WWF's Conservation Committee.

CHAPTER THIRTY-ONE

BURUNDI AGAIN

It must have been in February 1988 when Amir Rahemtullah telephoned from Bujumbura and asked if I knew an ivory buyer called White from either South Africa or Zimbabwe. Such a man was in Burundi wanting to buy all the ivory in the country, which at that time might be over 100 tons.

My first thought had been that this must be WWF's secret operator penetrating the trade. To get the inside track there was no better way than actually trading. After all, I had done it myself. I also felt whoever White was, he was certainly doing it in style: he wasn't just penetrating the business, he was taking it over!

Not long afterward I passed through Harare and asked around. Several sources independently came up with the same name - White. He had made quite a name for himself in the Rhodesian war as a Selous Scout. It was clear from the way people talked of him that he was tough and a bit of a 'skellum' (rascal) whose war record elicited uniform praise. Once Zimbabwe had self-rule, White had left the country to serve the South African security forces. All my informants thought he could well be the man in Burundi: he had the nerve to go for a scam on that scale. One informant came up tops. He had recently talked to White who not only told him that he had bought ivory in Burundi, but produced a photograph of himself standing by a Johannesburg registered Toyota Land Cruiser in Bujumbura to prove it. It had taken little effort to find out who White was, but then Africa is a small place for palefaces.

When Jacques Berney and I arrived in Burundi, Amir Rahemtullah told us more about White. He had arrived in Burundi on 6th August 1987 in a Trans Afrique Hercules (registered S9 NAF and piloted by a Captain Deang). On 8th August the 'plane had departed - ostensibly for Dubai, but in fact for Dar-es-Salaam - with a ten ton load of ivory.

In September White had again reappeared in Burundi accompanied by an Israeli/Zimbabwean, and who had earlier appeared in Mogadishu interested in the ivory that had been there in 1985. Their plans to buy all the ivory in Burundi were thrown awry by the coup as all exports had been frozen by government.

* * * *

The new Burundi President, Pierre Buyoya, was as much a dictator as Jean-Baptiste Bagaya, whom he had deposed in September 1987. Yet, he did lead the country to democratic elections in 1993 and was sensitive to international opinion. Burundi had been severely criticised over reneging on its first commitment to CITES and he had grasped the fact that western opinion gave the matter considerable weight.

Thus on 5th November 1987 the President banned further imports or exports of ivory or rhino horn with immediate effect, trapping an alleged 109 tons of ivory in the country on that date. The new Government would not allow this volume to be exported until Burundi had formally joined CITES, and all fences with the treaty's Secretariat were mended. The gravity of the ivory issue in the Government's eyes was apparent from the red carpet treatment accorded to Jacques Berney and myself when we arrived on 29th March 1988. There was a new Director-General of INCN - a Mr André Niyokindi – who had replaced Audace Kabayanda and for the next three days we met with a battery of officials: the Director of Customs, the Deputy Director of External Trade, the Minister of External Trade, Industry & Handicraft, the Minister of Finances, the Secretary-General and Deputy Secretary-General to the President and the Director-General for Europe, North America and the International Organisations of the Ministry of External Relations and Co-operation. For so small a country there was no shortage of impressive official titles.

Jacques' message for Burundi was uncompromisingly pessimistic. Burundi must join CITES immediately. Its ivory should be registered and a copy of the register deposited with the Secretariat. Even then, there was no immediate prospect of other Parties to the Convention allowing imports of any ivory from Burundi. This would remain the position indefinitely or at least until the Parties were convinced that Burundi had not only acceded to the Treaty, but proved that it was reliable. Jacques suggested that Burundi should send a delegation to a WWF-sponsored meeting to discuss ivory that was to be held on 9th April (little more than a week later) in Lusaka, Zambia. It should put its case to that meeting and persuade it to recommend to CITES that Burundi be allowed to dispose of the ivory now held in the country.

Jacques also suggested that things would move much faster if the Government confiscated all the ivory in the country, sold it and devoted the money received to conservation or, better still, gave the money to the CITES Secretariat! That last suggestion was unfortunate as it swept the Secretariat off a moral plateau to being yet another of many conservation bodies grubbing for money.

It was impossible not to grin at the deftness with which the Burundians seized the moral high ground. They understood, so they said, the Secretariat's want for money. Ethically they were unable to accept Jacques' advice. It was precisely because of the old regime's disrespect for law that it had been ousted. The new regime was committed to uphold law and whether one liked it or not, those holding ivory in Burundi did so legally under Burundi Law. For the new Government to now confiscate that ivory from its owners without compensation would negate the coup's rationale.

The Burundians were acutely aware that the international conservation movements did not so much want justice and an equitable solution as vengeance and punishment for those who had gone back on Burundi's earlier commitment to join CITES. Yet as they said, we *have* punished the previous regime: what greater punishment could we have inflicted than a *coup d'état*? For the CITES countries and their conservationists to punish the new Government for the sins of its predecessor was illogical and unjust. Their case was well argued.

Berney, bound by the CITES position, had little latitude to negotiate or strike bargains with the Burundi Government. The reality of the situation was *if* there were 109 tons of ivory trapped in Burundi, they had a market value of around $13 million. In poverty-ridden Africa it was ridiculous to expect a commodity of this value to remain untouched while wealthy nations bickered over its disposal. No-one appreciated better than the Burundians how the pressures to let the ivory go might sow discord between members of the President's immediate circle.

As WWF's representative I took a different view. I agreed there was no likelihood of any Party foreseeably issuing import permits for a second lot of Burundi ivory. It was even less likely if it meant the traders to whom it belonged would get a second windfall price increase. Yet I also saw the new Burundi Government's predicament. If the ivory sat in limbo within the country the $13 million dollars would cause endless problems.

I advised that WWF had no punitive wishes. Its first consideration was to stop the Burundi ivory traffic immediately and permanently. For as long as the ivory currently held in the country stayed, it would generate many malign influences and delay a clean ending. CITES procedures prevented any quick solution. The best way forward, therefore, was for the Burundi Government to force the ivory to be exported *before* acceding to CITES. Order the traders to sell it and have it out of the country within the next four weeks. Burundi would then be free to join CITES without an ivory issue to complicate or delay matters and also have no pre-convention stock.

Jacques agreed with this advice, but as a member of CITES' Secretariat could not publicly or officially advocate any trade in listed wildlife products between non-member states. Yet, between ourselves and informally with the government officials, he said WWF's suggestion was the obvious pragmatic solution and one he would not oppose - other than for formality's sake.

Between meetings with officials André Niyonkindi took us to see the 109 tons of ivory the Government said it had registered. It was difficult to accept there would be so large a quantity in the country. I had a crude rule of thumb that half a ton of tusks occupied approximately one cubic metre. If I could see the stocks and roughly measure the length, breadth and height of the area occupied, I was confident that I could estimate weights to within five per cent of what was there. Caspar Egebu (aka Gaspard Egebu Ndikumasabo) had been the front for the Malians Tanaba and Mamu, though this time the ivory was in his own premises. Of his stock of 24.5 tons, 21.7 tons had been 'seized' by the Customs Department as, so they said, it had entered Burundi after 5th November 1987. It was lax use of the word 'seize' as the ivory was still in Caspar's possession. Stored in standard shipping containers, it was easy to calculate volume and it matched what was claimed.

So, too, did the tusks in Jamal Nasser's house - 25 tons. Yet this was not the case with Tariq Bashir who claimed to have 59.7 tons but no matter how I paced out the rooms in which it was stored, I could not get a measure of more than 58 cubic metres, that is an estimate of about 29 tons or half the claimed volume. Jacques had spoken to Tariq at length. Perhaps it was this that led Jacques to enquire, waving a hand towards the stacked elephant tusks, whether Tariq did not have qualms of conscience about the slaughter it represented.

Tariq gave a measured reply. He referred to the Koran and belief that God had given man use of and control over all animals. He was, so he said, making use of elephants. In so doing, he provided much employment and enabled people in countries like Tanzania and Zambia to get food and goods that they otherwise could not have acquired. He felt that as many people had benefited he had wisely exercised the stewardship of animals that God had given man. Where elephants were concerned, he agreed that they might be overused and he was not sure where lines should be drawn, for the Koran also bids man be continent. Yet where ethics were concerned, and here his voice had taken a harder edge, he didn't think that Jacques or any white man was in a position to lecture him or anyone else for that matter. After all, it was white men who sold arms and ammunition purely for profit to both Iraq and Iran simultaneously in their eight year long war. Was that moral?

As Jacques said later, "He made good points." Tariq's arguments were sincere and he had obviously thought of ethics. Later it was confirmed, however, that as suspected, he did not have 59.7 tons of ivory. Government had initially weighed all stocks by loading them into trucks and then driving them onto a weighbridge. Tariq had paid the weighbridge official to record a higher weight than there actually was. The result was challenged. When his stock was re-weighed under supervision the fraud was discovered, and he was sentenced to four years in jail.

Here it is perhaps appropriate to make the point that all those involved in the Burundi ivory traffic had grounds to doubt western conservationist claims. When conservationists said elephants were in danger of extinction, many trading ivory did not believe them. If there was a general decline, how were the tens of thousands of tusks which they handled being produced? How could this be if there were so few elephants?

The strong strand of logic in this reasoning can only be set aside by carefully presenting all the evidence. In the light of history, many in Africa and Asia believed that white people lied. As Victorians knew what was best for the uncivilised, so the modern white conservationist has the same absolute God-directed sense of certainty when telling people of other races what conservation is all about. The supreme irony is that where large mammals are concerned, it is on white man's land that they have been conserved least. With modern technology it would be easy to reintroduce wolves, bears and boars to Britain from whence they were wiped out centuries back. Yet this will never be done. Such matters do not go unnoticed and erode belief in western statements about elephants, ivory and, no doubt, cabbages and kings.

Rightly or wrongly, neither the Burundi Government nor the traders accepted the advice to get rid of the ivory before joining CITES. They hung on to it believing that, sooner or later, the world would see sense. A decade later it had not and the 80 or so tons were a constant source of trouble. The Government did, however, accept Jacques Berney's advice and took the tusks into its own custody, for safety placing them in shipping containers that were held in an army barracks. As anyone with African experience would suspect, that might make the tusks free from interference by civilians, but what the army may have done with them is another matter.

As promised, I was allowed to take the next flight out of Burundi and returned to Nairobi with Jacques. I informed WWF of the position taken on its behalf, receiving a letter of approval, and my fee. That should have been the end of the Burundi business.

* * * *

In Nairobi Jacques had met up with the Secretary-General of CITES, Eugene Lapointe, and the CITES Ivory Officer, Joe Yovino, on their way to the Lusaka ivory meeting referred to earlier (which I had declined to attend) that would be chaired jointly by Prince Bernhard of the Netherlands and Prince Sadruddin Aga Khan. Nevertheless joining these CITES men for a social drink at the Nairobi Hilton one evening, discussion inevitably turned to ivory and the personalities connected with it. The name of the Israeli/Zimbabwean who I knew had been in Burundi with White came up and, never having met him, I had asked about him. All three present - Jacques, Eugene and Joe - had been complimentary. They enjoyed his visits to Lausanne because, and the comment was not wholly facetious, he always took them out to a good restaurant. More seriously, they felt he was an excellent man and the most co-operative and helpful of ivory traders. Take the Mozambique ivory quota, for example, he had given them the inside track on that, greatly relieving their minds on the issue.

The Mozambique ivory quotas for 1986 and 1987 had been less than 1,000 tusks while that for 1988 had leapt to 19,000 tusks. The reason, the Israeli had assured the Secretariat as he personally knew about it, was that the Frelimo government had seized this stock in the civil war then raging with the Renamo rebels. It was a one-off event, with nothing phoney about it. Thus assured, the Secretariat had passed the explanation to enquiring parties with all the authority those with the inside track habitually impart. It went a long way towards dampening conservationist curiosity and criticism.

In a nutshell, the Israeli was one hell of a good guy. He even checked first with CITES before doing business in a country to make sure that it was kosher. As I recall, it was Eugene who had made this point and then gone on to say that the Israeli had asked them about the wisdom of buying ivory in Burundi, from which CITES had dissuaded him. He had even asked them how to get to Burundi.

At that the penny dropped. What I had been hearing was somehow too good to be true. Asking how to get to Burundi was the lie that showed through. Why all these questions about Burundi when in fact he had been there with White in September 1987 *before* asking the questions? The following hypothesis sprang to mind.

Ivory could be bought in Burundi at less than fifty per cent of the market price. If this could be laundered through another country's CITES quota it would sell for one hundred per cent of market price and there would be big profits. The Mozambique quota increasing from less than 1,000 tusks to 19,000 tusks was an opening through which Burundi ivory could be directed. The Israeli had got close to the CITES Secretariat, convinced its members that the ivory had been captured by the Government from rebels in Mozambique's civil war. The Secretariat then passed on the story which, coming from so reputable a source, had suppressed further investigation or unease on the part of other CITES Parties. The ten tons flown out of Burundi on 8th August 1987 had been a trial shipment. From Dar-es-Salaam it had gone to Beira. The story that it was going to Dubai was simply to throw out a false trail. That the ivory had suddenly become available in Beira I knew because Peter Becker had bought some of it. Not knowing the background in detail, he was curious about the the otherwise inexplicable presence in Beira of Zaire basin forest elephant tusks, though they confirmed its origins to me.

The eighty or so tons of ivory stuck in Burundi offered a wonderful opportunity. The international price was edging towards $200 per kilo. With a Mozambican export permit, the value of the stock in Burundi would be around $16 million. Available in Burundi for around fifty per cent of market value plus a discount for bulk purchase, this would have given a gross margin of at least $8 million. In summary, Mozambique had been set up as an outlet for half-price Burundi ivory. This would be flown from Burundi via Dar-es-Salaam to Beira in Mozambique. The Israeli, being too clever by half, had blown it by asking the men in Lausanne how one got to Burundi - having already been there. I told Eugene, Joe and Jacques they were being duped. As soon as I reminded Jacques that Amir Rahemtullah had seen the Israeli in Burundi in the previous September, he confirmed it to the other two. The Mozambique quota was obviously set up to cater for Burundi ivory. If I was right, the 19,000 tusks which Mozambique claimed to have seized from rebels would *not yet* be in Mozambique, but still in Burundi. I bet that if they demanded to see the seizures the Secretariat officials would be able to confirm this and expose the fraud.

If protocol had been followed, once their suspicions had been aroused, Eugene Lapointe should have informed the CITES parties concerned - Mozambique and Tanzania (as the ivory would be passing through Tanzania) - as well as the Standing Committee (the body that runs CITES between the Meetings of the parties). To have done so would have immediately been leaked back to the parties concerned and not only blown the project to move Burundi ivory on Mozambican documents, but also lost any opportunity to capture the ivory.

We talked the matter over and decided that Joe Yovino must check whether the 19,000 tusks did in fact exist in Mozambique. If they really had been seized in military operations then they would be in Maputo the capital. As CITES Ivory Officer, he had every right to inspect them. As he was going from Nairobi to Zimbabwe, it would be easy for him to take a side trip to Maputo and see the tusks for himself. If they were not there, then the story about the tusks being spoils of war was disproved. It was agreed there and then that Joe should immediately visit Maputo and ask to see the tusks.

This he did, but he never saw them because 'the Storeman with the keys was out of town - apparently on leave.' Claiming that the only keys to goods worth $16 million would have been in the hands of someone out of town on leave was preposterous, to put it mildly, and, alone, ground to assume the 19,000 tusks were not there. Yet Joe was not the sort of man to challenge so ludicrous a claim. Accepting it at face value, he had resumed his interrupted itinerary. His appearance and enquiry may have warned those in the racket that the Secretariat was suspicious. On the other hand, his meek departure may have redressed this: after all, he had not been very determined.

Eugene Lapointe agreed to delay making formal representations as called for by protocol, while I set a trap. My thinking was simple. No African country would knowingly allow tons of ivory from Burundi to pass through its territory. After all, Burundi's traffic in its neighbours' assets was uniformly disliked. Whether driven by legal and ethical motives, or just for financial gain, all would seize such ivory and the 'plane carrying it. Losing both aircraft and ivory would make the owners' eyes water even if they escaped jail and additional fines. We needed to alert the appropriate authorities in the likely transit countries when the ivory left Burundi, and then just let Africa's processes take their course.

Who were the appropriate authorities? By CITES' rules each country has a designated 'management authority' - usually the national conservation organ. In Tanzania this was the Game Division and Tanzania was the most likely transit country. Yet given Tanzania's track record of not stopping ivory going into Burundi, I doubted that informing the Game Division would do other than speed the ivory on its way. The correct person to inform was surely the President himself?

I approached Julius Nyerere through his *éminence grise*, Joan Wickens (a Fabian from Britain, who was a behind-the-scenes advisor and sounding board all through Nyerere's career), gave a background to what had been going on with ivory vis-à-vis Tanzania, my supposition about the Burundi-Mozambique deal, and the need to intercept it when and if it passed through Dar-es-Salaam. She responded, giving me a name and 'phone number to contact in Dar-es-Salaam.

It was possible that alternative routes through Zambia and Zimbabwe might be used. Thus I approached the then President Kenneth Kaunda through Dr Richard Bell and likewise was given a contact in Lusaka. Rowan Martin of the Zimbabwe Department of

National Parks & Wildlife Management alerted Zimbabwe's President via its Central Intelligence Organisation (CIO) but remained my contact.

The final touch was to arrange for journalists and television people to arrive on the scene immediately after or actually during the ivory seizure wherever it happened, to make it that much more difficult for the owners to buy their way out of their predicament. Thus I approached Mohamed Amin, a friend who ran the regional television arm of Viznews, Mike Wooldridge of the BBC and Robert Powell of Reuters.

The trap set, we waited for Amir Rahemtullah to telephone that the ivory was on its way. His intelligence was that it would move before the end of April 1988, that is within four weeks of our plan being laid.

* * * *

We naturally speculated on who the brains behind the Burundi-Mozambique ivory heist might be. There had to be millions of dollars behind White's offer, and it seemed unlikely that he, personally, had such cash. Possibly the Israeli could bring other resources to bear, but for a hard-nosed trading concern, the risks were still substantial. Even the underworld might think twice, given the many possibilities for things to go wrong.

I suspected the plan had been hatched in South Africa's security/intelligence community. With wonderful irony, ivory originating from Mozambique's Renamo guerrillas had helped to pay for the munitions South Africa supplied, went to Burundi where South Africa would help extricate that Government from its CITES embarrassment of having eighty tons that could not be exported, then with South African help the ivory would go back to Mozambique – this time into the Government's hands and profit the officials involved handsomely. The same ivory would thus have helped South Africa (i) control Renamo rebels, (ii) do the Burundi President a favour, (iii) put the opposing Mozambique Government under obligation to it, and (iv) along the way, turned in a nice profit. Having all parties, regardless of their differences or opposition to one another, dancing to the same tune was after all the hallmark of South African intelligence. I put this to Robert Powell who was entertained briefly by my imagination, but too canny to buy the hypothesis at face value. If I was right, the evidence would become apparent in due course. Meanwhile we waited.

By the end of May nothing had happened and Robert methodically went through my evidence. I think he really began to feel he was on to something when he found that the traffic log for Dar-es-Salaam airport for 8th August 1987 had no evidence of the aircraft carrying ten tons of ivory arriving from Burundi. That page of the register had been torn out. This could have been by coincidence but we thought otherwise: someone had removed it to cover tracks. It could have been done around the time of the shipment as a precautionary measure against the aircraft's route being followed, but it seemed more prudent to assume it was recent and a response to the trap. If that was so, then there would be no transfer from Burundi to Mozambique. And as historical confirmation, Mozambique never did export the 19,000 tusks it said it had seized from its opponents in the civil war.

We felt that we would soon know whether the force behind the Burundi-Mozambique scam were elements of the South African system or a purely private venture. Our reasoning: if the initiator was the South African Government, its reaction to the project's cover being blown would differ from a private entrepreneur's. The Government appreciated how strongly the western public reacts to poaching in Africa and spin doctors would try to neutralise the potentially damaging story of South African involvement with ivory in Burundi. If private operators only had set it up, nothing would happen. Those involved would merely take note and lie low.

To jerk the chain a little, Robert put out an article on Burundi through the Reuters' wire service. It included a sentence referring to the well-known ivory trader, Mr White, having recently been in the country. To professionals this seemingly innocuous statement would confirm an interest in White's activities and nudge them towards damage control if South African Government agencies were involved.

* * * *

In June 1988, Alan Root, film maker and very old friend, came to me with an odd story. A man had arrived at television film company Survival Anglia Limited's London offices, met Mike Hay the senior executive present and said he had been a major player in Africa's ivory dealings for several years. His conscience had caught up with him and he wished to make amends. He wanted his tale made into a documentary film in which he would expose all. Before Alan finished I had interjected that the man had to be White and that he would have said that the people he worked for were not South African. I was right.

Was this the expected damage control? The story would direct blame away from South Africa. Coming as and when it did, White's appearance in Survival Anglia's offices seemed too much for coincidence. Later in the year Mike Hay himself visited Kenya and we talked at length about White's proposal.

Survival Anglia could not make a documentary of it: the libel prospects alone ruled that out. White's story had to be told as fiction. That being so, he had brought it to the attention of David Puttnam who was very interested. With films such as *The Killing Fields* behind him and reputed to have a strong social conscience, David was the person Mike thought best suited to make White's tale into a film.

In early 1989 I met Mike Hay and David Puttnam. David seemed genuine and during our meeting I outlined a much wider background to Burundi and ivory into which White's story fitted. This brought the Wata into it and what was happening to them. It led to a second meeting and I was sufficiently convinced of Puttnam's interest that I agreed both to get a comprehensive background on White and to write up a story on ivory for him.

The opportunity to put across what really went on in ivory in a new dimension seemed worth doing. Visions of fortune if not fame no doubt played their part, though I do not recall being particularly excited by them. As we parted, David had stressed the need to make the story topical, up-to-date and to flesh out the picture on White.

* * * *

White was well-known to journalists who took a serious interest in South African affairs. That he was fairly prominent in the realm of security and intelligence gathering was immediately apparent. Among various positions he held, one was as a director of a company called Longreach of which an even more prominent Director was Craig, a name to be reckoned with in the world of South African intelligence and apartheid national security, he was one of its 'heavies'.

David Puttnam was interested in White's veracity. Was his claim to have been smitten by conscience true? His connections with Craig and, according to his admirers, his strong sense of loyalty, were at odds with him confessing anything that might compromise his colleagues and employers. Could he spill the beans on ivory without compromising his activities in other fields? Why had he avoided reference to his work and relationship to the South African security system, when originally selling his tale to Mike Hay? If his approach had been truly independent, it ran considerable risk of bringing up activities that could embarrass his erstwhile employers. At the time it was not as if he was estranged from them: to the contrary he and they were still very much together in harness. The obvious alternative interpretation was that his approach to Survival Anglia/Puttnam was in fact officially approved and that White's film proposal was the damage limitation and misinformation exercise that Robert and I had expected to hide the plan to market Burundi tusks on Mozambique permits.

The politics of conservation indicate little real concern for elephants.
(Photo Peter Davey)

CHAPTER THIRTY-TWO

SOMETIMES YOU GET LUCKY

Sometimes you get lucky. In July 1989 I was driving from Francistown to Gaborone in Botswana with a South African friend and talking about cabbages, kings and hardly surprisingly, ivory. A man of many parts, he traded widely about Africa and knew Peter Becker well. Because commerce with South Africa was technically banned by many African states, and South Africa was subject to many international sanctions, business between South Africa and the rest of Africa tended to be confidential and one didn't pry. Nevertheless I had never connected this person with ivory or known that he was involved in it – in Burundi or elsewhere. Perhaps I should have been suspicious when he accompanied Peter Becker to buy ivory in Beira – part of the trial ten-ton shipment of ivory from Burundi that contained Zaire basin tusks. Indeed, in retrospect, he may have been the person who put Peter on to it.

We talked about South Africa's obvious deep involvement in ivory in other countries. Assuming I knew more than I did he let several cats that I had been looking for out of their bags. His agent, an Italian called Dominico, notified him on 2nd June 1988 that he was negotiating to get 169 tons of ivory out of Burundi from 'Jamala', Tariq Bashir and someone called 'Hassan'. As he could not produce CITES permits, my friend had not been interested. He went on to say that shortly thereafter two well-known representatives from the South African security system had together come to his offices and offered the Burundi ivory *with Mozambique permits*. What to that point had been hazy supposition could now be seen in a much clearer light. My intuition had been right.

It never rains but it pours, as the saying goes, and one stroke of luck was complemented by another. Some time after my Botswana trip, on a flight between Addis Ababa and Nairobi, Robert Powell had sat next to a jovial Italian called Dominico who was on his way back from Dubai to South Africa. His business was selling oil in eastern Zaire, Burundi, Rwanda, Zanzibar and mainland Tanzania - all countries that were bankrupt and in which a commodity like oil might be paid for in kind rather than cash. Robert, alerted by the word Burundi, steered the conversation toward ivory. Presumably because he appeared well informed, Dominico revealed that he, too, knew quite a bit about it. He had, he said, some sixty-four tons presently in Burundi. The ban on exporting was no particular problem. The tusks were being crudely carved, flown to Dubai, then sent by sea containers to China where the carvings could be finished or reworked. The pay-off to the Burundi Government had been $2 million. It described the situation that Jacques Berney and I had foreseen if CITES' Parties forced the ivory to 'stay' in the country: it was too valuable and if there was no legal way out, then an illegal one would be opened, subverting Burundi's good intentions.

For an irresistible rake-off, Burundians were bound to allow the ivory out. And to cover their backs they would allow more ivory in so that, if accused of not having held all the 1988 tusks they took away from the traders, they could throw open their containers and say, "See, there are the eighty tons". As foretold, the eighty tons had become an entrepôt in themselves.

Dominico also made other interesting comments. He knew all the men in the Longreach company and others connected with the South African security field.

This was obviously the Dominico my friend had employed as agent in Burundi. He certainly knew something about ivory in that country and the names he had come up with included those we already knew from other sources were connected with it. I sensed what he told Robert was a goulash of recent and not so recent information. Dissolve the big talk and there was still more than a whiff of truth. My impression was of a lesser man than he made out; an employee rather than an employer who talked big to impress someone whom he thought was *au fait* with the same field.

Yet, taking the bottom line, he certainly knew the people of whom he spoke. As I say, sometimes you get lucky. My trip in Botswana and Robert's on an aeroplane in the Horn of Africa produced unexpected confirmation of points upon which, till then, we had only speculated.

* * * *

Earlier in 1989 I had been approached by two friends - Rowan Martin in Zimbabwe and Richard Bell in Zambia - with virtually the same question. Did I know a Colonel Ian Crooke who, in cahoots with the founder of the British SAS Regiment, Colonel David Stirling, was undertaking covert anti-poaching work in Africa? In 1988 at a meeting in Lusaka, they had been approached by John Hanks and asked to co-operate with a secret WWF-sponsored programme to investigate the ivory and rhino horn trades, but had heard nothing more. Both had now been approached by Crooke and asked to provide him with intelligence on poaching as well as any information on corruption in their respective countries. He told both that where suspects could not be brought to court he and his team of ex-SAS men would assassinate them. He never used that word directly, but he left neither in any doubt that that is what he meant. The project was financed by David Stirling and 'concerned conservationists', but was operated out of South Africa. Was this the WWF-sponsored programme?

I advised both Rowan and Richard to have nothing to do with Crooke. Technically, he had asked them to spy. The possibility could not be discounted that his operation was known to both Zimbabwe's and Zambia's intelligence authorities, who may have been aware of his contacts with both Martin and Bell. Indeed, if the contacts had been noted and they did not inform their respective employers, it might well count against them, even if they did nothing for Crooke. I advised both to report Crooke's approach to ensure that they were in the clear. Later it transpired this advice was not misplaced as Zimbabwe's Central Intelligence Organisation was very much aware of Crooke's project and of the contacts he had made in Zimbabwe.

Over lunch one day I told Robert Powell about Crooke. We speculated that Crooke and his men may have been the professionals that Hanks had referred to earlier and that WWF could well be behind the project. We wondered how Crooke's team would handle the likes of White and Craig. For the South Africans to allow an alien mercenary group to operate out of South Africa, particularly when the apparent focus of its attention was something in which the South African security people were themselves deeply involved, some deal had to have been made. Intuitively we both felt that there must be more to this project than met the eye. The plan to kill outside the law was simple terrorism. That a bunch of ex-soldiers acting as covert vigilantes should execute 'poachers' incensed Robert's journalistic ethics and he decided to write an exposé. His enquiries stimulated an immediate reaction. David Stirling telephoned him late at night and was threatening, suggesting that Robert must be in the pay of ivory traffickers, saying that any article released by Reuters would endanger the lives of operatives in the field and that he knew I was behind the exposé and would be attended to. That he could pull strings was obvious as Robert's article was held up for nearly three weeks and Reuters' head office in London tried hard to get him to withdraw it. When it was released on 8th June 1989, the *Daily Mail* carried another announcing that David Stirling had commenced an anti-poaching operation in Africa using his own money. Clearly it was intended to take the sting out of Powell's report. It also announced that a well-known conservationist was about to be exposed for involvement in the illicit ivory business. In view of what Stirling had said about attending to me, we assumed it was to me that this referred.

At no time had Powell said or written that I was in any way involved with his article, yet from the start Stirling insisted that I was. Neither Rowan Martin nor Richard Bell had told Crooke or any of his men that I had advised them against collaborating. Even though most people involved with ivory and elephants would have known of my interest, assuming I put Robert up to exposing Crooke was, on its own, a leap into the dark. In fact, I had not made the suggestion. While I brought what was going on to Robert's attention, my initial reaction to exposing it had been mildly against, wondering what good it would do. Instinct told me that what Stirling said was in some way connected with Burundi ivory moving to Mozambique or with the related enquiries I had been making at David Puttnam's behest. If, as I thought had to be the case, there was some connection between Stirling, Crooke and the South African authorities, then trying to destroy such reputation as I might have before I revealed anything in public made good sense.

* * * *

The articles in *The Mail on Sunday* and *The Sunday Times* both came out after Powell's exposé in 1989 and were obviously what Stirling had been referring to in his conversations with Robert. It was a bad error to assume they would shut me up because until that point, as the link between White/Craig and Crooke/Stirling had yet to be confirmed, my interest in Stirling's and Crooke's project had been rather lukewarm. Forced to defend myself, it now became both hot and focused. My situation regarding CITES may have been the apparent point of Revill's article, but no one in the know believed this for a moment. Here Stirling had made another mistake because when examining the ivory trade down the years, I had always made sure that a circle of friends knew what I was up to. It would have been foolhardy to have done otherwise. Thus I always had impartial individuals able to affirm what I did had been planned in advance and was within the law.

The cost of bringing libel actions against my defamers was far beyond my means. Instead I employed professional investigators to make enquiries on my behalf. Robert Powell dropped out of the picture as he was posted from Nairobi to Lisbon, but Stephen Ellis of *Africa Confidential* took his place. What then unravelled would take a book in itself to detail and I shall present only a summary here.

* * * *

Stirling and Crooke called the company through which they were contracted to do anti-poaching work KAS. It was staffed mainly but not entirely by ex-SAS men. The South African project was code-named 'Operation Lock'. KAS was only one of many companies for hire in the realm of 'international security' staffed by a freemasonry of ex-SAS soldiers. The firm I employed was one of them: my thinking - set a thief to catch a thief.

The company's eventual report included much about David Stirling taken from a book that I could have read in many public libraries! With regard to KAS's presence in Africa the investigators wrote –

> *"It now appears through two sources in Southern Africa, that both the funding and the initiative come from the Worldwide Fund for Nature ... This is meant to be highly secret...*
>
> *"KAS were introduced to Craig ... Crooke has said that .... he has proved very helpful and useful to KAS ....*
>
> *"The KAS operation has a wide range of influential supporters* [here they listed eminent personalities linked with the conservation movement].
>
> *"Stirling's links with the press are extremely good. James Adams, the Defence Correspondent of* The Sunday Times, *states that Stirling consulted him as to what should be done about the Reuters article about KAS.*
>
> *"Brian Jackman of* The Sunday Times ... *was also the chief contact of the Environmental Investigation Agency, when* The Sunday Times *was about to run an attack on Parker, based on their findings. Eventually* The Sunday Times *got cold feet, but Jackman was allowed to pass the information to the Mail on Sunday, who sent Jo Revill out to trap Parker and run the story. Material was passed back to* The Sunday Times *by arrangement.*
>
> *"Mail newspapers were also responsible for the KAS 'whitewash' that snookered Robert Powell's Reuter wire piece. That article was, on the word of the journalist who was accredited for it, written by the instruction of the Foreign Editor, who supplied the necessary facts.*
>
> *"The attack upon Parker by* The Sunday Times *and the* Mail *was deliberate and co-ordinated ..."*

These extracts were in keeping with much of what I felt about KAS and the smears in the press. Yet there were angles that did not gel. A penultimate draft of the report differed in its conclusions from the final copy.

> *"You are a target of KAS. You need to be absolutely clear <u>why</u>, and take clear legal advice. You also need to be absolutely clear as to the problem of taking on the conservation establishment, a British war hero and two powerful newspapers to court, or being taken to court by them."*

The last paragraph seemed a thinly veiled threat to back off. Yet when I took the warning as a challenge, it was removed from the final report. I sensed 'my' investigators were trying to marry interests of which mine were only one set, and that they wanted me to back off and let sleepig dogs lie. However, as the dog had bitten me I wanted to know more, so I paid the bill and continued alone.

* * * *

As KAS was at the core of the Lock scandal it was worth learning a little about it. Was it set up, as had been implied, to specifically undertake anti-poaching work in Africa? A company - K. A. S. Ltd, with company number 1701389 - was formed in the U.K. in 1983, four years before anyone was contracted to do anti-poaching work; its directors were a G. A. Goreham with ninety-nine shares and an R. M. Peacey with one share. The company's assets were £100 and as of 31st March 1989 it had not traded, and it looked like a shell set up to be used at a later date.

On 28th July 1989, three weeks after Powell's article, Ken Edwards - who we knew was one of Crooke's men – changed K. A. S. Limited's name to Ratebold Limited. On the same day a company called Ratebold Ltd, with company number 2350101, changed its name to K. A. S. Ltd. Presumably as a result of Robert's interest, this was to throw investigators off the trail. If they were not thorough and merely checked what K.A.S. returns had been filed with the Registrar of Companies in Cardiff, they would have acquired Ratebold's record.

If K. A. S. Ltd had not traded since its inception, was there another company? A little more pushing and out popped KAS Enterprises Ltd which was registered on 8th April 1986 as a private company numbered 2007790 with a nominal capital of £100. The issued capital had been two shares. The principals were David Sturling, director and Peter K. Skinner, company secretary. Cleverly misspelling 'Stirling' as 'Sturling' could be passed off as typist error, but it would throw out a computer search for the name Stirling. Half of the capital was owned by Ian W. T. Crooke and half by David Stirling, the registered address was David Stirling's at 22 South Audley Street in London and the company business was given as 'management consultants'. The company was active from inception, but its annual returns showed little in the way of tangible assets and a substantial debt.

Later KAS International Ltd, KAS Proprietary Ltd and Kilo Alpha Services Ltd were spoken of, though I did not bother to verify whether they existed because it was KAS Enterprises Ltd to which Hanks of WWF had given the anti-poaching contract which developed into Operation Lock. The cluster of similar names confused as, I assume, it was supposed to. Whatever the concept behind KAS in its various forms, it appeared first in 1983, four years before any involvement in African anti-poaching work, and obviously its owners did not want anyone prying into its affairs.

On 18th March 1990 London's *The Sunday Times* published an article alleging that KAS had undertaken investigative work concerning Richard Branson's night-club 'Heaven'. On 17th June 1990 *The Sunday Times* 'Insight' team exposed a KAS rôle in what was referred to as the 'Car Parks Scandal'. National Car Parks Ltd, Britain's biggest car parking firm, was upset by the inroads being made on its market by a newer company, Europarks Ltd. KAS was contracted in 1988 by National Car Parks to spy on Europarks. In the process it had stolen material from Europarks and illegally tapped its

Once David Stirling's company KAS Ltd was under scrutiny, he tried to throw us off the scent by switching KAS Ltd's name with another company called Ratebold Ltd.

telephones. One can see why Stirling and Crooke were upset when Powell's 1989 article put a spotlight on KAS.

Regardless of involvement in 'honourable' anti-poaching work, switching names in the U. K. Companies Register, the cluster of similar names and a contract for industrial spying suggested an investigative scavenger, to whom anti-poaching suddenly appeared as an opportunity. The car parks scandal showed a less wholesome side of Stirling than the war hero portrayed in the media.

I tried to confront Stirling and Crooke face to face. If they genuinely believed I was the evil genius behind the ivory trade, perhaps I could save them a lot of money. Contacting them through Jack Barrah, an old Kenya Game Department colleague who knew both, I offered to meet them in London at a time and place of their choice. They not only declined to meet, but rejected the offer with a threat. As relayed by Jack Barrah, it was along the lines, 'while we have nothing against you and have never sought to harm you, don't rock the boat because we know too much about your past.' As there was nothing in that past to warrant such a threat, it seemed they and not I had something to conceal and thought I knew what it must be. Their anti-poaching 'work' was a front and they wanted to avoid any probing for their real objectives.

Inaccurate and amateur it may have been, yet being under surveillance was intensely annoying. The same source of the foregoing information gave me the minutes of a KAS Board Meeting held in London on 2nd April 1989 chaired by David Stirling. It was attended by John Hanks of WWF, Ian Crooke, Ken Edwards and one Evelyn le Cheyne (who called herself David Stirling's political and intelligence advisor). Minute 11 shows that KAS had a covert contract with Iain Douglas-Hamilton, of elephant fame, but that he needed 'control attention'.

Douglas-Hamilton is said to have hired Kroll Associates as covert investigators focusing on East Africa. Kroll's London office was then run by Arish Turle – another ex-SAS officer. Seemingly need to 'control' Douglas-Hamilton arose from the possibility of parallel investigations getting lines crossed.

In the same minute it also states that I, too, required 'control attention'. As the minute predates Robert Powell's exposure of KAS, it vindicated suspicion that their interest in me was not initiated by his article.

Once in Southern Africa, KAS had been pretty half-hearted about being secret. Its first major project was overt, not covert. At the invitation of the South African Administrator of South-West Africa (shortly to be Namibia), Crooke said KAS trained a Namibian anti-poaching force for which the Administrator provided some two million rands (at the time equivalent to $800,000) worth of military equipment. They recruited heavily among ex-members of the 'Koevoet' battalion, a crack counter-insurgency unit, that had just been disbanded in the run-up to Namibia's independence. This project was decidedly odd.

Namibian conservation authorities were better able and more suited to train an anti-poaching force than a bunch of Limeys with no African bush-sense. The Namibians' only defect was not having enough money to employ the manpower or buy the equipment needed. Logically, if poaching was on the rise the solution was to have beefed them up with more money, more men and more equipment. For their own government to bring in a bunch of strangers made no sense. In any case, Crooke and his men were in Namibia for too short a time to have trained a crack anti-poaching unit.

A more logical explanation is that they never intended training an anti-poaching unit, but were creating a force of British officered right-wing muscle under a conservation screen, while the South African forces were withdrawing and a left-leaning independent administration taking over. The plan was aborted as abruptly as it was conceived. KAS also offered training for men from Mozambique under the guise of 'anti-poaching'. Yet, as explained by defecting recruits, the purpose was indeed developing right-wing muscle rather than gamekeeping.

According to Crooke, South Africa needed his team's help because their own manpower was overstretched. Yet just as KAS arrived in South Africa, the South African Police was establishing its own special unit, the Endangered Species Protection Unit (ESPU) to tackle poaching thereby denying the lack of manpower. The need for KAS help was nonsense. If there was one thing at which the South Africans were past masters, it was training people in counter-insurgency (i.e. anti-poaching). Further, there were many South African security firms who would have taken on this work at the drop of a hat. The ineptness of Crooke's ideas and plans that struck both Bell and Rowan Martin so forcibly, was apparent at all levels. Perhaps the most ironical was Crooke's code name: Mother Teresa.

* * * *

Establishing who had contracted KAS took time. Initially Hanks denied any WWF involvement. So did Stirling and Crooke, though Stirling conceded that 'they may know about us'. Yet on 16th November 1987 Hanks had written to Ian Crooke of KAS Enterprises Ltd, confirming that he had funding, would pay £15,000 upon signing of a contract and a further £5,000 on receipt of an acceptable report by 29th February 1988. In the letter he wrote:–

> *"I would like to confirm that this whole operation should NOT be regarded as a WWF funded activity. I can also confirm that none of the WWF staff know of this project nor will they be informed of any of the activities until the project has been completed and the operators have the left the field."*

It is this upon which Crooke and Stirling based their denials of WWF involvement. Under press pressure, Hanks conceded that he and only he in WWF had known about KAS. With further pressure he then disclosed that Prince Bernhard of the Netherlands had funded the project and that only they were involved, but neither took part in the project or its management. The minutes of the London KAS meeting held in April proved this a lie. Hanks was present and prominent in making project management decisions.

Hanks' claim that only he and Prince Bernhard knew of the programme was destroyed by an in-house document[152] circulated to WWF's Conservation Committee in which he had written –

> *"professional investigations ... funded by external sources, were initiated in November 1987."*

He apparently forgot having told Richard Bell and Rowan Martin at a meeting in Lusaka in April 1988 that WWF was setting up a secret project with which he hoped they would collaborate. He forgot having written to me saying that investigators had been employed. He must also have hoped that no-one dug up the fact that Dr Esmond Bradley Martin, WWF's rhino trade consultant, had at WWF's behest briefed KAS in London in January 1988, further proving that knowledge of the project had gone beyond WWF office precincts.

In 1988 Hanks went to Lausanne to the CITES offices and asked the Secretariat to help WWF in its project[153], telling Chris Huxley, *"We* have heavy money and have engaged professional investigators in London to look into the rhino horn and ivory trade in Africa." WWF wanted Huxley to go to London, meet the investigators and brief them with all the

---

[152] Hanks, 1987. *Ibid.*
[153] Huxley C., pers. comm.

Secretariat's information on illegal dealers, poachers, corruption with governments, etc. Huxley could not do this without clearance from his superiors. The Secretariat, as a United Nations Agency, was privy to privileged government and inter-government information that could only be divulged to the Convention's parties.

Huxley was denied permission to brief the investigators, but allowed to write them a report providing that he accepted any blame that might arise for passing privileged information to WWF. He gave this undertaking, wrote a report and detailed the names of all those who he knew to have been involved with or suspected of involvement with the illegal trades in rhino horn and ivory.

While he had stressed the obvious need for secrecy, Hanks had clearly stated to Huxley that he approached the CITES Secretariat officially on WWF's behalf. Given the two organisations' overlapping fields of interest, the close proximity of their headquarters' and frequent social and official contacts between their respective personnel, it is improbable that he lied about this because the risks of exposure were far too great. The evidence from CITES shows WWF Headquarters in Gland was well aware of the covert operation. On one point, however, WWF needs defence: neither Hanks, nor anyone else in the WWF network in Europe that I am aware, ever gave indication of knowing that Crooke and his KAS operatives had spoken of assassination. One must assume that they were unaware and were not party to the KAS statements.

* * * *

Recalling how Hanks had kept Prince Sadruddin ignorant of my original proposal, I wondered whether the same was happening with Operation Lock where Prince Philip, now WWF's President, was concerned. If not, and if the project backfired as I thought likely, the political consequences might go beyond embarrassing WWF. The least I could do was to pass on the information and I gave the basics to a friend in Nairobi with the right connections, requesting it be passed to Buckingham Palace. No response was expected or needed.

A second opportunity to ensure Prince Philip was aware presented itself and I took it. Summarised, my message was that deliberately or in ignorance, WWF was involved in activities for which it was not equipped and which no charity should undertake. Among other things, and albeit minor, it had broken the United Nations arms embargo against South Africa by providing sophisticated night-sights for authorities in Natal. More serious, the threat to assassinate poachers could have far-reaching political implications. Aware that Prince Philip might not be inclined to take what I said seriously, I suggested he chose an Ombudsman in whom both he and I had confidence to listen to what I had to say. I emphasised the evidence was incomplete, but strong enough to warrant attention. If the Ombudsman decided it was not worth pursuing, there the matter would rest. If it was, then it would be upon the Prince to decide what to do.

The response was frigid: either present facts, names, dates, places, times and corroborating evidence or drop the issue. The logic didn't quite follow: if I had all the facts an Ombudsman would be unnecessary. Later a third opportunity through the offices of Dr David Jones of the London Zoological Society presented itself: this time I was asked to address my worries to Trustees of WWF(U.K.) and we met in the Zoological Society's offices in Regent's Park. I gave sufficient fact to establish my view, but held an ace or two in reserve. The general response had been why have you been making these enquiries; what is your objective; what are you getting out of it? Not one person present expressed any interest in what the evidence meant for conservation, or its implications for WWF. Nevertheless, Jones was sufficiently impressed by what he heard to pass it on to Buckingham Palace. On 4th September 1990 he wrote[154]:–

> "...I spoke further to the Palace and, as a result, had a very useful discussion with Prince Philip, in which Charles de Haes and John Nash also took part. That was followed a day later by a full meeting of the Executive Committee of WWF International at Buckingham Palace, which I also attended, and at which I summarised that part of your story which it was useful for them to hear. I also had individual meetings with Charles de Haes, others on the Executive Committee, and with John Hanks who came over from South Africa for a couple of days for this purpose [Hanks had returned to South Africa not long after he became a focus of media interest].
>
> "The outcome of all this appears to be that although de Haes and one or two other senior members of staff in Switzerland were aware of general discussions going on between Hanks and the various interested individuals and groups relating to the need to break the poaching rings, they did not have any direct knowledge of specific arrangements which Hanks made on his own. Even those only seem to go as far as linking up the private donor with KAS. Once that link was made, it seems that Hanks took no part in making the detailed arrangements and, although he was probably aware of their general progress, he deliberately kept at arm's length from much of it. My feeling is that some further detail was probably known by the then Director of the South African Nature Foundation, but that Gland itself was not aware of such specific activities going on in Southern Africa.
>
> "I talked again to Prince Philip about the whole matter and also to an independent assessor, who was brought in to look at the various questions being raised and to look at any relevant paperwork on file in Gland. The feeling currently is therefore that WWF, in the form of John Hanks only, made certain arrangements between a funding party and an organisation who have, it would seem, had some recent successes in the anti-poaching field in Southern Africa. Even if, looking at it with hindsight, making such arrangements was not politically wise, it was certainly done with the best of intentions . . . "

This looked like a whitewash job if ever there was one. If Hanks attending a KAS Board Meeting, or approaching both the CITES Secretariat and individuals in Africa for their co-operation in WWF's name was 'taking no part in the detailed arrangements', then moons are blue. A final point contradicting the claim that the only person involved with Operation Lock in WWF Switzerland was John Hanks, was provided by John Ledger, of the South African Endangered Wildlife Trust (EWT) who clearly knew about Lock. After Hanks had returned to South Africa, the South African Nature Foundation (SANF) which

---

[154] Jones D., 1990. Letter DMJ/JK/5994 of 4th September to I. Parker.

was later renamed WWF (S.A.) bought rhino horns for Crooke from Natal and Namibia. This excited suspicion and when a South African journalist, De Wet Potgieter, started enquiries, Ledger telephoned WWF in Switzerland to warn them. As Hanks was in South Africa, the call must have been directed to and taken by someone else, proving WWF Switzerland obviously knew both its South African chapter and Lock were, for whatever reasons, buying horn.

The Duke of Edinburgh did appoint an assessor – Lord Benson – to look into matters, but as he only examined what WWF presented to him, and did not approach WWF's critics or assess their evidence, Benson's rôle was equally half-hearted. After David Jones' meetings in Buckingham Palace I concluded Prince Philip knew more about Operation Lock and the risks it posed than I had originally believed.

Protestations that the whole business revolved about 'honourable' anti-poaching activities and was well intentioned, were again undermined by Hanks. As a *quid pro quo* for the information Huxley had provided, he had agreed to keep the CITES Secretariat informed of progress. Before Powell's article on KAS appeared, he telephoned Huxley in considerable excitement saying the report was to hand and its contents were 'incredible' and 'dynamite'. Huxley, intrigued, asked when CITES could have its copy and Hanks replied not until cleared with his superiors – yet again confirming that more than he were involved.

Huxley reacted saying, "You can't tell me it's dynamite and leave me in suspense without a clue of what is in it!"

Hanks had replied, "It holds evidence and proof of South Africa's programme of political destabilisation."

The Secretariat never got a copy of the report. After several reminders, Hanks said that he was not permitted to release it.

* * * *

On 7th October 1994, Mr Justice M. E. Kumleben, a South African Judge of Appeal, was appointed by President Mandela to be Chairman and sole member of a Commission of Enquiry into the alleged smuggling of and illegal trade in ivory and rhinoceros horn in South Africa. His findings were presented to President Mandela in January 1996[155]. Unfortunately the Commission's terms of reference restricted it to South Africa. Yet this notwithstanding, the Commission took considerable interest in Operation Lock, confirming that WWF (S.A.) assisted Operation Lock in certain transactions. It had paid the Natal Parks Board R250,000 (at that time worth ± $89,300) for fifty rhino horns. It also bought sixty Namibian horns for R150,000. The facts were not disputed. The Commission called Hanks to give evidence about WWF International's rôle and Justice Kumleben wrote of his testimony:–

"I must say that I do not find this explanation convincing."

---

[155] Commission of Enquiry Into the Alleged Smuggling of and Illegal Trade in Ivory and Rhinoceros Horn in South Africa. Report of the Chairman Mr Justice M. E. Kumleben, Judge of Appeal, January 1996. Typescript.

Unable to subpoena witnesses from Europe, he concluded –

"In sum, on the evidence available to this Commission, one must conclude that Lock was not a WWF venture but that the latter cannot contend that it had no knowledge of Lock or was totally divorced from it."

Had Bell, Rowan Martin, Esmond Bradley Martin and Chris Huxley testified, and had the minutes of KAS meetings been made available to the commission, that conclusion would not have stood. With regard to Crooke's claims to have done sterling work in Namibia Justice Kumleben wrote –

"The reports [Crooke's to sponsors] *refer, for instance, to Lock members lobbying in various countries, including Namibia, as a result of which enactments were amended to impose stiffer penalties for poaching and smuggling. The Commission knows that the impetus for such amendments in Namibia came from another source. In any event it is difficult to visualise how, acting incognito, they could have helped in this regard. Both reports appear 'padded'.."*

Proof that Crooke and his men planned to kill was presented to the Commission by Lt-Colonel Pieter Lategaan of the South African police who testified –

"Mr Chairman, they [Crooke & his men] *had all sorts of weird ideas of how to deal with people. They just thought that a 'shoot to kill' policy would also work ..."*

This testimony was accepted. With regard to the 110 rhino horns bought for Operation Lock by WWF (S.A.) the Commission concluded –

"In the light of these facts - or rather the paucity thereof - one may well question whether the rhino horn received was used for the intended purpose."

There never has been a public accounting of what happened to them. One of the more interesting documents submitted in evidence (Exhibit Q) came from a Mr Mike Richards who was contracted by Crooke to help in Operation Lock along with his company R. & TG Consultants International (Pty) Ltd. Mike Richards (alias Harry Stevens) was working at the same time for South Africa's military intelligence. By his own word, he took the contract offered by Crooke to gather intelligence. Certain aspects of Exhibit Q are of such relevance to some of the questions that I had been asking that they are worth quoting at length –

"There is a possible problem and there always has been ... of [Operation Lock] *running into S. A. D. F.* [South African Defence Force] *operations.*
"This point is recognised by the investigation team [Operation Lock] *and at the beginning of the operation in February 1989 a decision was taken to avoid*

any possible contact with SADF personnel involved in such activities [handling ivory and rhino horn].

"A second problem exists whereby information may be received of dealers... [who] are actually permanent force SADF members. The policy upon the discovery of such information, is to deal with it internally and to pass it on to the actual command structure of the SADF personnel concerned and to take great care that such information does not fall into the hands of the media or other departments.

"Another problem which was recognised at the beginning was the activities of the South African backed Renamo and Unita which have large-scale rhino horn, ivory and other endangered species smuggling routes in operation. Once again, a decision was taken to avoid possible confrontation in this area as far as possible."

Here we have it from an operative under contract to KAS that the Company agreed to avoid South African Defence Force involvement with illicit rhino horn and ivory. It answered my question of how any serious investigator could avoid running into the illegal traffic run by South Africa's armed forces. The excerpts from Mike Richards in Exhibit Q of the Commission's evidence were explicit: Crooke made a deal to ignore the largest element in South Africa's illegal trade. It could not have been otherwise as without such a commitment, he would never have been allowed into South Africa.

Exhibit Q indicates what Stirling and Crooke were really after:–

"The next advantage [of co-operating with Operation Lock] is the possible co-operation which has been offered by Ian Crooke concerning the monitoring of anti-South African bodies which are situated overseas."

Crooke actually volunteered to collect intelligence – that is, spy – for South Africa, bearing out my warnings to Rowan Martin and Richard Bell. It also raises an interesting but as yet unexplained event in London. Powell's article released by Reuters was picked up and run by the London *Evening Standard* and *Africa Confidential*. Stirling threatened to sue both papers and Reuters itself. All three published public apologies to him and made undisclosed payments to his 'conservation fund'. What particularly irked Stirling was Powell quoting a Zimbabwe Minister's feeling that Operation Lock was part of South Africa's destabilisation programme. As disclosed by the Kumleben Commission, this statement was on target. The Reuters article was accurate and most definitely not libellous. Every point in it could be substantiated. Reuters' legal experts had vetted and passed it.

What pressure was placed on the three media organs to recant, apologise and pay into Stirling's 'conservation fund', when no real ground for doing so existed?

The Kumleben Commission was not the only exposure of Operation Lock. The journalist – De Wet Potgieter – who had links with the South African intelligence realm, became closely involved with Operation Lock and subsequently wrote a book[156] which

---

[156] De Wet Potgieter, 1995. *Contraband: South Africa and the International Trade in Ivory and Rhino Horn*. Queillerie Publishers. Cape Town.

further documents its connections with the South African intelligence milieu. Robert Powell did interview Ian Crooke and during that interview asked if he had had any contact with Craig. Crooke told Powell that Craig had been 'useful' to him, making the connection between KAS and Craig.

My own investigations threw up information that did not figure in either the Kumleben Commission's findings or in Potgieter's book. At least three South African conservation bodies – WWF (S.A.) already mentioned, the Rhino & Elephant Foundation and the Endangered Wildlife Trust - collaborated with Crooke in setting up and running Operation Lock. It was definitely not a secret in South African conservation circles.

So much for what Hanks so disparagingly referred to as 'the conspiracy theory.' There definitely was a conspiracy and it was not of my making!

* * * *

David Puttnam commissioned me to follow up on the tale told by White, stressing the need to keep it current. That I did, even though the story unravelled uncontrollably away from what may have been expected. Least anticipated, perhaps, was the participation of venerable conservation bodies and respected conservationists in what, in the kindest terms, could be interpreted as sleazy activities. Yet what Puttnam initiated exposed aspects of ivory's international influence that might otherwise never have seen the light of day. Sadly, his interest evaporated very quickly after the Revill exposé. Keeping the story current as he had asked me to, proved costly. Yet I suspect that had he known at the outset that the web would enmesh WWF and even its ennobled officers, his interest would never have been kindled. I am not certain that mine would either.

CHAPTER THIRTY-THREE

THE DREAM TEAM

Public reaction to Operation Lock's exposure was muted. Involvement of a British war hero and WWF in a sleazy vigilante project, ostensibly to stop poaching, but which bought rhino horn and offered to collect intelligence for apartheid South Africa during its last days made few headlines. The British press swept the whole business under the carpet out of sight. Yet I did not let the matter go so easily.

True, my attitude may have been influenced by a wish to right what I perceive as wrongs from KAS's attempts to 'control' me, though I don't think so. Stirling said I was motivated to damage WWF which is untrue. From *Ebur* days I have made no bones about its incompetence, an opinion of which it was well aware. Even so, we had worked together and I did not try to damage it. For the most part I avoided it and when I did co-operate, it was at John Hanks' request to build a new relationship.

At one level the events that culminated in Operation Lock bore out Gafaar Busaid's observation that ivory leads into the corridors of power. A web of economic and political influence, reaching from far corners of Africa's wildernesses, was manipulated by the continent's most powerful intelligence apparatus and spread to enmesh the highest in Britain and Holland, and compromise that civilised citadel - WWF.

Driven back defensively by each successive exposure, the tattered and untenable position that WWF held until public interest evaporated was that Operation Lock originated in Switzerland as a genuine but misguided conservation attempt. This view was, very charitably, accepted by the South African Kumleben Commission. I am not so certain and think it could all have begun in South Africa.

When he testified before Judge Kumleben, John Hanks was shown evidence of Crooke's deal with the South African intelligence people and seemed deeply shocked. The Commission accepted he had been duped and had been unaware of both SADF involvement in illegal rhino horn and ivory trafficking. The shock could also have arisen from the exposure of evidence that he thought would never become public.

From my lips to his ears in Switzerland, Hanks knew that SADF was a major player. Without trying to verify my evidence, he implied I lied and talked disparagingly of conspiracy theory. If he held my information is such low esteem, what was he doing paying for me to come to Switzerand to advise WWF on strategy?

Yet another person gave Hanks evidence of South African activities. Huxley, officially in the name of the CITES Secretariat, did so, with trade statistics proving far greater exports than could be accounted for from within the country and expressing frustration over the South African authorities' failure to reduce such exports.

One particular case irked Huxley. A consignment of rhino horn and hide (misdeclared as something else) was sent by post to Hong Kong where it had been seized and found to have come from South Africa's most sought-after illicit trafficker, a man with excellent links to the SADF. The consignment was redirected by post back to this man, but not before the South African authorities were notified the shipment was on its way. As a result, the man was caught red-handed in South Africa with rhino horns, rhino hide and proof that he had both exported them in the first place and then received them on return. What should have been an open-and-shut case leading to imprisonment got him no more than a slap on the wrist. Thus from CITES data Hanks knew of failure on South Africa's part to enforce the law. And if Hanks had ever wondered about the reasons, he himself had first told Huxley that the first KAS report revealed South African destabilization tactics.

Then there was that fateful KAS Board Meeting on 2nd April 1989 which Hanks attended. Minute 3.4 was in reference to work in Botswana and Swaziland and included the following excerpt –

> *"A valuable contact has been made in Swazi who would be willing to trade traffic information* [in rhino horn & ivory] *against observations made of terrorist activity."*

Here it is stated that, in his presence, Crooke was proposing to swap information on the illegal trophy trade in return for political/military intelligence on terrorists of the South African ANC and PAC movements. Minute 3.5 records that Crooke was put in contact with 'C' (Craig) by Evelyn le Cheyne. Hanks must have been aware that 'C' was not the local scout-master! No wonder he was shocked when this sort of information appeared before the Kumleben Commission.

* * * *

The evidence that culminated in the Kumleben Commission bore out predictions I had been making since at least the mid-1970s, but which were ridiculed or ignored by conservationists. Originally there was rejection of the material I eventually recorded in *Ebur*. As I became more involved outside East Africa, so the South African situation became progressively more obvious. However the predictions and speculations I made were not based solely on the modern evidence of that time. I routinely tried to place it in the context of both a historical backcloth and current politics. It seemed to make little sense for ivory, elephants and their conservation to be considered on their own.

Standing back from events in the late 1980s and the 1990s, the broad historical tapestry against which they took place gives useful perspective. South Africans of Dutch ancestry are proud of their conservation history. Some of the steps they took to conserve wild animals are given in Chapter Four, many at the time were in advance of conservation measures being taken elsewhere. Yet as the Boers spread and settled the land, game disappeared in their wake.

Trek Boers shot a great deal and both marksmanship and hunting prowess were machismo aspects of their culture. They rued the game's going, but perversely boasted about the volume of wildlife shot. It had parallels in the machismo culture of the American west: of gun fighters, cowboys, covered wagons, cattle barons, settling the prairies and, bringing bison to the brink of extinction.

In both cowboy and trekker culture and less blatantly with the white settlers taking land in Kenya, dominating and taming wilderness gave pride and shooting out the game was sadly part and parcel of the process. As the British Army found in two Boer Wars, the hunting which was part of their life made Boers marksmen. So formidable indeed that bright red uniforms which had served well enough for centuries quickly gave way to cryptic khaki that blended with the landscape.

In highland Kenya the first white settlers found plains animals roughly comparable with the modern Serengeti situation. Yet in under three decades most of those animals were gone.

Old timers said the game had been shot out: "We shot thousands of animals – we lived off them."

In their own minds they had indeed shot out the game. Yet pressed for details a different picture emerged. How long did a buck last the family? The answer was usually several days. Did they shoot every day? No, of course not, most people shot once or twice a week. How often did they have big battues and drives to really get rid of animals? Not often ... such events could be recalled to the year. 'In 1925 we had a big zebra shoot' was the sort of answer I got. Gradually by sanding down the subconscious machismo myth, it was obvious that while settlers shot a lot, it was only a lot relative to other people's shooting, but not relative to the array of animals available. It is unlikely the settlers ever took off what the thousands of animals in the highlands could have theoretically sustained. The settlers' 'we shot thousands' was a figure of speech. What was actually shot should have made little impression on the game hordes.

In Kenya, as in South Africa and the American west, what got rid of the animals was settlement. Constant disturbance, competition with domestic stock, grazing lost to cultivation and so forth and eventually fences. Like the Kenya settlers, the Boers themselves were at least partly to blame for the game-killer reputation that haunted them, even though not wholly true. The belief in Afrikaner culpability was inflated by the ill will between them and English-speaking South Africans. Just as Francophobia caused the British to call syphilis the 'French' disease, through Boerophobia the English attributed bad things to the Boers, among them the disappearance of game, selectively forgetting that men such as Cornwallis Harris[157], outstanding for his butchery, were British. The Boer Wars, which were the outcome of this general ill-will, were fought at a time when conserving game was catching on generally. In the circumstances it was not surprising that Boer attributes of marksmanship and being hunters were held against them: hence game killers. The first white Kenya settlers from South Africa included both Rooineks (British) and Boers and they brought their prejudices with them. As a fledgling game warden I was warned to be on the lookout for 'bloody Boer biltong hunters'.

My historical point is this. Having been so criticised for being game killers, Afrikaners are aware of the emotion the issue stirs in the British breast. Better than most, Afrikaners knew how damaging the label 'poacher' was. The accusation 'game killers' burned all the more deeply because they made strict game laws centuries before the British settled in Africa.

* * * *

Turn now to western feelings for animals. Millions of people across Europe and North America sit daily in front of television sets and watch violence. Not only do they see it *verité* in endless war and crime coverage, but as if this were not enough, Hollywood makes it up in

---

[157] Cornwallis Harris W., 1838. *A Narrative of an Expedition into Southern Africa During the Years 1836 and 1837.* American Mission Press, Bombay.

ever more gory and unrealistic detail. Yet this violence is only acceptable between humans. Showing an animal being hurt is *verboten* and generates outrage. In western culture human distress, pain and maiming, both fictional and real, are entertainment: animal distress is not. Human distress becomes boring. Endless footage of the world's poor starving, in refugee camps, or fleeing fighting creates fatigue: animal distress does not. Appealing to charity for animals can run and run. We need not go into the psychology of this seeming paradox, other than to note that it exists. People making films, running television stations or producing newspapers are very aware of it, as are all in public relations work - the art of propaganda.

Turn back to South Africa: the core of apartheid philosophy was belief that Africans were incapable of civilised governance. This was why they were denied votes and governmental responsibility. Apartheid's principal public relations thrust was to convince the world's 'civilised' people of this 'truth'. And what more subtle way of saying it than providing as much evidence as possible of African poaching?

Remember, conservation is apolitical - or so they say. Written and screened footage on poaching is not consciously appreciated for its virulent political content. Yet instance after instance about poaching, animal decline and animal massacre in country after African country had an irresistible subliminal message. Africans cannot conserve. And given the sense that conserving is the essence of being civilised, few issues sustain the idea that Africans are not fit to govern than continuous conservation failures. As anti-black propaganda the fate of animals in Africa was powerful stuff. The subjects of governance or competence need never be mentioned: if the evidence of animal distress is kept before them, white viewers will make all the desired deductions for themselves consciously or subconsciously. Knowing this, when South Africa was going to great lengths to mould western thinking on African matters, it would have been odd if its planners had not made use of it.

In the context of South African history this was so obvious that, given the South African regime's needs and competence, such a propaganda peach had to have been picked. Positing such an idea then looking for evidence to fit it is a highway to finding reds under all beds if not restrained. There may have been no deliberate South African attempts to keep African animal distress in the western public's eye. As western journalists and film-makers picked it up so automatically as and when they come across it, maybe there was no need for a South African initiative. The work was done for them by others. The political message they wanted put across went out with every African poaching case publicised. All films recording, or articles written about, conservation failures in Africa subliminally and coincidentally supported the apartheid message. If South Africans were influencing this field, all that was needed was a touch or two here, a suggestion there, impossible to detect. It would have been extremely difficult in such circumstances for me to ever have proved my case.

Yet, if I was right, the mirror image of what I felt might produce the evidence I sought. The South Africans would have been exceptionally sensitive to the negative outcome attending any exposure of their government's role in trafficking illicit ivory and rhino horn. Here we do have evidence that this was so. Again from Exhibit Q in the Kumleben Commission's findings in which Mike Richards said –

> "The last and **most spoken about problem** [my use of bold] *was the possibility of giving South Africa bad international publicity if the media were to take the information and put it across to the world that the South African government is*

tolerating the smuggling of endangered specie [sic] *and wild-life products as part of the destabilisation process of its neighbouring states.*

"*This point has received much attention at the liaison and command level and an early decision was taken to closely co-ordinate all investigation actions* [by Operation Lock] *with the South African authorities to, as far as possible, avoid such a repercussion which would have a serious detrimental effect, not only on the host country, being South Africa, but also on the British subjects involved in this investigation.*"

Here is proof from the realm of South African intelligence of sensitivity to bad publicity from conservation failure. It was not the only evidence. I had previously been offered $5,000 to tell all that I knew about official South African involvement in the ivory trade. Ostensibly the offer came from conservation authorities, but when challenged it was admitted that it came from the Ministry of Foreign Affairs whose interest had nothing to do with conservation and everything to do with the national image. My information was wanted in order to give it an appropriate spin.

Held in general opprobrium over apartheid, the South African regime did not want to jeopardise one of their few publicity aces - their successful conservation versus black Africa's failures. Here was ground to discredit me and deter me from enquiring after those behind the Burundi project to move ivory through Mozambique, or seeking the truth about Stirling and Crooke's pseudo-anti-poaching venture.

* * * *

The foregoing illustrates how initial deduction from very general background knowledge, would shortly become supported by much firmer evidence. Aware of the dangers of circularity, this support drew me back to the question, did WWF think up Operation Lock, or was it dreamed up in South Africa, which then used WWF as a tool in a greater play? Again I gave my mind free reign.

From a South African view, WWF was an ideal vehicle for conveying influence. First, there was the oft-stated ideal that conservation is outside politics. While baloney, it is nonetheless generally accepted. Thus whereas commercial and political contact with apartheid South Africa was internationally discouraged, dialogue over conservation was not. An international body headed by the world's social and commercial elite and devoted to unimpeachable, philanthropic, apolitical conservation was good cover for contact with altogether different motives and well worth penetrating and controlling. Had this been done? Could it *not* have been done?

Was there evidence to support this hypothesis? If so, who was the team? An ardent supporter and source of WWF funds was Anton Rupert, the patriot Afrikaner tobacco magnate and multi-millionaire. His rise from running a dry-cleaning business to millionaire tobacconist gave him the wealth to be very influential. Anton Rupert's donations to WWF gave him contact with its luminaries like Prince Bernhard of the Netherlands or the British Duke of Edinburgh.

Charles de Haes was Secretary-General of WWF for some two decades. He came from a stable of bright young executives in Anton Rupert's empire. Although there is no evidence and I make no suggestion that he ever acted other than in WWF's sole interests, he was well placed to press whatever South African case he cared to. Had I been looking for someone to represent South Africa's interest at WWF, Charles is the sort of man I would have chosen. Presentable, urbane, self-confident, as mentally agile as a wagon load of vervet monkeys and to use the vernacular, not short on balls. In the same laudatory tone Afrikaners reserved for Jan Smuts, Charles de Haes was *slim* (sharp).

And to top off my team, a man who was presentable, *au fait* with the biological jargon used in conservation, not obviously South African but beholden to South Africa, a man like Professor John Hanks would have been useful. The title and also being an academic zoologist exuded the right aura. The pomposity and upper crust English accent would ameliorate South African connections.

A South African tribunal linked WWF(S.A.), in which Rupert had been very influential, with Operation Lock (putting up the money for buying rhino horn). Before that, Hanks admitted setting up Operation Lock. As Operation Lock did work for apartheid South Africa's intelligence apparatus, two members of my hypothetical dream team were linked, no matter how tenuously, with it. That my dream team just happened to be where I would like to have placed it can only be coincidence. I merely state this to illustrate how my mind was working. I have no evidence that Rupert's, de Haes' or Hanks' actions concerning WWF were undertaken other than with its best interests at heart and more than this should not be read into my words.

But if I had been running South African propaganda and having penetrated WWF with my dream team, how would I have used WWF? Without any intervention, African conservation failure was working for me anyway, so all I need do would be to publicise results. This posed no problem as stressing failures was part of the process of appealing for western donations: 'Give to help stop poaching' brings in more than 'Give to boost something working well.' What is working well doesn't need help. So naturally my team would be instructed to downplay African conservation successes.

Furthermore, every WWF donation to conservation projects would be publicised, but the team would be told that no donation should be large enough to ensure success. Every project has a threshold minimum funding if it is to succeed. If it be $100, only provide $50, if it be $1 million, only provide $200,000. Spread available funds widely in this way and it will look as though the team is doing great things. A gormless public would not appreciate that by deliberately failing to cross the essential threshold, you would ensure general failure. If challenged the team could waffle plausibly that 'every little bit helps' and 'our rôle is to provide seed money'.

The donating public, being trusting, does not appreciate that 'every little bit helps' is incomplete without the rider, 'providing that it helps cross the rubicon necessary for success.' Where there is small chance of attaining this threshold, then all that 'every little bit' will ensure is a bigger ultimate loss. It is, after all, the bankrupt's lament.

How has WWF actually done relative to my hypothetical strategy? Look at its publicity and compare the evidence of failures to the paucity of publicity on success. A newcomer reviewing the record could be excused thinking there were no successes. This is in fact not true. There have been conservation successes up and down the continent. And as success is the *only* sound basis for progress, the absence of analysis, mention and congratulation stands out. In law enforcement, for example, why did WWF never analyse where and why anti-poaching efforts succeeded, before spending money on law enforcement that failed? Why were the Woodley/Sheldrick successes never identified? Why did WWF not establish the financial threshold for success in projects before spending money? When it was known that the key to rhino conservation was concentrating animals in small defensible areas, why did WWF pour money into huge holes like the Luangwa and Zambesi Valleys in quantities that were never going to fill the gap? That it is certainly arguable that WWF's actual strategies achieved what I would have wanted my dream team to have achieved is, of course, wonderful coincidence. A study by Simon Anstey[158] where rhino were concerned bears out much of the foregoing and all that can be said in WWF's favour is that it asked Anstey to make the analysis.

Whilst the foregoing may the ramblings of a cynic, it is also logical deduction construed from events and evidence. Given the recurrence of coincidence after coincidence, I do not think that speculating along the lines I pursued is unreasonable. Unwittingly no doubt, in my view WWF did propagate a political message favourable to apartheid South Africa. If it was consciously done, Niccolò Machiavelli would have given it his highest rating.

And yet in the cold reflective light of today, I doubt that there was a master plan and if events panned out to South Africa's advantage, it was as much through happenstance as from any other reason. It was the bungling and amateurishness of everything to do with Operation Lock that, in the end, stood out. What argues more convincingly than any other factor for events being driven by coincidence and chance, is that the incompetence that stands out so starkly is not unique to WWF or its particular players, but characterises a great deal of conservation.

Already Operation Lock is nearly forgotten. It matters little, for events have been overtaken and swept aside by history. If the incident has relevance now, it is to show conservation is neither apolitical nor a sphere in which human endeavour is purer than in any other. What should not be lost sight of though, is that Lock used conservation as a screen for ulterior motives. Just because it happened then does not mean that conservation today has become any more immune to such manipulation.

---

[158] Anstey S., 1987. *An Evaluation of WWF Rhino Projects 1961-1987*. MSc Thesis. University College, London.

Conflict between elephants and humans is very much ignored by western news and entertainment media: yet Kenya's papers carry such frequent records that it warrants little more than minor headlines.

CHAPTER THIRTY-FOUR

THERE'S SOMETHING WRONG WITH CONSERVATION

In the previous chapter I stated that incompetence characterised not only WWF, but also conservation generally. Perhaps I use the wrong word. I use it to cover a magnificent capacity to ignore evidence. This sweeping generalisation is nowhere more apparent than in the international perceptions of what has been happening to elephants and rhinos for the past century. Conservationists maintain with a religious passion, that from the early 1970s until today, a huge rise in the price of both ivory and rhino horn triggered such hunting pressures upon both species that both declined precipitously in numbers. Implicit is the assumption that before the early 1970s neither elephants nor rhinos were in decline, which is not true. The evidence that both have been declining over the past century is documented clearly. Proof that it has little to do with the prices of either ivory or rhino horn is incontrovertible. That this evidence has been so persistently ignored as it has, suggests either that something is very wrong with conservation, or that I do not understand what it is all about.

* * * *

Africa's people have increased at between two and three per cent annually for most of the twentieth century. This set the rate at which they have taken up new land, which is also the general rate at which elephants have lost range. This is no coincidence: human increase and elephant decrease are two poles of the same phenomoenen.

Before the game management scheme or elephant culling ever started, I had shot several hundred elephants on control, participating in the process of displacement. In less than ten years I saw elephants abandon the best part of 5,000 square kilometres (2,000 square miles) in areas where I had been a warden. I had been both a party to and a witness of the process.

Then there was the sheer scale of elephant control undertaken by the Game Departments of Kenya, Uganda and Tanganyika. Uganda had the best records going back to 1890 when Lugard was consolidating the British East Africa Company's rule. Among his first edicts he claimed all ivory. Referring to King Kasangama of Toro, Lugard wrote[159] –

> "I made a treaty with him ... elephant were not to be shot without permission, and were the monopoly of the Company."

When the British Government took over from the Company, it kept this monopoly. In doing so it rendered itself morally if not legally responsible for controlling elephants. Until then, white men were ignorant of the fierce interaction between farmers and elephants that characterised agriculture generally across Africa. They had been unaware that much ivory entering trade came from animals killed in defence of crops or in revenge for damage they had done. Controlling this interaction proved a prodigious undertaking and completely beyond the Uganda Government's powers. In 1912 it not only handed the problem back to

---

[159] Lugard F. D., 1893. *The Rise of Our East African Empire*. Blackwood, Edinburgh.

the natives, but also issued them rifles and ammunition with which to kill elephants. Between 1917 and 1921 it doled out 20,328 rounds of ammunition to those who had received rifles. Although the natives were supposed to hand over the ivory from the elephants they killed, it is unlikely that they surrendered all their tusks.

The policy did not resolve the problem of too many elephants. Dissatisfaction with this arrangement led in 1918 to District Commissioners being empowered to employ white elephant hunters whom over the next four years or so killed 3,992 elephants. This still did not resolve the problem. In 1922 a committee was set up to advise on further steps. It recommended the employment of salaried Europeans (as civil servants) to kill elephants. In addition it advised that the country be divided into elephant and non-elephant areas. In the latter *elephants would be vermin and shot on sight*[160].

The Government said it did not have the funds to go on employing salaried hunters so issued special licences for twenty elephants at a time. Only 385 were shot on such licences and they, too, failed to solve the acute conflict between elephants and farmers. Hardly surprisingly the licence holders spent their time looking for big tuskers and not necessarily shooting elephants in the vicinity of crops. The Governor stopped the programme and decreed that shooters should be full-time employees and the money be found to engage them. Not for the first time a government found funds that it had previously said did not exist. At the same time C. F. M. Swynnerton, the zoologist of tsetse fame in Tanganyika, assessed the situation. His report[161] included the following –

> *"A small number of salaried European shooters should be employed for the protection of the 'cultivated areas'. They should be reliable, energetic and experienced local men, working on the one hand in consultation with the Administrative Officers, and on the other in close touch with the natives and the herd* [of elephants]. *Each should be given a large district and each should shoot only in direct protection of the 'cultivated areas'... "*

This recommendation resulted in 1924 in the Uganda Elephant Control Department under Captain Keith Caldwell, late of the Kenya Game Department. In 1925 his post was taken over by Captain Charles Pitman who held it for the next quarter century. And though the department's name was changed to the Uganda Game Department within a year or so of its formation, elephant killing remained its principal occupation until Idi Amin's lads undertook their terminal control work after 1971.

From what we now know of elephant densities relative to rainfall and the area of the country where they were found, there were likely to have been around 250,000 elephants in Uganda in 1925. Between 1925 and 1959[162] the Uganda Game Department recorded the deaths of 46,583 elephants, most being shot by its own staff. In the next decade a further 10,000 elephant were taken officially to bring the Government score to around 57,000. In 1969 there are unlikely to have been more than 50,000 surviving. The balance of 150,000 or so unaccounted for went through illegal hunting combined with reproductive failure

---

[160] Swynnerton. *Ibid.*
[161] Swynnerton. *Ibid.*
[162] Brooks A.C. & Buss I.O., 1962. *Past and Present Status of the Elephant in Uganda.* Journal of Wildlife Management, Vol 26.

brought about by disturbance, displacement and the severe stresses related to compression into ever smaller ranges (as was described in the research by Laws, Parker & Johnstone[163]). Between 1925 and 1959 elephant range contracted from around 70% of Uganda's land surface to 11% and had dropped to about 8% in 1969 – most of which was in national parks, or game and forest reserves.

In little over a quarter of a century, human distribution changed from being islands in a sea of elephants to one where elephants occurred as shrinking islands in a sea of humans and there were virtually none left except in areas where human settlement was barred. The loss of virtually all elephants outside protected areas was the outcome of deliberate policy, officially instituted and caused, not of trade, but by competing with man for space.

In Tanganyika the same was happening with the Game Department alone killing over 100,000 elephants between 1930 and 1959. All their tusks entered trade, but trade was *not* the primary cause of the elephant deaths.

While the Uganda and Tanganyika Game Departments were preoccupied with killing elephants, they had neither the resources nor incentive to stop other people killing them or selling ivory illegally – though many old-time rangers and wardens dispute this. I contend from what I learned about the illicit traffic that all through these colonial years the local people killed more elephants than the government elephant hunters. This was certainly so with the 'poachers' in eastern Kenya.

One has only to consider the number of rifles licensed in Uganda relative to the number of game licences issued to see that it happened there too. Between 1951 and 1958 an average of 1,968 rifles were licensed annually against an average of only 463 game licences issued. It is naïve to think that the 1,505 licensed rifles unmatched by game licences were not used. And when an amnesty for illicitly held ivory was announced in 1970, from whence came the eighty tons that were handed in? Obviously a large, unlawful trade existed.

Similarly in Tanganyika, when one appreciates both the greater number of firearms in addition to a far bigger disparity between gun and game licences issued, official statistics for elephant deaths were considerably underestimated.

Massive elephant mortality was a feature throughout the colonial era. It was not new to the post-colonial period or to the 1970s, but had been rising steadily through the century. The annals of Fauna & Flora International (formerly the Fauna Preservation Society and before that, when founded in 1903 The Society for the Preservation of Wild Fauna of the Empire), have correspondence on the subject. An outburst by Dr A. H. B. Kirkman of the London University Animal Welfare Society on 7th March 1938 was typical. In a lecture at University College he stated –

> *"Africa is killing off the elephant... Soon it will have none left... Elephants were being killed off at the rate of 36,500 a year, greatly in excess of their reproductive capacity. The British game wardens – nominally the protectors of wild animals – could only show a satisfactory credit balance by slaughtering the animals... Thousands of African natives were working their own sweet will on Africa's unique fauna with guns and every imaginable and diabolical form of trap."*

---

[163] Laws *et al. Ibid.*

I do not know where Kirkman got his figures, but he was right in saying that thousands of elephants were being killed annually. Tusks imported into America, Europe and Asia proved that. He was wrong saying that this was done so that governments could make money. The elephants were shot primarily because they conflicted with farming and expanding humanity: that they made money was a bonus.

East African ivory exports outlined the trend. Between 1950 and 1970 (the last decade of colonial gamekeeping and the first decade of independence approximately), they rose at an average of 5·9% a year. In East Africa we knew that the main reason elephants were eliminated between 1925 and 1970 was human increase. We *know* that human increase continued uninterrupted after 1970 and into the present. We *know* that if East African ivory exports continued to rise at over 5% annually, the 220 tons of 1970 would exceed 350 tons by 1980: which they did. We *know*, too, that this could not go on and, for example, in 1970 I forecast the collapse of the Uganda elephant population: which then happened. Looking back, the predictions were consistently borne out by subsequent events.

Until 1970 or thereabouts there was little dissent about why elephant had declined and were declining. Their range had been and was being taken away from them. That, above all others reasons, was why national parks were formed: we *knew* that elephants and other large mammals had to have land. People created the parks primarily to provide such land and to block the process of displacement. Ivory was a by-product of the process, though until I looked at the trade statistics, no one had used its production to measure the human/elephant interaction.

Several had, however, measured the loss of elephant range: Alan Brooks and Irvin Buss mapped the contraction in Uganda[164] while Don and Joyce Stewart had done the same in Kenya[165]. Gerry Swynnerton (son of the renowned C.F.M. Swynnerton) had done so in Tanganyika as had Jonathan Kingdon who summarised all this work in his monumental *East African Mammals: An Atlas of Speciation*[166]. Loss of range was the critical issue. An animal or plant community can be reduced in numbers but through ability to multiply can bounce back. It happens naturally all the time. Droughts are endemic to savannah Africa and periodically reduce animal populations – sometimes drastically – but when they are over, surviving animals respond to better conditions by increasing rapidly *providing that they have the space in which to do so*. By eroding space decline becomes irreversible.

These simple facts are irrefutable. Yet conservationist and, even more, media reaction to evidence that post-1970s ivory production was predictable from pre-1970s history has been aggressive scorn. Supported by the 'scientists' financed by western sources, they set the earlier knowledge aside and replaced it with a new hypothesis. In the 1970s the price of ivory climbed steadily from between $5 and $6 per kilo 1969/1970 to around $75 per kilo in 1979 and on upward until over $200 at the end of the 1980s. Coincidentally, in the 1970s widespread corruption and declining standards of governance were manifest across independent Africa. Conservationists and media said these factors explained the rising

---

[164] Brooks A. C. & Buss I. O. Ibid

[165] Stewart D. R. M. & J., 1963. *The Distribution of Some Large Mammals in Kenya*. Journal of the East African Natural History Society & Coryndon Museum XXIV (3) (107).

[166] Kingdon J., 1971. *East African Mammals: An Atlas of Speciation*. Vol. IIIB. Academic Press. London.

ivory exports, which mirrored elephant deaths. Their solution: break the mirror (stop trade) and the deaths will stop! Lewis Caroll would have loved the logic.

For all that corruption was rampant, it was a red herring that turned attention away from the wealth of evidence on the fundamental cause of wildlife decline. Corruption's greatest influence was not so much on how ivory was acquired as on how it was disposed of. The real cause of decline that conservationists so arbitrarily set aside was as valid post-1970 as at any time before. Human increase continued exponentially. Elephant loss of range kept pace. People were replacing elephants to whom it made no difference whether the bullets that blew their brains asunder were fired by colonial wardens who said they loved them, or Amin's soldiers who did not: what counted was their deaths.

The price-of-ivory/massive-corruption theorists ignored the colonial records. The control record in Tanganyika and Uganda (minimally killing 157,000 elephants in two decades), of declaring elephants vermin, the documented loss of range; all this was swept under the carpet. And the great weakness of arguing that the output of ivory was a response to rising price was exposed by its corollary. If production rose in response to rising prices, logically it should have fallen when prices dropped. That never happened. The price of ivory fell by 69% in the Great Depression of 1929-1933 and, in both real or relative terms kept falling for nigh on forty years until 1969. Yet throughout the four decades of declining price in either absolute or relative terms, ivory production kept rising *at the same exponential rate that it rose after 1970*. If further proof is needed this was repeated in the 1980s. As I have already reported, for the hunter in the field, the price for ivory that moved through Burundi (over 50% of all ivory exports out of Africa) fell by 75%. Yet it had no impact upon the volume produced. In a nutshell, the pattern of production (that is of elephant deaths) was independent of whether prices for ivory were falling or rising.

Elephant decline was driven by a biological process – displacement of wild species by the increase of humans and their domesticates. If economics were involved it is as a product of and subordinate to this biological process. In the words of Graeme Caughley and Anne Gunn[167] –

> "*The increase in ivory production from 1970 reflected a seamless extension of a long-term exponential trend in production effort that had been in place from at least 1950, that trend being perturbed only minimally by fluctuations in the ivory price. Exponential trends can cause unpleasant surprises, nothing much appearing to happen for a long time and then the state of the system suddenly changes ... The debate on slowing then stopping the ivory trade has, in the 1980s and 1990s, hijacked elephant conservation.*"

While corruption in the Kenya Game Department was epic, as I have illustrated, it was made to *seem* far worse by *assuming* that prior to 1970 the Department was effective – which it never was. The 'Great Defection' of the Kenya Game Department and the WCMD which succeeded it was serious where good governance was concerned and gave unique insight into what was going on generally in Kenya. Yet the inexorable loss of square metres

---

[167] Caughley G. & Gunn A., 1996. *Conservation Biology in Theory and Practice*. Blackwell Science. Cambridge, Mass.

The Uganda kob produced no trophy of value, but had lost 89% of its 1925 range by 1975.
(Photo Peter Davey)

that had been going on across the country throughout the century was far more important to the elephant population (and all other large wild mammals). Elephant decline had been exponential at least since 1925 and what happened after 1970 was merely extension towards the logical end point of a process that had been happening over the previous fifty years. The phenomenon was as apparent in colonial times as it was during African governance, only no one saw it for what it was.

Rhino decline was prominent in the colonial era. Among several authors who wrote about it at the turn of the century, in the 1910-1 and 1911-2 Report of the Game Warden for the British East Africa Protectorate he commented, "This animal is, I think, undoubtedly decreasing." The last black rhino recorded between Lake Turkana (Rudolf) and the Uganda border was in 1917[168]. George Adamson reported their disappearance from the rest of Turkana country by the end of the 1930s, pointing out that it was a recent loss as he could still find their crystallised urine on rocks where they had deposited it to mark their territories. By 1932 they were gone from the Mau Forests and between Solai and Baringo where they had formerly been abundant. Declining rhinos was a constant refrain through George's departmental correspondence – as much in the early years 1939-45 as later.

He was not the only one. The Administrator of the NFD during the early war years - Reece - also complained about rhino decline. And of course Jonathan Kingdon summed it all, showing that black rhino had lost 82% of their East African range between 1925 and 1970 and that the greater part of this loss had happened before 1950[169].

The loss of black rhino in the eastern half of the African continent shows a distinct pattern: as if the rhino 'carpet' has been rolled up from north to south. They were gone from the central Sudan and all but the extreme south of Ethiopia before the turn of the 20th century. They were gone from northernmost Kenya and all but southern Somalia by 1920. They were all but gone from the rest of Kenya's NFD and southern Somalia by the mid-1960s: the process was reflected in Somalia's Customs statistics for rhino horn exports between the late 1940s and 1960.

There is something distinctly odd in the way this tide rolled southward, regardless of the prevailing government: independent in Ethiopia, colonial in Southern Sudan, Uganda, northern Kenya and Somalia; independent capitalist in central/southern Kenya; independent Marxist in Tanzania and socialist Zambia; and then progressively south through the war-torn Angola, Mozambique, then Zimbabwe and now South Africa. The

---

[168] Rayne H., 1923. *The Ivory Raiders*. William Heinemann, London.
[169] Kingdon. *Ibid.*

process took place regardless of government. I offer no explanation for this pattern and merely point out that it existed, was obvious to anyone who looked at the data, and was far advanced decades before 1970. As with ivory, the decline went on whether the price of rhino horn was rising or falling.

Yet as with elephants, conservation scientists ignored these facts and were supported by the press. Wilful exclusion of the truth on this scale is tantamount to lying. The essence of what they told was a tale of cops and robbers in which price increases post-1970 seduced feckless Africans into raping their wild resources.

The 'kongoni' or Coke's hartebeests have neither horns nor teeth of value that were traded, but have still lost space and are irrecoverably fewer than they were twenty years ago.
(Photo Peter Davey)

Particularly telling was how they also avoided referring to the declines of other species which did not produce valuable, money-earning trophies. Jonathan Kingdon[170] presented changes in the distributions of thirty large mammal species. All had declined through the half century before 1970. Extent of range loss did not relate to value of trophy. True, elephants and rhinos were second and third in the extent of their range loss, but the animal which had lost more of its range than any other (89%), the Uganda Kob, produced neither horns nor a hide of value. Leopards, whose pelts ranked along with ivory and rhino horn in value, were at the bottom of the loss table. They had lost less of their original range than all but two other species, despite the value of their pelts. This point was confirmed in 1987 when CITES consultants noted that trade in leopard skins had negligible influence on leopard numbers[171]. Animals whose habitats were arid and which could exist without free water for months at a time (lesser kudu, oryx and gerenuk for example), had lost less of their range than those whose habitats were better suited for agriculture.

What stands out from the wealth of published evidence is that the declines of elephants, rhinos and large mammals generally in East Africa have everything to do with human increase and displacement by people, and relatively little to do with the value of their products. These declines have been going on throughout the past century and suggesting that they are post-1970 phenomena associated with African governance was downright dishonest. Critically, in this respect, the declines were far advanced and some species such

---

[170] Kingdon. *Ibid.*
[171] Martin R.B. & de Meulenaer T., 1987. *The Status of Leopard Panthera Pardus in Sub-Saharan Africa.* CITES, Lausanne.

as black rhino had actually lost most of their former 1900 ranges during the colonial era. While probity and corruption may have influenced what was happening, they were so secondary that the trends were the same regardless of governance. Just as unscrupulous, was the failure to report that ivory production held to its upward trend regardless of whether prices were rising or falling.

* * * *

When IUCN/WWF/NYZS published the results of their survey of the elephant's 1978 status in Africa, they claimed it derived from the best available contemporary information. It had eighty-one informants on a continent that had a population at the time in excess of four hundred million black people of which at least hundreds of thousands lived in proximity to elephants. Yet only six listed informants were black. The balance were mainly white and of them a substantial proportion were transients or relative newcomers to Africa. Yet the authors were never challenged over their claim to have pooled "*the best available contemporary information*" and to have "*...relied on hundreds of informants from all corners of Africa.*"

In fact, given an elephant range of over seven million square kilometres that must have included some millions of people, the sources of information were so paltry as to verge on the ridiculous. That all sources were not tapped is apparent from the case of Jonathan Kingdon: an academic of unimpeachable scientific integrity.

Brought up in Tanganyika; a hunter and naturalist throughout his youth and fluent in Kiswahili, Kingdon had direct access to Africans including peasant hunters all over Kenya, Uganda and Tanzania. Living among the animals, they knew infinitely more about local distributions than did transient white research boffins. Coupling this huge reservoir of knowledge to, in his case, truly hundreds of resident farmers, rangers, wardens and hunters as well as scientists, his was the most comprehensive collection of data ever compiled on East African fauna. Kingdon was never asked for this information, so he took the initiative and offered all his elephant data *gratis* to the WWF/IUCN/NYZS study. It was spurned. The same happened to his data on rhinos, which was equally outstanding.

In another IUCN example, its Antelope Specialist Group described the status of Africa's seventy or so antelope species. Informants were again a handful of whites. The millions of indigenous black people who assuredly knew something about the antelopes amongst whom they dwelt (they were poaching them, weren't they?) contributed infinitesimally – mainly because no one had asked or, as likely, could not ask them because those collecting the data could not speak vernaculars.

The *African Elephant Action Plan* and the Antelope Specialist Group perpetuate the vision of Africans so benighted that they could not describe the whereabouts of the animals that shared their environment. And if this is not enough, there is the case of the Zimbabwe Department of National Parks & Wildlife Management.

After the unilateral declaration of independence in 1965, the Rhodesian Department was by far the best-organised and competent in Africa, oriented towards hands-on, practical results in the field. Confident and competent, it was doing all the things other African Departments wanted to do, but seldom got beyond debating. This was what newly independent Zimbabwe inherited. It had scientists and wardens who had *done* and *did* rather than merely met and talked, who were examples for the rest of Africa to copy, and exactly what conservationists said they wanted for the rest of Africa.

Yet conservation NGOs aware of this were soon approaching Zimbabwean wardens and researchers with attractive offers to take employment elsewhere. No-one blamed the wardens and researchers for giving in to this seduction and leaving Zimbabwe's meanly paid Department of National Parks and Wildlife Management. Thus western-based conservation bodies ironically prattled about developing indigenous competence, while actively dismembering Africa's best wildlife department, poaching its experienced staff for their own projects in other countries.

The Customs statistics between 1940 and 1960 documented the loss of Italian Somaliland's rhino.
(Photo Peter Davey)

Challenged on this, WWF retorted that staff left the Zimbabwe Department of their own accord because they were not paid enough, which is true. Yet the solution was surely to get them better terms of service, keep them *in situ* and keep the Zimbabwe department functioning. If the problem was conditions of service, the solution was rectifying those terms. Instead, Africa's most competent conservation department, technically better than any of the overseas agencies peddling conservation to Africa, was not held together, but pulled apart.

* * * *

The writer Ray Bonner[172] attempted to explain the paradox of conservationists failing to conserve. In his view, which he supported with considerable evidence, most international conservation organisations were more interested in raising money for themselves than for animals. I believe this explains some of what I have observed. Reporting an inevitable, unavoidable biological process of people replacing animals would not have loosened the strings of charity, whereas the tale of cops and robbers in which conservationists placed themselves with the 'good guys' did. Being able to employ the best men for the job on 'their' projects overrode any interest in keeping the continent's best conservation unit going. Self-interest ruled.

There was so clear a connection between expanding human settlement and declining wild species' ranges that one did not have to introduce the value of trophies to explain the

---

172 Bonner R., 1993. *At The hand of Man*. Alfred A. Knopf. New York.

trend. Yet why it was not publicised was summed by two television film producers, with whom I worked: one from the British BBC and the other from the American NBC. Independently, both said, "But the public is not interested in population issues."

Therein lies the nub: only tell stories that the public wants to hear because it will only buy, literally and metaphorically, what it wants. Cops and robbers, the goodies and the baddies, Robin Hood and the Sheriff of Nottingham are themes that sell and sell and sell and have done down the millennia. They fit what the faith wishes to believe. Human increase is a non-story: after all, if you ask *who* is too many there is the dreadful possibility that someone will say '*You*'. The only time I ever got a vigorous response over human increase was in Australia when I brought the Pope into it. Ray Bonner was right: making money selects what those who depend on charity and the media propagate. Yet while it is very important to them, I believe that it only partially explains why conservationists and media are at such odds with truth.

CHAPTER THIRTY-FIVE

WHAT I TELL YOU THREE TIMES IS TRUE

The root of the word 'conserve' derives from the Latin *servare* to keep. Conservation is preserving what has been and still is. To be conservative is to favour what has worked, is proven or what one is used to and to be wary of change. It logically follows that one might expect conservationists to be conservative. Yet those proclaiming themselves as such today seldom see themselves conservative politically, or in any other way. To the contrary, it is more usual for them to be politically liberal and, excepting subject, to speak and act very much as Marxists did twenty-five years ago. The absence of conservative behaviour among conservationists suggests that they have hijacked the word, rather as homosexuals stole 'gay', and have given it an alternative meaning.

The motives behind conserving and being conservative where farmers and landowners generally are concerned, has a basic evolutionary common sense: of staying with what you know works and looking after the environment with which you are familiar. The unknown, of which you have no experience and therefore lack expertise to handle, may be dangerous. It is in this sense that it was first applied and used in the context of preserving wildlife. For centuries and in many places, landowners were both conservative and conserved as a consequence. It was always more a rural than urban activity and derived from an outlook and attitude rather than a cause or ideology, and driven by common sense, pragmatism and flexibility. Environmental events are so fundamentally unpredictable and variable and reflect random chance and chaos so consistently that it cannot be otherwise. Consequently successful land management, of which conserving wildlife is an aspect, is a process of give and take and not amenable to dogma and dicta.

Italy's *fascisti* espoused law and order, out of which evolved fascism's disregard for law. Communal welfare was the germ of what became harsh and illiberal communism. The laudable goal of wealth for all led to socialism's destructive economics. Fascism, communism and socialism were rooted in excellent intentions and common sense. Yet once their good ideas were codified into rules and dogma that precluded alternative options and explanations, they became pernicious, socially divisive and eventually self-defeating. That, it seems, is what is happening in conservation. It is becoming conservationism (aka environmentalism), is losing common sense, pragmatism, flexibility and becoming doctrinaire, with its gurus believing that it is the only 'true way' forward. It has emerged as a belief, faith or ideology (which are the same thing) that is righteous and righteousness has always been defence against reason.

The world is currently experiencing a wave of fundamentalism that is readily apparent in all the great religions. Doctrine, dogma and fundamentalism are all facets of the same psychological phenomenon. Characteristically, doctrine becomes accepted through repetition and is embodied in prayers and mantras. Priests and gurus ask their followers to 'say after me,' the other side of which coin – so neatly captured by Lewis Carroll – being 'What I tell you three times is true.' And if you don't believe me, listen to those in the green, environmentalist movements. Their statements are endlessly repetitive. Ecosystems are always *fragile* (whereas in reality they are extraordinarily robust), species are usually *endangered*, biodiversity is *essential* for human survival: this long catechism is acquired by rote very much as the Lord's Prayer, Koranic verses and Hindu mantras are learned. Industrialised, urban people divorced from the pragmatic give and take that dominates rural life, overwhelmingly drive this conservation fundamentalism.

Ray Bonner was shocked by the hostile reception of his eminently logical book *At the Hand of Man*.[173] He failed to perceive this arose, not from pointing out conservationist mercenary policies, but from challenging a faith - conservationism.

* * * *

The manner in which conservationism had proceeded is not unique. Causes attract parasites. David Livingstone was a classic example. In the mid-18th century *the* cause of the day was the anti-slavery movement. The passions and rhetoric of anti-slavers then and conservationists now are disturbingly similar. For all that both may have solid philosophical foundations, the vehemence of expression in both has the same touch of fundamentalist madness to it. It is this that leaves them open to abuses.

Today the young western biologist appeals for funds to go forth into the Third World, preach conservationism, rescue endangered species and live on a shoestring. Then, dedicated young people qualified as preachers or doctors, or both, begged for funds and went out into the uncivilised world to stop slavery (i.e. poaching). Organisations sprang up to support them: for example, the Church Missionary Society (CMS) and the Universities' Missionary Society (UMS) were analogues to WWF, IUCN and similar bodies. Both then and now those at the top of such bodies lived very nicely, thank you.

David Livingstone took religious orders, acquired a medical degree and went forth as a Christian missionary. It is as a missionary and philanthropist that he is still revered. Victorian Britain lionised him. Unquestionably intrepid and a strong personality who gave western science new geographic records, he made Europe aware of the chaos that ivory gathering and slaving were causing in the African interior. Yet there was a huge difference between the public perception of himself he cultivated and the man he really was. Even more important, he contributed powerfully to Victorian Europe's badly flawed concept of benighted Africa.

Livingstone may have trained as both minister and doctor, but from his own records, in his entire career he only converted one person (Chief Sechele of the Bakololo) to Christianity and his record of healings is paltry. He never stayed long enough in one place to be theologically or medically effective. Perforce he started out preaching the Word under the man who became his father-in-law, Dr Robert Moffat, in the north of what is now South Africa and in modern Botswana. A thoughtful medical study by Ransford Oliver[174] gives insight into his personality. From the outset he bent even seemingly trivial facts in every situation to his advantage. Thus between 1841 and 1853 he travelled a great deal by wagon and these trips have come down in Livingstonian lore as prodigious feats. For the most part, however, they would have been relaxed and pleasant experiences. As Oliver points out,

---

[173] Bonner. *Ibid.*
[174] Oliver R., 1977. *David Livingstone, A Reassessment With Particular Reference to His Psyche.* Ph. D. Thesis. University of Rhodesia (Faculty of Arts).

although they entailed occasional anxiety about obtaining water, ox-wagon journeys were a great deal more comfortable than modern travel through the Kalahari by Land Rover.

Livingstone's first journey north from Kuruman where he was stationed under Dr Moffat began with Rogers Edwards on the 18th October 1841. Oliver observed:–

> "Livingstone implied in his correspondence that it was he who had initiated the expedition ... but here he was being a little less than frank; Edwards had planned it long before Livingstone had reached Kuruman, and the newcomer merely volunteered to accompany him ... "

David Livingstone: intrepid, unpleasant, besotted with fame and afflicted with cyclothymia.

Shortly thereafter, in August 1843 Edwards and Livingstone again went to set up a new mission station at Mabotsa among the Bakgatla and again Oliver was struck by Livingstone's weakness:–

> "... Livingstone wrote a report on setting up the Mabotsa Mission. Most unfortunately an extract from this report was published by the London Missionary Society (in the Missionary Magazine & Chronicle VIII (50)) during April 1844. Ambitious for personal renown, Livingstone had given remarkably little credit to Edwards for his share of the work (it was a lapse which was repeated when he came to write Missionary Travels) and the Chronicle made things worse with a gushing statement concerning the initiative of 'our intrepid missionary, Mr Livingstone,' without saying a word about his colleague."

He made his first big pitch for fame outside of the missionary field over the discovery of Lake Ngami. In this case he stole fame from William Cotton Oswell. Hearing the natives speak of a lake and swamp to the north of the Kalahari, Oswell determined to go there. As a wealthy, sporting bachelor, he had the means and assembled a party of his friend Mungo Murray, his servant George Flemming and a trader called Wilson. Late in the planning he invited Livingstone to accompany them as a guest. They found the lake, but it was Livingstone who seized credit for the discovery, writing it up for the Royal Geographical

Society. In the write-up he paid due tribute to Oswell's financial contribution (he had funded the expedition entirely), but never suggested that he deserved the credit of discovery.

Fortunately for Livingstone, Oswell took no offence over the theft of his glory. The missionary knew when he was on to a good thing and persuaded Oswell to finance three trips north from Ngami to the Chobe and Zambezi rivers in 1849, 1850 and 1851. In letters to his father-in-law and his superior, Freeman, of the London Missionary Society he blocked public perception of Oswell's role in these ventures. Indeed, instead of acknowledging his own indebtedness, he reversed their roles and made it seem that he was the initiator and drive behind the projects:–

> "Oswell was excessively anxious that I should promise to let him accompany me next year, but I declined, yet I don't know how to get quit of him."

In 1853, now knowing the way, Livingstone went alone. He bought land for a mission at Chonwane from Sechele of the Bakololo in exchange for a gun, some ammunition and some beads. As supplying the natives with guns was widely frowned on, he concealed the fact, writing in Missionary Travels[175] that he had bought the land with £5 worth of goods. His own correspondence reveals that not only did he act as Moffat's agent in selling guns and powder to the Bakwena, but was quite prepared to repair their muskets for payment, and even sold them some arms directly[176].

Going to the Chobe and Zambezi rivers Livingstone implied that he went among natives who had never before seen white men. This was untrue. The Bakololo had only recently left South Africa because of the Difaqane dispersals. Livingstone's father-in-law had written about them and they certainly knew white men. Thus when Livingstone went north, he went neither among total strangers nor among people who knew nothing of the outside world. And he was not the first white man in the area.

First he met Caetano Jose Ferreira and Norberto Pedro de Seña Machado who had been on the upper Zambezi for a while before him and were talking of going on to Mozambique. Livingstone refers to them scathingly as half-caste slavers. They were neither, but well-to-do Portuguese farmers from Angola exploring for the fun of it.

Next he met Antonio Francisco da Silva Porto, a somewhat better known Portuguese explorer who was also a farmer from Bihe in Angola, who had first reached the Zambezi in 1848 and in 1853, was intent on traversing the continent to Mozambique. While da Silva Porto fell ill and never made his crossing, several of his men went all the way to Zanzibar then returned to Angola by sea via the Cape.

And then there was Laszlo Mágyar, a cultured Hungarian whose journals establish him as one of the great unknown white explorers of the 19th century. He is not well known because he did not publish widely. Yet he, too, had been on the upper Zambezi a year before Livingstone. Mágyar described the route between Linyati on the upper Chobe and the Atlantic to the Bakololo and they, in turn, passed it to Livingstone who used it as the guide for his trek to Luanda. Mágyar tried to meet Livingstone, but the great man refused

---

[175] Livingstone D., 1857. *Missionary Travels and Researches in South Africa*. John Murray Ltd, London.
[176] Oliver. *Ibid*.

any contact. Churlish behaviour, but understandable in the light of how Laszlo's achievements diminished Livingstone's own.

More important, Livingstone never gave credit to three Arabs (Swahilis?) he met in da Silva Porto's camp who told him much about Lakes Tanganyika and Malawi. They had walked across the continent from the Indian Ocean to the Atlantic via Lake Tanganyika, Katanga and Luanda and were now on their way home[177]. Later, he met another Arab whom he referred to as 'Ben Habib' who arrived among the Bakololo from the east. Tellingly, Habib undertook to guide a Bakololo ivory caravan to Luanda, which the man could scarcely have done had he not made the trip before.

Referring to the Batoka near Victoria Falls, Livingstone[178] wrote that they were so ignorant that they did not know the value of ivory. Yet they and the Bakololo wore cotton cloth and carried muskets obtained from trading tusks and slaves with the outside world. As proved by the Arab/Swahilis he met, this commerce was not southwards through South Africa, but both eastward to the Indian Ocean and westward to the Atlantic. However, in the lands from where they had recently arrived, the Bakololo would certainly have known about demand for ivory and the white men who carried it southwards.

The more he travelled, the more Livingstone realised that wherever he went, it was in Portuguese and Arab footsteps. All he could safely claim was that he was the first Briton to have crossed the continent. The archaeological record now sustains this picture; indeed it proves people in the upper Zambezi valley had been in contact with the east coast for over a thousand years.

Because the Portuguese were Europeans and might diminish Livingstone's achievements by describing their own, his attitude towards them changed from friendly to venomously critical. Lambasting them for condoning slavery (at a time when it was still legal in the United States), he threw a screen across their achievements that might lessen his glory as an explorer. He projected himself as the great anti-slaver because this was the cause of the day assuring him popularity and funds. Yet he was less critical of the Arabs and Swahili whose hospitality he accepted in the field while they were actually acquiring slaves. There is also no evidence that he tried to dissuade them from slaving.

Ransford Oliver explained Livingstone's behaviour as the product of cyclothymia which, in extreme cases, becomes a form of madness. Whether he was in fact psychotic cannot be determined at this distance.

All the foregoing suffers the ill effects of reductionism. Livingstone was a great man - at least in determination. He advanced western science and he wandered on an epic scale. He was nonetheless an altogether lesser person than he made himself out to be. On its own that might be trivial historically; what is of greater import is how people saw Africa as a consequence of his self-publicity.

Above all other individuals, David Livingstone was responsible for nineteenth-century Victorian perceptions of Africans as benighted primitives without a history. He deliberately suppressed information to the contrary and got away with it because it was what Imperial Britain wanted to believe: even more, it was what the Christian churches - the Faith - wanted

---

[177] For a compendium of this evidence see Main M., 1990. *Zambezi - Journey of a River.* Southern Books (Pty) Ltd. Halfway House. South Africa.

[178] Livingstone. *Ibid.*

to believe. The vision of pagans in utter darkness and ignorance and under the unmentionable yoke of slavery, isolated, shrouded in ignorance, exploited by evil men, thirsting for God's Word and above all, European civilisation with its industrial goods was just what was wanted. It went a long way to creating the ethical environment in which expanding the Empire became self-justifying. The material motive may well have been commerce and power, but this was difficult to justify morally. Loving thy neighbour, justice and fair dealing were difficult to equate with profit and territoriality. If dressed as philanthropy, of bringing light to benighted savagery, of being *good* and *righteous* then, as it was, the cause would be firmly seized.

Yet Livingstone could not have warped history as he did, alone. To a degree he was a cypher in the hands of a public who wanted a hero and was not too particular about details of the heroism. He was not responsible for the intensely Anglocentric outlook which prevailed while he lived, nor for the Eurocentric ambience within which it existed, but he parasitised both shamelessly and sculpted the truth while he lived. Livingstone's defects were apparent to many of his colleagues. Those who saw through him kept quiet and continued to do so after he died. Missionaries and the media, reaping the material benefits, built him into a cult figure so successfully that for more than a century after he died, we were taught what a wonderful person he had been. He was an extremely successful cause manipulator whose legacy was the great misconception of Africans benighted beyond reality which was then taught as fact across the British Empire. It played no small part in setting our colonial attitudes towards them and in the difficulties thereby generated.

There are parallels to Livingstone in modern African conservation. It is particularly obvious where elephants and ivory are concerned. There is the same basic common sense in conserving them that there was in stopping slavery. There is similar international interest. And like the mission societies there are the conservationist societies: both producing a few 'heroic' individuals who today, as Bonner observed, were profiting all the way to the bank. Like Livingstone, they are not too concerned with truth and accuracy. So long as the cash flows in, does it really matter what causes elephant decline?

* * * *

Conservationism dogma are obvious. Take 'the balance of nature' or in more recent idiom 'ecological equilibrium', for example. Both imply conditions in which, excepting slight population fluctuations, life within an area stays more or less the same over long periods: that a state of stasis can prevail. Maintaining such stasis is a conservationist goal and best seen in the idea that nothing, but nothing, should become extinct.

Yet there is no stasis in the cosmos. Everything from the outermost dimensions of the universe to the smallest particles of matter constantly changes. Physics, astronomy, chemistry, geology, meteorology and indeed all science affirm this. Yesterday will never be repeated; today is like no other; tomorrow will always be different from what has been. Darwin saw how this applied to life with palaeontology being the study of evolution past and ecology its study in the present. Stasis does not exist. Change may not have a constant pace yet it never stops.

If DNA illustrates the uniqueness of every individual, so, too, it proves that no two generations are the same. The difference between generations may be hard to perceive in outward features or gross internal anatomy, yet when it comes to the genetic inheritances of, say, resistance to disease, they can be profound. When influenza and measles were introduced to South Sea people who had not experienced them previously, large proportions of their populations died. However, in the space of a generation or two the genes of those who did not survive were eliminated and the survivors had changed into a population resistant to colds and measles. Outwardly they looked the same, but genetically and in ways still beyond analysis, where influenza and measles were concerned, the surviving populations were profoundly different. Such changes at the 'invisible' molecular level can be fast and obviously go on all the time.

While lack of stasis in Nature is indisputable, a craving for it seems intrinsic to being human. The Garden of Eden and Heaven are ideals where time - which is change - did or does not exist. Perhaps it is no more than simple, intuitive, hope for eternal youth and rebellion against death: a relentless, inescapable and dreaded process. Whatever the reason for craving it, stasis only exists as an unattainable concept in the human mind. Philosophically, it contradicts the principle of evolution - which is constant change. To pronounce a dogma sanctifying the balance of nature, or ecological equilibrium and giving high moral value to preserving nature as it was or is at a point or period in time, flies in the face of common sense.

Yet another illustration gaining wide acceptance is that human welfare depends on preserving the world's 'biodiversity' - the term meaning all forms of life. This, like claiming that there is a balance of nature, is nonsense. Much life is inimical to our welfare. All Africa would rejoice if the plasmodia and anopheline mosquitoes which respectively cause and carry malaria became extinct. Black rats seriously lower the quality of human life. It has never been proved that *any* other single species is actually essential to human existence. Dodos certainly were not, nor were Steller's sea cows.

Some species are very useful to us. Others we like to have around but are neutral in the material sense. Yet others are wholly injurious. Many have no fixed status. Elephants as sources of pure joy to Africa's visiting tourist are worse than smallpox to the peasant whose crops they have just destroyed. The world's biodiversity in the holistic sense is obviously not necessary to human welfare. Indeed a strong counter argument exists. Human welfare is measured by the control we exercise over our environments and the degree of our success in eliminating life inimical to, or which competes with us. Some of the most awful prospects facing humanity - HIV and Ebola viruses for example - came upon us unexpectedly and unwanted out of Nature's living cornucopia. Arguing that biodiversity *per se* should be preserved so we do not lose chemicals and cures yet to be discovered, must surely be countered by the evidence that it may as equally be a source of further horrible diseases. What other eboloid sicknesses lurk in rain forest shadows?

The dogmas that have replaced the common sense of conserving are powerful clues to the behavioural rôle of conservationism. What counts is how people behave and think. The rules arbitrate the ephemera of 'right' and 'wrong'. Dissenting views are not permissible and savagely attacked. Facts fitting the formula are acceptable, but as Galileo found out, any which buck the prevailing explanations - regardless of their verity - must be suppressed. Just as the major religions need the presence of poverty to sustain their shapes, so conservationism needs plants and animals near extinction as the material for fund-raising.

Understanding conservationism as a faith or ideology explains its apparent failures. When I observed that WWF routinely failed to establish minimal monetary thresholds for success in its projects, I assumed a businessman would view this disregard as incompetence. I did not appreciate how it is characteristic of religions. They proceed identically. Churches dole out money to the poor, certain that it will not stop them being poor: it is the giving which matters. If, by miracle, secular governments could abolish poverty, ignorance and sickness and thereby the need to give alms, educate the ignorant, help the sick - all means to righteousness (loveliest of feelings) - they would cripple all extant religions. Indeed, it is probably no coincidence that the rise of wealth, scientific knowledge and state medical services in western industrial societies, has been paralleled by declining religious power and influence. It is when put in the context of a faith that the monstrous incompetence and inefficiency of CITES come into focus. What matters is not so much successful law enforcement as behaving correctly and having the right attitudes.

As with faiths in general, conservationism has a corona of fundamentalist fanatics. Fundamentalism is the bedrock of terrorism. Today there are Animal Rightists; Greenpeace; the Environmental Investigation Agency (EIA); The Humane Society of the United States and WWF, who may vary in the degrees of their fanaticism. Animal Rightists planted bombs across Britain, Greenpeace used a vessel to ram a whaling ship: quite literally an act of war. And when Greenpeace became too tame for its crazier fringe, splinter groups like the EIA were generated, for all the world like Trotskyites coming off mainstream communism to enforce what they thought was 'right'. One can but note that, in an age when the death penalty is being removed from many countries' statutes, it is seen as most appropriate for African (but not European) poaching. Several hundred poachers have been killed in East and Southern Africa since 1975, and hardly stirred comment from bodies like Amnesty International who make such a fuss about the death penalty elsewhere.

A more recent case of setting aside the norms of law are in the 'Lusaka Agreement on Co-operative Enforcement Operations Directed at Illegal Trade in Wild Fauna & Flora'. Adopted on 8th September 1994 by Ethiopia, Kenya, Lesotho, South Africa, Swaziland, Tanzania, Uganda and Zambia, it was an inter-government arrangement to prevent illicit international trade in wildlife products. In the Kumleben Commission's words it would –

> "... function, as it were, as a regional Interpol within the territorial boundaries of the member countries."

On the face of it, as sensible as Interpol itself, it was not the outcome of *professional* analysis. No one explained why existing law enforcement agencies needed to be augmented by a force with far more power than presently given to conventional police. Other than for the South African Police, no police forces were represented. As with CITES in Africa, it arose from the false assumption that new law was warranted because extant laws were not enforced. Yet failure to enforce laws because of corruption and inadequate funding is no case for new law. No corruption and adequate funding are what are necessary. Without them new laws will be implemented no better than the old laws.

At the meeting in September 1994, a committee was established to carry the concept forward. The idea was 'bought' by the United Nations Environment Programme (UNEP) which from then on provided secretarial and administrative support, giving the Lusaka Agreement the *gravitas* of a United Nations-backed enterprise.

At the second expert group meeting held at the UNEP headquarters in Nairobi UNEP provided twelve (27%) of the forty-five participants. Three were EIA 'operators' (L. A. Carter, Ros Reeve and Heena Patel). It is germane that Judge Kumleben said of the EIA –

Be honest: who sheds a tear for the departed dodo?

> "(i) The assertions of the EIA in their reports were a major source of discrediting certain traders, the police and State officials in the RSA.
> (ii) The EIA, and its employees contributing to these reports, were unable to provide the Commission with evidence to substantiate these damaging allegations, despite invitations to do so."

He was too kind. I believe the EIA ranks among the least acceptable faces of conservationism. Seeing ends justifying means, its faith in 'being right' sets aside need to comply with legal norms. Typically, Carter, Reeve and Patel, gave their addresses as Lusaka Agreement Co-ordinating Secretariat, Environmental Law & Institutions Programme Activity Centre, UNEP, Nairobi. By offering to serve on the co-ordinating committee set up in Lusaka, they got a UN address. The tactics are classic 'penetration'.

Section 10 of the Lusaka Agreement read –

> "10. For the purposes of paragraph 8 of this Article, the Director, Field Officers and Intelligence Office of the Task Force shall enjoy, in connection with their official duties and strictly within the limits of their official capacities, the following privileges and immunities –
>
> (a) immunity from arrest, detention, search and seizure and legal process of any kind in respect of words spoken or written and all acts performed by them; they shall continue to be so immune after the completion of their functions as officials of the Task Force;
> (b) inviolability of all official papers, documents and equipment;
> (c) exemption from all visa requirements and entry restrictions;

> (d) *protection of free communication to and from the headquarters of the task force;*
>
> (e) *exemption from currency or exchange restrictions as is accorded to the representatives of foreign governments or temporary official missions; such other privileges and immunities as may be determined by the Governing Council."*

'Strictly within the limits of their official capacities' is undefined. When, say, a South African Task Force member came secretly into Kenya, under whose rules would he or she operate - South Africa's or Kenya's? Task Force operatives would have a degree of freedom 'in connection with their duties' that would cover libel, slander, harassment or, for that matter, execution. No other law enforcement agency could examine the Task Force's files. Its officers would be free of immigration formalities and able to move money at their discretion. And on top of this, the Governing Council could grant them yet further freedoms.

True, if Section 10 constituted *carte blanche,* Section 11 toned down it down slightly for it says [my underlining] –

> *". . The Governing Council shall have the right and the duty to waive the immunity of any official in any case where, <u>in the opinion of the Governing Council</u>, the immunity would impede the course of justice and it can be waived <u>without prejudice to the interests of the Task Force</u>."*

The immunities granted under Section 10 could only be exercised at the Governing Council's discretion. Yet recall how such discretion was abused in Kenya through Chief Game Warden's Permits. As I hope I have shown, the conservation record in Africa has been heavily influenced by people who did *not* abide by law. Slipped on to the statutes by harnessing the prestige of UNEP, ironically part of the United Nations established to safeguard human rights, the Lusaka Agreement is extraordinarily dangerous where those rights are concerned. In the hands of conservationism's fundamentalist fanatics it could unleash unpleasant consequences in the name of law.

* * * *

Simple, commonsense conservation has been perverted into conservationism just as the cornerstone truths of fascism, communism and socialism were hi-jacked and distorted into the negative causes that they became. And as with them, conservationism has all the ingredients for its own destruction. The dictum, thou shalt not hunt, now in vogue in Britain is characteristic. Many in the Third or Underdeveloped world (which is where wild flora and fauna are) grow tired of being lectured by the West (which is where the plants and animals are not) on the need to conserve. The worms will turn and, as is always the case when people react, unnecessary loss will follow.

CHAPTER THIRTY-SIX

THE TYRANNY OF FREEDOM

In my own mind I have reached a point at which I think I understand a little about conservation, conservationism, elephants and ivory. Across the years I was driven to try and place them in the context of life in general. If conservation is to succeed in any degree, it has to fit into the context of African governance where it is among the least urgent of all responsibilities. While the imperial system had many shortcomings and in the light of present philosophies may have been fundamentally wrong, in the Anglophone Africa in which I grew up, it produced good governance. For me, the ivory trail exposed just what a tyranny freedom has been to Africa.

Events in my *alma mater* - the Kenya Game Department - were not exceptions, but part of the main. The shambles it became reflected governance in both Kenya and elsewhere in Africa. As a wag whose identity is lost to history said, independence arrived on the slogan "one man, one vote, once." Quoting from John Reader's *Africa*[179] –

> "More than seventy coups occurred in thirty-two of Africa's independent nations between the assassination of Sylvanus Olympio [first President of Togo] in 1963 and the overthrow of Zaire's President Mobutu in 1997. Of the [30] states represented at the inaugural session of the OAU, only five (Cameroon, Ivory Coast, Senegal, Tanzania and Tunisia) had not suffered a violent change of government by 1997."

Between 1960 and 1998, only three Heads of State were voted out of office (Aden Abdullah Osman Daar of Somalia in 1967, Kenneth Kaunda of Zambia in 1991 and Hastings Banda in 1994) and only three voluntarily relinquished office (Leopold Senghor of Senegal in 1980, Julius Nyerere of Tanzania in 1985 and Botswana's Quett Masire in the 1990s). The acquisition of freedom unleashed tyranny across the continent, one product of which, among others, are over 5.1 million refugees in Africa south of the Sahara.

Kenya settlers opposed independence for Kenya not so much on ideological grounds (most were ideologically barren) as on experience. They felt that the transfer of power would greatly reduce standards of governance and thereby reduce the chances of being able to exist in the manner to which they had become accustomed. Corruption and inefficiency would prevail. At the time, we were condemned as bigoted racists - which we may have been - but it was a *non sequitur* to then argue that our experience was invalid. Events since have proved our fears were well grounded. It is no longer an issue of opinion about what *would* happen, but of cold historical fact that *did* happen. Conservation is one among many fields affected by massive mismanagement, an example of which was published by Kenya's Auditor General[180]. He listed the following queries (which the Government never answered) for 1995/6 over official expenditure –

---

[179] Reader J., 1997. *Africa: A Biography of a Continent*. Hamish Hamilton. London.
[180] The Controller & Auditor General's Report for 1995/6. Kenya Government Printer, Nairobi.

	US$
Unconstitutional spending (i.e. barred by the Constitution)	257,354,400
Goods & services paid for but not delivered	4,597,521
Diversion of public funds to private use	221,986
Over-invoicing	3,909,383
Withheld from the public exchequer	17,767,900
Wasteful expenditure	167,763,400
Unsupported (i.e. unexplained) expenditure	<u>213,544,700</u>
all of which totalled	**<u>665,159,290</u>**

These facts bear on the survival of elephants. Anyone conserving them must do so in the context of this reality. I quote Kenya because the country still publishes documents like the *Auditor General's Annual Report*. There are equally startling cases of inefficiency and corruption elsewhere across the continent, but the data are not so available. All in all, African governance in the closing decades of the twentieth century has not been encouraging and one wonders how freedom unleashed such tyranny.

Once I felt it must be due to African defects. The idea had a fine sense of the circular about it. Africans were not ready for self-government therefore there would be chaos; the chaos which did follow proved that Africans were not ready for self-government. Communism's demise threw a different light on events.

One cannot argue that the countries of eastern Europe were not ready to govern themselves: they have done so for as long as Britain. Yet the sudden collapse of their centralised Marxist regimes produced much in common with post-colonial Africa. In both cases authoritarian governments were suddenly replaced with different systems in which neither communist apparatchiks nor the young, rising Africans had the experience or training to run. In both cases, money was the only practical alternative to sustain or acquire status and shield against the wide insecurity of the changed circumstances. The similarities in post-colonial Africa and post-Communist Russia and Eastern Europe destroy assumption that what we have seen of African governance is something uniquely African. They appeared earlier in the French Revolution and the factor shared between eighteenth-century France, twentieth-century Eastern Europe and Russia and post-colonial Africa was abrupt changes in the manner of government. In due course things should settle as they have through history: if they are allowed to, that is.

* * * *

Will Africa be allowed to settle down? Initially, the Cold War interfered. Now the West participates in Africa's affairs at a level that was unthinkable in 1975. Consequently its politicians have been forced to accept pluralism. The World Bank and the IMF insist upon market-oriented economies and corruption being controlled. African behaviour is now criticised in a manner impossible during the Cold War. Over a quarter of a century ago I wrote in *Ebur* –

> *"If we want the rule of law, as perceived by westerners (and to which Africans in public pay lip-service) to prevail in Africa, then westerners must involve themselves in African affairs."*

I did so thinking that westerners would never involve themselves in African affairs, or certainly not within a century of having so recently pulled out of it. I was (again) wrong. The West's rising involvement in Africa's affairs has been on a scale beyond my wildest imaginings. Yet I am not sure it is for the better. Instead of letting its own system of order evolve out of the chaos brought about by the abrupt change from colonialism to independence, there seems a very real danger that western interference will impede that evolution and prolong the chaos.

As a mirror reflecting greater spheres in sharp focus, conservationism highlights the developed world dictating to Africa. Conservation policy is still laid down by white men. It comes across crudely when the animal loving organizations in the United States tell Kenya that if sport-hunting is reintroduced, it will campaign against American tourists coming to Kenya. Just as crude is United States' support for banning the export of Africa's ivory with the proviso that it should not apply to American citizens' taking tusks as sporting trophies. Closely allied to this is the US Fish & Wildlife Services' insistence that *it* be the judge of whether African countries are managing their elephants effectively or not, before imports of wildlife products from Africa are allowed into the USA. The European Union has a similar philosophy. If pointed out that this contravenes an independent country's freedom to manage its own affairs, the reaction is, 'tough, but that is how things are.' Yet despite the imposed measures (which are still without regard for the consequences of human increase), elephants go on getting fewer and fewer.

Much in conservationism in Africa is repeated on a grander scale in the extraordinary phenomenon of aid from the industrialised nations. This vast business, documented by Graham Hancock[181], is perhaps the most spectacular show in town. By any rules of fair play and probity, it has become a scam of almost unbelievable dimensions: a business that in 1989 was spending over $60,000,000,000 a year. Case after pathetic case can be dismantled and, in the light of its stated objective, shown to be incredibly inefficient, usually misguided and often corrupt. Hancock pointed out that aid does not do what it says it sets out to do. Like the money given for conservation, much, if not most, remunerates the aid managers.

Hancock refers to decades of development turning tenacious survivors into helpless dependants. Conservation aid donors achieved just that with the Zimbabwe Department of National Parks & Wildlife Management. All the inconsistencies of the aid scene are manifest in conservationism and for the same reasons. Instead of catering for refugees there are animals that have to be rescued from poachers and endless experts' meetings to mop up the available funds.

Hancock raged at squandering of $60 billion annually, massive failure to achieve objectives and the principal beneficiaries being the donors themselves. He assumed the purpose of aid was to help poor nations: this hid the obvious. Helping the recipients was *never* the primary purpose. Spending $60 billion, year in and year out, *must* have a positive result as failure on such a scale would be unsustainable.

---

[181] Hancock G., 1989. *The Lords of Poverty*. MacMillan, London.

I think that there is an alternative to Hancock's hypothesis of waste. At the most basic level there is innate human behaviour. Humans are a social species, as committed to hierarchies as they are to the shapes of their faces. Individuals are condemned by their genes to establish their personal niches in social hierarchies. Nations, being agglomerations of individuals, exhibit the same behaviour. Concepts of egalitarianism, fair play and equality are merely lubricants for both personal and international affairs: they may help wheels turn but have no influence on the direction those wheels take. They cannot release us from our behavioural commitment to hierarchies.

Graham Hancock did not perceive the behavioural reality of the picture he painted. The bigger a hierarchy the further apart its two poles. Thus as populations expand and the poles become further apart - the rich will inevitably be richer and the poor poorer. Moralists rant against this, but as I said, morals are lubricants. If you doubt this, think for a moment of those you know as cleverer, richer, not as clever or poorer than yourself, and my point is made. That virtually every term we use to describe our fellows places them, directly or indirectly, on polarised scales relative to ourselves, makes it again. It's in the genes.

In the pattern of dominance and subordinancy innate to behavioural hierarchies, the initiatives do not derive solely from the dominant pole. Hierarchies are formed not solely by the stronger and cleverer animals forcing the less well endowed downwards to a nether pole. For the system to work economically in terms of the community's (species') benefit, there may be an innate drive to establish status, but the ability to accept subordinancy must be as integral to it as assuming dominance. Relations between the developed (dominant) and the undeveloped (subordinate) worlds reflect this and the two poles - dominant and subordinate - contribute equally to the pattern.

Cases, again from the conservation realm make my point. After seven years running a small cropping project for zebra and topi on the shores of Lake Rukwa in Tanzania and thereby becoming the world authority on this particular activity, the country's Game Division asked for an expatriate 'expert' to advise them on how they should proceed. I blew my top and remonstrated with them. What greater experts were there on cropping zebra and topi at Lake Rukwa than themselves? How could an 'expert' who knew nothing about Lake Rukwa or its environs, who had never managed either zebra or topi anywhere, tell people with seven years' practical experience of doing so on site what to do? There had been no argued response: just embarrassed glances, a muttered concession, "You are right", but the request for an expert was not withdrawn.

I had felt it might be a racial phenomenon until white South Africans - *the* world experts in African fauna management - recently went through similar self-abasement. White South Africans know more about managing their indigenous fauna than any people on this planet. They brought both their elephant and white rhinos back from the brink of extinction to the point where they lack room for more. These two cases stand high among the twentieth-century preservation success stories. South Africans need no advice on how to manage such animals and, what is more, they *know* that they need no advice. Pre-eminent in the world, *they* are best placed to tell others what to do.

Once apartheid was banished, their certitude went with it. In no time they, the real experts, were seeking opinion and advice from western conservation bodies. In Third World style, they had their hands out with a professionalism that implied they had been brought up begging and, humiliatingly, were prepared to accept aid with strings.

Tanzanians were the world's authorities on managing topi and zebra.
(Photo Peter Davey)

Specifically they accepted funds with conditions that they never cull, crop or kill wild animals on the land that the money was donated to buy. At issue is not being either pro- or anti-culling, but accepting money - $5 million - with strings attached. The end of apartheid should theoretically have had no bearing on the unchallengeable competence of South Africans to determine conservation policy. Forgetting that they introduced game laws three centuries before North America coined the term national park, they threw away their pre-eminence like an old shirt. Now they grovel before immensely ignorant western conservation lobbies with all the genetic expertise of a dog rolling over on its back when confronted by a dominant. And when I have remonstrated with them for this appalling performance, damn me if I haven't got the same reaction as from the Tanzanians over Rukwa, more than twenty years ago.

I have already mentioned IUCN's assumption that Africans know nothing about the elephants and antelopes amongst whom they dwell. IUCN's arrogance may stand out, but far more inexplicable is Africans allowing IUCN to get away with it. If Africans do not deny being as benighted as made out, or that IUCN's findings are not the sum of human knowledge on elephants or antelopes, then by default they are equally, if not more responsible for their own negative images. If they don't object, then no-one else is going to do it on their behalf.

Where Africa is concerned, it seems that a psychology of dependence rules. The end of the colonial era did not end outside dominance. The continent is awash with aid workers from Europe, American and Japan, spending hundreds of millions of dollars annually. Take those counting and administering refugees, distributing food to them, keeping the camp records and inoculating them: *nothing* that they do is technically complicated or industrialised. Why are *any* Europeans, Americans, Australians or New Zealanders involved? Africans could easily run the whole refugee show themselves. So why don't they do it? Hierarchies and behavioural ranking explains this psychological mindset of subordinancy that is otherwise difficult to understand.

There is of course evidence from the other hierarchical pole of dominance. The United States is the planet's top dog. Its citizens' passion for freedom of information sometimes produces evidence that 'in the national interest' might better be kept under wraps. In 1947 George Kennan, first Director of the United States' State Department's Policy Planning Staff outlined the country's basic policy in a secret memorandum[182], an excerpt of which is worth quoting –

> "... we have about 50% of the world's wealth but only 6% of its population ... Our real task in the coming period is to devise a pattern of relationships which will permit us to maintain this position of disparity without positive detriment to our national security. To do so we will have to dispense with all sentimentality and day dreaming ...
>
> "... We should dispense with the aspiration to 'be liked' or to be regarded as the repository of a high-minded international altruism. We should stop putting ourselves in the position of being our brothers' keeper and refrain from offering moral and ideological advice. We should cease to talk about vague - and for the Far East - unreal objectives such as human rights, the raising of living standards, and democratization. The day is not far off when we are going to have to deal in straight power concepts. The less we are then hampered by idealistic slogans, the better."

That is the truth. The critical words are this *position of disparity* which translate into position in the hierarchy. Tony Duff, when British High Commissioner in Kenya, said as much when he allowed that his primary duty was to protect Britain's investment in Kenya. The issue is not one of morality so much as reality and it is difficult to see how matters could be otherwise. All through recorded history *realpolitik* has been driven by crude reality, with ideals never being more than grease for the movements. 'You gotta do what you gotta do,' as the saying goes.

In this light the $60 billion aid doled out annually that so upset Graham Hancock, actually makes very good sense. Not only does most of it go back to the donors directly or indirectly but what remains with the recipients is intrinsic to the Third World's debt of no less than 39% of its gross domestic product. This amount in 1987-88[183] – $1,270,000,000,000,000 – is an outstanding success in the Kennan wish for top dog staying top dog. The amount owed outlines the 'position of disparity' that he recommended and international aid serves its purpose. The behavioural phenomena of dominance and subordinancy go a long way towards explaining the double standards at CITES and international relations generally.

Coming from a colonial background I am in a strange quandary of seeing the new colonialism rampant and not approving of it in either principle or practice. In a very real sense the trap of debt into which Africa has cast itself represents a tyranny that was never apparent when today's lenders said, "Here ... take your freedom." I am not alone among African-born whites in finding myself between the two stools of my Caucasian heritage and African environment. Time after time I am disappointed or shamed when Africans accept help or ask for it where I would have refused it and opted for independence. This is

---

[182] Bonner R., 1987. *Waltzing With A Dictator*. Times Books. New York
[183] Hancock. *Ibid.*

nonetheless the way things are. And if one wishes to influence the destiny of any animals or plants, then this is the ballpark in which play takes place: as much in conservation matters as in anything else.

* * * *

Then there are the media. From personal experience of the press, I am biased against it. London's the *Mail on Sunday* and *Sunday Times* both tried to discredit me, knowing that that I did not have the means to retaliate. When Dorothy Ravenscroft refused me any right of reply I wished her all the evil in the world. Reuters' trying to stifle Robert Powell's exposure of Operation Lock, though the matter was petty, spoiled my belief in press impartiality. Notwithstanding that I have several good journalist friends, my experience destroyed such respect as I may once have had for the media.

Curiously, disrespect for the media is widely shared wherever I have been - Africa, Europe, Asia and Australasia and for the same reasons. Inaccurate reporting, the claim of a 'right to know' and imputing that individuals have a duty to divulge information to journalists and, soaring over all, journalistic disrespect for privacy. Underpinning these reasons is the sense that the press has an unearned and resented power over both the public and the individual. At issue is uncertainty over whether the media created the present western mind-set, or are they its inevitable product?

Freedom of the press evolved as a concept when journalists were held in high repute and editors had a sense of when *not* to publish. They restrained themselves from printing when it would cause distress without public benefit. This restraint is now rare and no distinction is made between titillating curiosity - which is insatiable - and 'the right of the public to know'. The right to know, unbalanced by other considerations, causes needless pain. The only justification for invading personal privacy is if a greater public good is served - and satiating curiosity is not such a good.

Television and the papers they fuel are overwhelmingly commercial and for profit they publicise what sells. Sensation sells, personalities sell, controversy sells and embarrassment sells. Selling the normal condition is not easy, which is why crises are magnified and contrived. Like the international aid industry, the media thrive off the same diet of crisis and sensation which, *inter alia*, perpetuates the image of benighted Africa. The way in which they both fuel and benefit from conservationist fundamentalism is a case in point.

The late Mohammed Amin, himself a cameraman and photo-journalist, described a vignette that revolted him beyond description at the time. Mo' had just exposed the great Ethiopian famine of Mengistu's years. Scenting a bonanza, aid and media people poured into the country. They wanted the most gruesome scenes, the better to support their stories. A small enclosure surrounded by hessian sacking had been set up in the midst of a refugee camp: an indescribably sad situation. In it aid workers and press representatives could put their belongings and take a break. And it was in there within feet of people dying of starvation that Mo' listened to colleagues complaining that the only choice of sandwich had been smoked salmon.

That they could cold-bloodedly eat - or even think of eating - within feet of people dying of starvation and complain about the food's lack of variety, exposed a reality totally at odds with the sense of mission they presented to the world. Squeezing a tear in front of the cameras, these journalists could have taken conducted tours of Auschwitz operating at its peak, looking for the best angles, the appealing child about to die, without turning a hair. And as I write this their cameramen compete in Iraq for the most ghoulish sights, for all the world like vultures feasting on carrion, making the same point. Quite literally, journalists feast in the fullest hedonistic sense of that word on humanity's worst aspects – and profit from it. Mo' felt that the vignette revealed something about his profession which he preferred not to think about. As hardened as anyone else in the business, that scene shocked him to the core of his being. The film clip that he should have taken, but never did, was to have panned from a white journalist's face, cheeks bulging with salmon sandwich, to a dying Ethiopian, prostrate in the hot sun nearby, flies surging around the eyes so soon to close.

Having to sell news and profit from it establishes a fundamental conflict with reporting impartially. That such a conflict of principle has existed all along was recognised, but leaving it to be resolved by men of principle - which worked when principles counted - is no longer effective. Too many journalists are now unprincipled. Editors did present views which conflicted with their own, but that is now rare. Rupert Murdoch blew it out of the water recently when he attempted to stop one of his companies from publishing a book by the last British Governor of Hong Kong - Chris Patten - because Patten's ideas conflicted with Murdoch's financial interests in China. Therein lies what is possibly the nub of the press issue: personal influence. Murdoch owns so large a proportion of the world's media that he may quite literally be the most influential individual on earth. That he can pull strings behind the scenes to stifle ideas or information which conflict with what he wishes people to know doesn't sit well with concepts of democracy.

Journalists have far more influence over our lives than the average politician. They demand freedom of the press (which is also freedom for themselves) claiming to exercise some sacred trust on the public's behalf. They say one does not have to read what they write or watch what they film, knowing we cannot escape them. Unlike a computer, one cannot purge the human mind of knowledge. What passes into our brains through any one of our five senses cannot be expunged. We may suppress it in our conscious minds, but there is little we can do about its lodgement in the subconscious - which holds the greater part of the memory bank. Thus, I may not want to buy a paper or read what has been written, but my eye crosses a headline many paces distant and, quite outside my control, it lodges in my head. This power to insert material into other minds is awesome. As I found out painfully, there are situations in which the individual can gain no redress for wrongs inflicted by journalists.

The media at large are neither subject to public selection nor special qualification. Given their power over people and society, this is at odds with the principles of democracy which sets the rules of the current games we play. If we have choice over politicians who govern us, there is surely an equally strong - if not stronger - case for a selective and elective process for controlling the media which influences us so very much more profoundly?

Freedom of the press evolved around a core of common sense, balance and continence. With time these controls have dissolved. Driven ever harder in the quest for profit, press freedom has escaped society's control. We have not quite yet reached dictatorship by the Fourth Estate, but its power is so great and exercised in so unprincipled a manner that society would be sensible to control it through an elective harness. Until this happens, the press will go on exercising as virulent a tyranny over freedom as there has ever been.

CHAPTER THIRTY-SEVEN

FAREWELL THE PLEISTOCENE

Several erudite friends have read this text and among their comments, one cropped up repeatedly: having criticised conservation so broadly, surely I must have something to put in place of what I so disparage? The truth is that, in the holistic sense it is thought of today, I do not believe conservation can or will work. Perhaps it would be more accurate to say that the African megafauna is beyond conserving.

* * * *

In 1925 one could walk from Lake Turkana in northern Kenya to the Rovuma River on Tanganyika's southern border and, with relatively little deviation, be in elephant country all the way. In 1950 I might have done so, but the route would have involved large deviations from a straight line. By 1975, it was no longer possible because the elephant ranges were islands separated by large tracts in which the species no longer existed. The same general pattern is apparent on an east-west axis from say Mombasa to Rwanda or from Dar-es-Salaam to Lake Tanganyika. It applies to black rhino and zebra too, as well as many other large species. Where, a scant hundred years ago, they had general distributions that extended from the Sahel nearly to the Cape or from the Atlantic to the Indian Ocean, now they only occur in shrinking, discontinuous islands, separated by broad swathes of Africa devoid of big animals.

The trend of what was happening was perfectly apparent from the dawn of the colonial era. The perceived remedy was to create 'sanctuaries' – which was done. Today there is a common perception that these were not enough. This is yet another case in which conservationist thinking arises against the available evidence. Ignored is that in terms of relative governance, more was being done to conserve Africa's large fauna before 1900 than is the case after 2000. Before 1900, some 23% of what is now Kenya had been declared game reserve: a status in those days denoting absolute sanctuary. Today, we use the term national park for the same notion, but national parks are less than 5% of Kenya. If one is honest and removed the 20,000 square kilometres of the Tsavo Parks from this total as it has been already extensively used by cattle, absolute sanctuaries only comprise 1.4% of the country. Proud claims of increased park areas in recent times sound like progress towards progressively more land being set aside until, that is, one appreciates what happened to the far larger game reserves. Sleight of tongue and semantic use of 'national park' hide earlier words connoting the same thing. Through a century and more of supposedly rising 'conservation consciousness' sanctuaries together with both animal distributions and numbers have been progressively becoming less and continue to do so.

Why, is readily explained. No species can increase numerically except at the cost of other species sharing the same environment. People increase through expanding the extent of their 'slave' domesticated species. The sum of increasing people and their domesticated plants and animals has to be at the cost of life that is 'wild'. It cannot be otherwise and what has happened and is happening is as simple as that. As human increase is predicted to continue, so wildlife will decline in symmetry.

* * * *

Island biogeographic theory is compelling. Let me simplify it here. It is to do with space and scale: the size of an island determines (a) the number of species it will hold, (b) the number of individuals it can contain and (c) their chances of survival over time. The larger the island, the more species and the more individuals it can hold and the longer the species' chances of survival will be. The concept arose from observing life on oceanic islands, but is equally applicable to islands howsoever they are created. Africa's national parks are all becoming islands to which the theory will apply. If it is correct, their existence as islands must result in an inevitable and irreversible trend towards fewer species within them, each with a rising vulnerability to becoming extinct. They can never be true samples of the greater systems of which they were part.

When one considers that the 'ecosystem' within which many of Africa's large mammals occurred only a century ago exceeded 15,000,000 square kilometres, even the largest modern sanctuary is less than one per cent of what so recently was. Having reduced continental Africa's fauna and flora to fragmented and shrinking park islands, if we now leave them alone to let nature take its course, its immutable laws will redesign these faunas and floras to suit the space available. Possibly management by humans might give direction to this trend, but that is all.

Elephants make the point clearly. No elephant population yet studied has been stable or 'in balance' with its environment. Every one has been either increasing or decreasing at differing rates. As there is no evidence of natural population stasis from all the research undertaken on them, it is reasonable to assume that population instability is intrinsic to elephants – at least at population levels. Chronically unstable, they were one of the late Pleistocene's great evolutionary engines.

A century ago, Africa held thousands of elephant populations, each a cell in a continental mosaic, and all at different points in interminable cycles of increase and decrease. Hence my transect from Lake Turkana to the Rovuma being within the mosaic all the way. If, from time to time, one cell declined to zero, soon enough the area it had occupied would have been reinfected from other cells. This chronic instability could only have evolved where a hugeness of scale was available to cater for it. The continental mosaic of population cells, the interminable, irregular cycles, and overlapping elephant ranges, were as much part of 'elephanity' as the individual's size or longevity or trunk or tusks. It was only vast scale that gave the species some semblance of stability: taking Africa as a whole, the pluses may have roughly matched the minuses. Without that scale the species' instability is what stands out, and the smaller the scale, the greater that instability. Elephants need continents, not islands – even if they are called national parks. Academic geographers in their recognition of theoretical land scape scales have come closest to the scales that apply to elephants.

As park managers the length and breadth of Africa have been learning, elephants introduce chaos because no park is big enough to hold more than a population cell or two – at most several. Confined within park islands, woodlands vanish and with them go their dependent animals. If the elephants then suffer a population crash – as happened in

Elephants and Tsavo spectacularly illustrated island biogeographic theory in the 1960s and 1970s.

Kenya's Tsavo – woodland may return. But it will be deficient when there are no longer any extraneous sources of plants and animals from which it can be recolonised. Through elephants, diminishment of life's variety within parks shows how the laws of island biogeography manifest themselves with surprising speed.

Where a scant three decades ago conservationists argued about whether to hold elephants at predetermined levels or let nature take its course, neither side considered island biogeographic theory or the prospect that something whose ecology was continental simply could not exist in islands. The problem would be complicated enough if what elephants did was confined to themselves, but their chronic instability passes through to the plants and animals around them. We have seen how they have converted forest into grassland and, through their absence, grassland into forest. Only humans have a comparable ability to inflict change on all other life about them. The presence of elephants enormously complicates the management of island parks. Indeed, while not widely talked about, a growing body believes that elephants should be excluded from many of them for precisely the same reasons that humans are kept out.

What we face is that just as a village would be unrepresentative of humanity, so one or two elephant population cells will be equally unrepresentative of elephanity. Indeed, one or two cells in a park is, in principle, just as artificial as having one or two elephants in a zoo enclosure. It is hardly ground for optimism.

* * * *

Strictly speaking, the Pleistocene is a geologic age that stretched from about ten thousand to 1·6 million years before the present. Being characterized by abundant

In terms of elephant ecology, a single elephant population in a national park is as incomplete a phenomenon as a single elephant in a zoo enclosure. (Photo Peter Davey)

mammals in great variety, some gigantic, gave rise to the terms Pleistocene megafauna and Pleistocene assemblages.

Pleistocene megafaunas once existed on all continents and on many of the world's larger islands. In the last quarter of the geologic Pleistocene, they started to decline. This first happened in temperate Eurasia and spread down the Indonesian archipelego and into Australasia. Next it happened in North and then South America. In North America it went in two distinct waves. First, and most extensively, it started when people arrived from Asia across the Bering Straits. A second phase followed humans from Europe penetrating the continental interior. The idea that human arrival caused the trends is supported by the events on islands such as Madagascar and New Zealand. It is, of course possible, that what caused the animals to disappear may also have caused humans to arrive, both being responses to some as yet undescribed effect: we tend to forget correlations are not necessarily causal.

Africa, where humans evolved, was the exception: a Pleistocene assemblage lasted in place into the twentieth century. On the face of it, this is the strongest case against the idea that man caused the Pleistocene declines. If people have been in Africa longer than elsewhere, why did the Pleistocene megafauna not disappear there first?

Consider this possibility. If people evolved within and as part of Africa's Pleistocene assemblage, they will not only have adapted to the life about them, but that life will equally have adapted to them. Edenesque harmony is unlikely to have featured and the human/nature relationship would have been rough. Not only did human mindsets evolve to fit the circumstances, but plasmodia that cause malaria, trypanosomes that produce sleeping sickness, together with HIV and Ebola-like viruses were all equally integral to it. Humans were part of the African Pleistocene assemblage of life in which its forms were adapted to one another, and like rabbits in their original habitats, we were no obvious threat to the general ecology.

Tourists may be as malign as a multitude of poachers in their multifarious influences upon animals and habitats.
(Photo Peter Davey).

Emigrating, humans went among animals and plants that were not used to them. Just as rabbits up-ended Australian ecologies, the arrival of humans radically changed flora and fauna in environments lacking Africa's resilience and checks upon their influence. How the declines and disappearances actually came about is not at all clear. Popularly, they are attributed to over-hunting by people. With what we know about the economics of hunting, of the theoretical reproductive capacities of most large mammals, and of the apparently low human densities, this is unlikely. There were seldom enough people to inflict continental disappearances. Human-caused negative influences are likely to have been both more subtle and multifarious.

Once away from Africa, the emigrant humans' descendants evolved in other ways to the people who had remained behind. By the time they returned to Africa they were as alien to its Pleistocence assemblages as rabbits were to Australia. Confronted by these humans who had evolved *ex situ*, the megafauna has collapsed faster than elsewhere. Against them, Africa's natural controls like malaria no longer served their purposes. One doesn't need to postulate technologies, economies or anything else: the presence of people unadapted to nature in Africa and *vice versa*, to which nature in Africa was no longer adapted, was enough.

A latter-day observation shows how unexpected the influences can be. Somehow conservationists have evolved an idea that they can enter Africa's ecosystems without ecological consquence. That because they 'love' wildlife (their term) they can go among the pristine animals as tourists, without leaving footprints, as it were. Yet they are wrong; for the footprints they leave are ecologically deeper than those any dinosaur left. Take a bejewelled American Belle Dame in her frail dotage, visiting Kenya's Mara Reserve. Forgetting the internal combustion engines, the burned hydrocarbons, the immense expenditure of energy and chain of people that placed her there; in her wrinkled self, she seems ecologically harmless. Yet animals watch her in her four-wheeled drive chariot, among them some that have learned that where such as she stop and look, there is something for them to look at too. Thus, for example, lion and hyaena watch tourists, scientists and photographers and are led to cheetahs, which they kill, particularly if immature.

Seemingly harmless aliens in minibuses induce change. So do scientists. Their careful, loving, research into the Serengeti's wild dogs ended in the death of every one they handled. Why? Well they don't really know - but they *loved* them, dammit, and they had degrees. The point: nothing - animate or inanimate - can enter an ecosystem without influence. It may seem miniscule, like the movement of a butterfly's wing over Tokyo that seeded the subsequent Oklahoma tornado, but its minute influence is nonetheless irredeemable The passage of a million tourists through Kenya's national parks may seem harmless, but it is a profound influence whose complexities are beyond us to describe. By such reasoning do I say that some single, simple influence, like hunting, was unlikely to be the sole cause of Pleistocene declines.

When the emigrants who had evolved through millennia ex-Africa re-entered the continent, they found the humans that were still part of the continent's Pleistocene assemblages different to themselves. This finding was often expressed in pejorative terms – 'primitive', 'backward', 'uncivilized', 'conservative', 'benighted' and 'poor' being among them. That such characteristics might be adaptations to the Africa of their evolution, has not been considered. Living in harmony with nature - a term connoting loving partnership - is no more than a drawing room hope. The reality of nature 'red in tooth and claw' encompasses predation, competition, parasitism and many other features that are distinctly uncomfortable. If such qualities be the price of Pleistocene assemblages, then who wants them? Certainly no developed nation has recreated or reconstituted their lost faunas - even in such small parts as they may be able to. The dimensions of the chasm between that bejewelled Belle Dame in an airconditioned Toyota Land Cruiser, peering through tinted lenses at the denizens of an African national park, and the cultivator contemplating his elephant-ravaged crops, is immeasurable. Suffice it that it reflects post-Pleistocene and Pleistocene reality

As part of the Pleistocene collapse and using the technological magic that the ex-Africans brought with them, the continent's people broke their bonds to the megafauna and underwent an unprecedented increase in numbers. Naturally it was counterbalanced by corresponding wild animal declines. Unspeakable thought, but in a nutshell, the disappearance of the great game benefited many people. Kids can go to school with no fear of coming across large dangerous animals on the way. It could be argued that not until the white man penetrated Africa's hinterland, were Africans freed from the thrall in which wildlife hitherto held them.

Pleistocene Africa survived for as long as its megafauna existed *despite* humans. This condition where big game still dominated landscapes, I saw across East Africa and far beyond with my own eyes. It ended that instant the animals only existed *because* of humans. Now, courtesy of *Homo sapiens* as is the case elsewhere, the animals are where we ordain they may be, as in national parks. In extreme irony such parks, made to preserve the great faunas, cannot be more than cages: nice cages maybe, but cages. Equally true, such parks are the Pleistocene's gravestones.

By chance, my life and those of the great documentors like the academic Jonathan Kingdon and the film-maker Alan Root, span the going of Africa's last Pleistocene assemblages. Elephants and the rest of the megafauna as we knew them had a grandeur that we longed to preserve but, without the space, it is impossible. Now only Antarctica remains. Species may survive in small refugia (the national parks) that, at best, will be living museums and no longer part of the African environment in any holistic sense. This is what we are now witnessing. So when I am asked if I have nothing to put in place of the edifices I have knocked down, the answer is no, not really. My endeavours to understand the ivory trade may have some historical interest, but little conservation value. CITES, WWF, IUCN and the bunny-huggers at large will play their games and believing what they wish to believe: after all, what they say three times must be true. If it gives them fun, why shouldn't they? It will not stop what is happening.

Sad? Indubitably; but with the going of the game, Africa is at long last joining the rest of the world in the Holocene; or should it, more aptly, be the Homocene?

Abakuna Gumundi. (Photo Tony Archer)

EPILOGUE

There were five of them, all with AK47 rifles. Their poor Kiswahili betrayed their origins in Somalia and all had bad faces. They had gone to old Kenga the Giriama's shop near Kasikini on the white sand under the tall, small-leafed *Brachystegia* trees. Shop was perhaps too proud a term for the rude, rough-thatched hut where he sold sugar, salt, cigarettes and such items – when the old man was not out of stock, that was. Their leader had asked for a box of matches. Kenga produced one and told him the price: three shillings.

Their leader had seemed surprised. "You want payment?" he asked.

"Yes," Kenga replied, whereon the Somali shot him three times in the chest.

"For an infidel, an unbeliever, that was payment enough. In fact it may have been too much, for bullets are not cheap. Yet we believers are not ungenerous."

His companions had laughed and they returned to the bush from whence they had emerged, leaving news of Kenga's death to spread by bush telegraph through the thinly scattered local community of mainly Wata people. It was the most recent in a long string of abuses and mindless atrocities inflicted by this particular gang. Like so many others, it had come into the area looking for elephants, but as its members were inept hunters, they had not prospered. As the elephants had become fewer and more difficult to find, they preyed with rising frequency on the local people. Being far from any administrative centre and on the fringes of the great Nyika wilderness, news of this did not reach the authorities. While the officials were always calling on the people to report the presence of such Somali 'poachers', the Wata preferred to keep officialdom at as great a distance as possible and retain their independence. Yet clearly something had to be done. The gang was small: surely they could deal with it?

Their conservative souls rebelling against action out of the ordinary, the possibility was discussed endlessly; each time increasing the certainty that eventually the Wata hunters would take their great bows and dreaded poisoned arrows and kill the aliens. If they were resolute, a dozen of them could take them as they slept. Two bowmen to each recumbent Somali were good odds and provided that they could get into position – each couple within six feet of its selected victim before the first arrow was loosed, they should kill all of them.

Perhaps they should first learn more about the Somalis? Like most of these gangsters, they might be grateful for a little assistance and welcome a local person as porter and guide? A few weeks with them should reveal its members' habits, their weaknesses and the best time to attack them? Discussion had reached this point when Kenga – old, inoffensive, well-liked Kenga – was murdered.

Abakuna was son of Gumundi of the Wata Gamado clan. His father had been Gumundi son of Abakuna who in turn was Abakuna son of Gumundi who was son of Abakuna who was son of Gumundi in a chain of elder sons that stretched beyond the horizons of Wata memory. It reached forward, too, through his son Gumundi and his grandson Abakuna. As a successful hunter, versed in healing and interpreting omens, Abakuna was looked up to by the people around Kasikini and had been foremost among those advocating direct action against the Somali gang. As tracker, gun bearer and great friend of a white professional hunter, he had travelled – as far as India even – and his view of the world was wider than his fellows'. He saw that the community must destroy the gang quickly.

Kenga had been a friend since boyhood and his senseless death triggered a response in Abakuna that, even as he experienced it, made him uneasy. He sent a young man to tell the police in distant Malindi: "Come – there is a Somali gang at Kasikini."

* * * *

The police came. An inspector, a corporal and five constables: they arrived reluctantly in a battered Land Rover. They did so because the President of Kenya had been outraged by the security forces' lack of success against shifta gangs poaching, raiding and ambushing with apparent impunity across eastern and northern Kenya. In his ire and with a politician's lack of precision he had said poachers must be shot wherever they were found. He meant, and everyone who heard him knew that he meant, that it was the Somali shifta, the bandits who themselves shot first, who he demanded be dealt with. That he had used the word 'poachers' for them was accepted in the same sense that they used it for themselves: everyone knew to whom he referred.

The policemen from Malindi were educated men, each with certificates to prove the levels of their learning. They appreciated good clothes, liked pointed, highly polished shoes and were all accomplished disco dancers. Cities and towns were the environments to which they were accustomed and to the bottom of their educated hearts, they loathed the bush. As far as they were concerned, the wilderness should be left to wild animals and poachers; the Somalis were welcome to it. Most of all, they dreaded meeting these bandits of whom everyone spoke. Fighting was simply not in their make up: the limit of their aggression was setting roadblocks and demanding bribes from luckless motorists. To be shot by wild men from the north was a dreadful prospect. Even worse, as one lay wounded, was the possibility of having one's testicles cut off and one's throat slit like a goat on some slaughter house floor.

The policemen came with a single intention: to get back to Malindi as fast as they could. Only the threat of transfer to Ileret in the desert north, the post furthest from a disco in all Kenya, persuaded them to get to Kasikini. Their plan was simple. They would go to the nearest waterhole half an hour's walk from Kenga's shop, make themselves as comfortable as possible in the horrible circumstances, wait several days, then return to town reporting an arduous but fruitless patrol.

Young Abakuna, grandson of the great elephant hunter who had summoned the police was unaware of this. He and two others had been away hunting for the past week. They had gone for meat for their families. A small black snake had crossed their path from right to left not long after leaving Kasikini. Being young and carefree, and the snake being quite small, they had paid it no heed. As now ordained, they were unlucky and failed to kill anything. On the way home they came across a half-grown *golja* – a wart-hog – which Wata laughingly referred to as their 'sheep' because they were so staple a source of meat. They watched as the hog came to the entrance of its burrow, stop, look around and then carefully reverse into it. The three young men grinned for now they would not go home empty-handed.

Creeping up quietly, young Abakuna drew his knife and crouched on the burrow roof just at the entrance. Ascertaining the burrow's direction, he gestured to his companions – Bonaya and Guyo – who positioned themselves above it and started a rhythmic stamping.

Above the thump thump of their feet came a muffled scrambling from underground and as the *golja* shot out, Abakuna stabbed it expertly through the neck.

Each young man carried a portion of wart-hog as they approached the waterhole where the police lay idle and bored. They were not silent; indeed they made a point of speaking loudly so that nobody, particularly aggressive Somalis, came upon them unawares and open fire on them. As part of their presumption they prayed that any Somalis would also wish to avoid a confrontation. Hearing the policemen the three young hunters cached their meat and approached cautiously. From cover they watched and wondered. Young Abakuna then whispered to his friends that he would ask the police what they wanted. The other two tried to dissuade him, but he was determined. He emerged from cover while his companions remained hidden but watching.

"For whom do you wait?" young Abakuna had called out, and the police were on their feet, weapons in hand, in an instant. Seeing him alone and unarmed, they bade him come closer.

"We wait for poachers," the inspector had replied as he and his men clustered round Abakuna. Hostile and suspicious, with some logic they had wondered what he had been doing in the wilderness.

"Maybe you are a poacher," a constable had said half in jest before another said, "He is. Indeed he is."

Abakuna had been careless, for there was a small stain of wart-hog blood on his loin-cloth. His accuser reached forward and touched the stain. "It is damp. This is fresh."

Bonaya and Guyo watched in horror as the policemen's interest changed from what they had clearly anticipated would be diversionary hassling. Young Abakuna was cursing himself but clearly was not too worried as he tried to brazen it out. "It is sheep's blood. I come from slaughtering a sheep." Presumably this was said for his hidden friends' entertainment.

Yet things went wrong. One of the policemen could track, and back-tracked Abakuna's trail. Bonaya and Guyo retreated keeping distance between them and the policemen. The police posse followed their tracker and after painfully slow progress came upon one of the hidden wart-hog haunches, together with a bow and a quiver of arrows.

"What is this?" asked the inspector.

"It is meat for my women and children," his prisoner replied with the simple directness that so characterised the Wata.

"Then you are a poacher," said the inspector, "and our President said that all poachers are to be shot."

"Indeed he did," young Abakuna replied, "but he referred to the Somalis, the men with guns who go to the national parks to shoot elephants and tourists."

Nothing more was said then, but the other policemen drew back. The inspector cocked his sub-machine gun and as the young man looked him in the eye, squeezed the trigger sending a single long burst through Abakuna's middle.

Bonaya and Guyo took to their heels. The broken body was left where it lay beside a piece of wart-hog and a broken bow. The police went home reporting that they had killed a dangerous 'poacher'.

Next day this good news was announced over the radio. And righteous white conservationists praised the police: "It's the only way with these bloody African poachers," they muttered, "have to get tough with them y'know."

* * * *

Cobras, unlike mambas, are creatures more of night than bright day. Now and again the hunter had seen one sunning itself on a termitarium whose funnels into the earth's bowels were close sanctuary. Bold but nervous, these snakes nevertheless preferred avoidance to confrontation and moved out of one's way. Though he held them in due respect, he did not fear them for what they were. It was as omens that they made him afraid. Abakuna's sick old heart lurched and he all but dropped his bow as a huge white-lipped cobra slithered slowly, almost languorously, across the path from right to left ahead of him. Tongue flicking, nearly black with a metallic sheen in the morning sun, it vanished into the undergrowth. That malign forces were abroad had never been less equivocal. Go home, do not hunt, and do whatever was necessary to placate them.

Even as he absorbed the message, the naturalist in him wondered at the manner of its telling. White-lipped cobras were not uncommon closer to the coast where it was wet and humid, but rare on the top of Dakabuko hill above Kasikini. More usual, were excitable, beautiful, copper-red spitting cobras and insolent, heavier, brown Egyptian cobras. Harbingers of bad news though they were, neither had ever conveyed so malignant a feeling as this snake from further east: its very blackness made the point.

As the initial shock wore off, the elder Abakuna did not retire as tradition prescribed. While not crossing the snake's trail, which would have been foolhardy, he turned aside and clambered up a huge rock that broke through the greenery near Dakabuko's highest point. Squatting on its flat summit, bow beside him, he looked out across the great Nyika plain that rolled over the horizons to south, west, and north. Westward, far away as distant lands beyond water, lay the faint blue hills of Ukambani – the land of the Kamba. Southwest, even fainter, rose the Taita mountains and beyond them, yet more distant still, Kilimanjaro's improbable cap. Here and there thrusting out of the arid Nyika sea were the small islets: Kulalu, Sala, Lali, Dakacha, Hoshingo whose sedimentary rocks held fossils that made white oil men's eyes glitter. The low outline of one of the more distant – Dakadima (the red hill) – looked for all the world as an ancient, abandoned aircraft carrier whose rusty red bows were tilting upwards as its stern settled beneath the sea of flatness. As nowhere else except, perhaps the great hinterlands of Australia, all was pervaded by senses of immense age, loneliness and, inexplicably, great sadness. For the small brown man on the rock it was as though his soul and the spirit of the land merged and he entered both the past and the future.

His people, the elephant people, unlike others never came from anywhere else. These hazy grey-blue lands of the Nyika below, were where they had always been – as had the *arba* (elephants) to which they were mystically joined, even more than Masai being one with their cattle. If there were no elephants then there would be no Wata, for one could not be without the other. His life was as an overlay across the vista.

Down there as a small boy, near where dark trees marked the Sabaki river, he had trapped brilliant blue rollers in the April rains. Flying northward in waves, these birds progressed from prominent perch to prominent perch. With his age mates he had caught and pinned locusts, kicking to get free, in a patch of bare ground beneath a well used perch. The struggling insect was irresistible to every roller arriving at the perch. Swooping down on the bait, unaware that the strategically placed twigs around it were smeared with a sticky bird lime made from resin and euphorbia latex, they were trapped. On a good day he and his mates might take twenty rollers, each heavy with fat to fuel its journey northward and, roasted over embers, was so delicious, that memory alone made him salivate.

Near a clump of *hadama* euphorbias he had arrowed his first elephant and, aged thirteen, became a man. Nearer the base of the hill he had hunted with Boru 'Magonzi' (the sheep), so named because he had once accepted a sheep for a tusk from old Ndibo Molu, the Kamba ivory buyer. He and Magonzi had been thoroughly chased by an angry *dadnaba* (a matriarch). Elephants have poor sight and they are all but blind to movement in front and along their flanks. Tucked in directly behind their wrinkled sterns one was invisible and, providing the wind was right and one was silent, it was possible to go up and touch them – though he knew of none who had actually done so. Yet in a fifteen degree arc either side of the blind area, elephant ability to pick up movement was surprisingly good and had cost many a hunter his life. On that day long ago he and Magonzi had been young and careless and the old she-elephant spotted them as they closed in. Wheeling and charging in a single movement, it had been a near-run thing when she abruptly broke off and left them breathless and all but helpless from laughter. With Magonzi one was always laughing.

Motionless and watching, memory piled upon memory. Things had started to go wrong when white men came, claiming all ivory and forbidding anyone to hunt elephants or any wild animals without their leave. The coastal Giriama had been so incensed by such infringement of their traditions that they took up arms against the government. Soldiers with rifles and machine guns put down the rebels armed with bows and arrows. As punishment, the Giriama had to pay a communal fine, each village contributing. When the Giriama said that they had no money, which they hadn't, the District Officer *Chambe* (Champion) said that they might, instead, pay in ivory. That was ironical because in no small part the war had arisen over the white men's demand that the Giriama stop trading tusks. Now they wanted them to pay in tusks. As observers the Wata were as confused as the Giriama, though some turned the situation to their advantage. The Giriama near Kisiki-cha-Mzungu got Boru Magonzi's uncle Kiribai to give them two tusks from a *tofa* to pay their fine. They had paid him in maize and other foods.

Later, unaware that the Wata were the elephant people, white men said they were killing so many elephants that there would be none left, which was ridiculous to men whose ancestors had hunted them since before memories began. Most men Abakuna knew had been sent to jail. Yet less than ten years after they said there were too few elephants because the Wata had hunted them, the same white men were saying that there were too many and that they must be made fewer. They killed them as elephants had never been killed. *Dadnaba, dudurucha, dubaro, ilbaad, korma arbai* – all elephants, it made no difference, they were driven by helicopters and shot in piles. Yet we, the Wata helped them do this, going even to Uganda for the money. Elephants should never have been treated thus, for if they are part of us, then we were part of them. There was no respect, no hymn sung at an elephant's death as it should be.

Then black men became the government and their greed for ivory was without measure. They treated us worse than the white men – unless we hunted for them. And then the Somalis came. Wild and speaking little Swahili. They shot anyone in their way – like Kenga, whom they killed for three shillings.

Misery bore down like a great physical weight on the small brown man, greatest of elephant hunters, immobile, high on Dakabuko. Because he had called the police, young Abakuna died ten days ago and with him went all prospect of old Abakuna's or his ancestors' rejuvenation or immortality. The chain connecting his ancestors' spirits to both present and future had broken. Henceforth their spirits and those of the elephants would only be heard as the wind moaned through the blue-grey Nyika adding to its ineffable loneliness. They would cry and sigh until the winds stopped blowing. From the snake he knew that this was the end. He descended and returned to Kasikini.

* * * *

They came quietly to Abakuna's simple home on the white sand beneath the tall, small-leafed *Brachystegia* at Kasikini and called him. Rising from where he had been sitting he faced them. They bade him sit again before them. They called snowy-haired Hagaiya, wife since his teens, and bade her sit too.

Pitiless, thin-lipped and hawk-faced, their leader drew a rifle round, burnished and brassy from his ragged pocket and asked in thick, awkward Kiswahili, if Abakuna knew what it was.

"It is a *risasi* (bullet)."

"Do you know where I got it?"

"No."

There was a long pause before the Somali said, "From a friend. From a friend in the police. From a friend in the police at Malindi. From a friend in the police at Malindi who came here to Kasikini last month and shot a Somali poacher – so the radio said." Dark, glittering eyes watched Abakuna's face for effect.

Abakuna had returned the look calmly, though he understood the Somali's inference: someone in the police had told the gang who had betrayed it. There was a pause before the man went on.

"We are your friends." It was like a puff-adder's hiss and Abakuna marvelled that such kind words could convey such evil. "You have a chest problem," and he had thumped his own to make the point. "This is medicine which you must try," and the Somali handed across a small dark bottle that contained a liquid. "Drink it," he commanded and standing back cocked his weapon with a loud noise – kra-kakata.

Abakuna knew he had to drink whatever was in the bottle or depart this life from a body shredded and disfigured by a hail of bullets, further disgracing ancestral spirits already embittered at being cast adrift. So he unscrewed the cap and put the bottle to his lips. Hagaiya flew at him, trying to wrench it away, but one of the gang stepped forward and brought his gun butt down on her head and she fell senseless.

Abakuna had touched her face gently and said, more to himself than his listeners, "She sleeps ... which is as well."

In one movement he swallowed the bottle's contents. For a moment nothing happened. Then it was as if he had been struck a great blow. He had gasped, his eyes started and rolled upward as he fell over onto his side. Two spasms wracked his body and Abakuna, son of Gumundi, the greatest of the Wata elephant men and the last of them became a voice in the Nyika winds. High overhead in the tall, small-leafed *Brachystegia* tree, an oriole sang his requiem in pure, liquid notes.

* * * *

Abakuna's murder is true and a metaphor for the Wata story. Theirs had been an ancient culture based on sustainable elephant use. Living in symbiosis they had what conservation aspires to a millennium or more before Livingstone set foot on Africa. Yet fellow humans could not let them or the elephants alone. No matter where they came from, whether as driven Somalis behind the shield of their faith, or righteous whites equally imbued by conservationism, the result was the same. In a nutshell we – all of us – bloody well blew it. If Abakuna's tale was a metaphor for the Wata, perhaps the Wata story is a metaphor too.

A metaphor.

INDEX
(People, tribes and companies only)

Abakuna Gumundi, 400-407
Abd Wak, 154, ,155, 157
Abdi Ibrahim, 154
Abdurrahman Mursaal, 156
Achieng A., 133, 171, 173, 174, 231, 233, 243
Adamson G., 50-59, 61, 63, 64, 71, 72, 74, 75, 109, 163, 167, 370
Aden Abdullah, 385
Adjuran, 150
Aga Khan, 329, 332, 336
Ah Pong, 265-267
Ahab King, 87
Ahmed Ibrahim Ghazi, 150
Ahmed Maghan, 155
Ahmed Y. Dualeh, 159
Akare Morijo, 177
Al'Masudi, 89
Ali Shiré Warsame, 275, 276
Ali Suleiman, 290, 303
Alpers E., 25
Amhara, 150
Anderson J.H., 38
Arabicho s/o Boru, 91
Archer A.L., 7, 231, 255, 278, 282
Arkle J., 240, 241
Auko Daudi Mabinda, 186
Aulihan, 154-157
Ayerst P., 87

Baali N, 173
Baker S., 207, 208
Baring Sir E., 75, 76, 78, 79, 83, 97, 102, 208
Barratt I., 240, 241
Barrah J., 348
Batiaan, 29
Bell Ginger, 112
Bell R.H.V., 79, 306,338, 344, 345, 350, 354, 355
Ben Habib, 379
Bisa, 203, 305
Blixen B. von, 112, 132
Bon Mariham, 57

Boni, 54, 57, 85, 88-90
Boone Daniel, 46
Boran, 150, 153, 157, 165
Borana, 153
Boru Badiva, 91
Boru Debasa, 110
Brown D.W.J., 7, 40, 73, 114
Bryant A., 147
Buechner H., 132
Burton R. 69

Caldwell K., 103, 366
Carroll L., 5, 375
Carter Judge M., 93, 94
Carter L. A. 382, 383
Casebeer R. L., 226
Caughley G., 229, 318, 369
Caughley J., 229
Champion (Chambe), 405
Cheong Pong. 266. 267
Chepkorio, 177
Chewa, 47
Child A., 76
Chonyi, 64
Cook J., 70
Corfe, 57
Corfield R., 152
Cornwallis Lord, 32
Covilha Pera da, 23, 24
Cowie M., 66, 94, 99, 129, 133, 138
Craig, 340, 341, 344-346, 356, 358
Crawshay R., 42
Crockett Davy, 46
Crooke I., 344-351, 353-358, 361
Cullen A., 99, 171, 172

Dabanani, 299, 327, 329
Dalton G., 163
Darwin C., 380
Davidson B., 89
De Haes C., 329, 352, 361, 362
De Meulenaer T., 371

De Souza E, 114
De Souza M., 74, 114, 115
Deang Capt., 333
Dedan Kimathi, 126
Deitrich K.H., 190
Delamere Lord, 32, 48
Dempster D.H.M., 224
Diaz B., 23
Digo, 64
Dominico, 343, 344
Donohew M., 70
Dorobo, 51, 54-56, 102
Douglas-Hamilton I., 256, 257, 325, 349
Douglas-Hamilton O., 325
Duff Sir A., 239, 246, 390
Duruma, 64
Dutch East India Co, 31
Dyer A., 7, 237-241, 247

Edwards K., 347, 349
Edwards R., 377
Egebu C., 290, 335
Egebu G., 335
Eliot Sir C., 31, 155
Elliott (DC Serenli), 156
Elliott R.T.E., 68-72, 75, 109, 134, 156, 162-165, 167, 168, 170-179, 182-186, 188, 190-196, 199, 220, 222, 240, 249, 253, 269, 273, 284
Ellis R., 325
Ellis S., 345
Engels, 13
Erskine F., 97
Europarks, 347

Ferreira C.J., 378
Fichat J., 31
Fichat S., 31, 32
Flemming G., 377
Foster R., 110, 111
Fox G., 69
Franklin D., 126
Fraser-Darling F, 101.
Freeman, 378
Friedlein T., 257, 258, 270, 278, 282, 293, 318

Frois (Governor), 25
Fuchs Sir V, 122.

Gafaar Elias Busaid, 213, 214, 357
Galileo, 381
Galma Dida, 157
Galogalo Kafonde, 111, 112
Gawa, 331
George, 295, 297, 298, 301, 302
Gethi B. 183, 186, 194
Giltrap C., 176
Giriama, 28, 64, 85, 110, 401, 405
Gishilian Leserewa, 173
Goldfinch G.H., 43
Gontier, 176
Gordon General, 151, 208
Goreham G.A., 347
Gosha, 155
Graham A., 125, 230
Grant H., 59
Gray J., 90
Grimwood I., 67, 71, 107, 115, 126, 127
Grogan E., 32
Guest W, 112.
Gunn A., 369
Gurreh, 150, 159

Haak J., 263
Hadza, 54
Hagaiya, 406
Haigh J.C., 255
Hale W., 40, 67-71, 73, 74, 100-102, 107, 114, 312
Hancock G., 387, 388, 390
Hanks J., 202, 204, 205, 327-332, 344, 347, 349-353, 356-358, 362
Hardinge A.,51
Harris C., 189, 190
Harris W.C., 359
Harvey W., 111, 112
Hasaballah, 213
Hassaco, 191
Hay M., 340, 341
Hemsted R.W., 93
Henderson M., 73

Herd R., 175
Herti, 154, 160
Hobley C.W, 84, 90.
Höhnel L. von, 29
Holman D., 94
Howell Tiger, 257, 258
Hunter J.A., 75
Hutu, 285, 186, 311
Huxley C., 7, 269, 273, 277, 285-288, 302, 303, 316, 322, 350, 351, 353, 354, 357, 358

IBEAC, 23, 29, 42
Idi Amin, 13, 124, 126, 219, 262, 269, 366
Il Tikiri, 54
Ilsley J., 259, 261, 263, 266, 267
Irwin J., 186, 190, 192-194
Ismail Mohammed, 207-209
Issak (Ishaq), 151

Jackman B., 325, 346
Jaluo, 195, 227
Jamal Nasser (Jamala), 290, 295, 297, 299, 303, 335, 343
Jarman P., 221, 222, 224
Jenkins P.R., 7, 93, 111, 118, 132, 145, 147
Jenner A.C.W, 155, 156.
Jeevanjees, 31
Joginder Singh Sokhi, 186
Jones Dr D., 351-353

Kabayanda A., 290, 293, 295, 296, 300, 301, 303, 334
Kabukoki Lolokoria, 173
Kalenjin, 39
Kamba, 12, 28, 58, 65, 72, 75, 77, 79, 82, 93-95. 110, 117, 131, 134-136, 140, 144, 145, 160, 227, 305, 404, 405
Kangane M., 177
Kapila, 179
Karanja Gaturu, 227
Kariithi G., 231
Kariuki J.M., 174 178, 179 242-244
KAS, 346-348, 350-358
KAS International, 347
KAS Proprietary, 347

Kasangama (King), 365
Kathuo Kagala, 110, 111
Kaunda (President), 165, 311, 338, 385
Kenga, 401, 402, 405
Kennan G., 390
Kenyatta J. President, 12, 39, 106, 126, 127, 134, 171, 183, 186, 190, 194, 226, 227, 231, 233, 234, 241-243, 246, 250
Kenyatta Margaret., 227, 241, 246, 250
Kenyatta Mama N., 227
Khalifa Abdulla, 208
Khoi, 259
Kikuyu, 28, 29, 32, 37-39, 97, 138, 233, 243, 305
Kilo Alpha Services, 347
Kimiti (General), 227
Joao (King of Portugal), 23
Kingdon J., 7, 368, 370-372, 398
Kingele J.J.M., 190
Kioko J., 143, 145
Kipling R., 147
Kiribai, 405
Kirimanya J., 177
Kirkman A.H.B., 367, 368
Kitchener (General), 208
Kroll Associates, 349
Kubai S., 188
Kulang Mabor I., 211
Kumleben M.E., 265, 353-358, 360, 382
Kunz F., 222

Lapointe E., 332, 336, 338
Lategaan P., 354
Laws R.M., 7, 122, 123, 132, 133, 202, 316, 367
Le Cheyne E., 349, 358
Leakey L., 29, 62
Leakey R., 106, 147
Lebesoi Lelisimon, 69
Ledger J., 352, 353
Leete D., 97
Lenin, 13
Lerumben Lenjatin, 168
Livingstone D., 12, 204, 311, 376-380, 407
Lolimiri Ekaran, 173

Longreach, 340, 344
Lord Cromer, 208
Lugard Lord F.D., 365

MacArthur C.G., 58, 63-65, 72, 94, 111, 112, 163
Machado M. P. de S., 378
Mad Mullah, 12, 151, 152, 155, 156
Mágyar L., 378
Mahamud M. Farah, 158
Mahdi, 151, 208
Main M., 379
Maitha, 110, 111, 327
Mandela N., 265, 266, 353
Manganja, 47
Marehan, 154, 156
Markham B., 112
Martin Dr Esmond, 350, 354
Martin R., 7, 46, 213, 214, 273, 275, 277, 302, 316, 338, 344, 345, 350, 354, 355, 371
Marx, 13
Masai, 27-30, 38, 45, 51, 54, 61, 67, 90, 154, 163, 244, 404
Matangana J-B., 290, 298-300
Matheka G.M., 233
Mathenge I.M., 227
Mbatia wa Gaterimu, 29
Mboya T., 233, 243
McCabe D., 75, 76, 186
McClinton P., 168, 169
MacDougall K., 156
McIlvaine R., 254-257
McKenzie B., 226
McOdoyo P.B., 186, 187, 234
Mengistu President, 391
Meredith M., 7, 241
Meru, 126, 154, 156, 157, 189
Mijikenda, 47, 85, 90
Minot F., 253, 254
Mizilikazi, 260
Mobutu President, 385
Moffat R., 376-378
Mohamed Ahmed, 208
Mohamed Aideed, 152
Mohamed Amin, 339, 391
Mohamed Gelle, 157

Mohammed A. Hassan, 151, 152
Mohammed Ali, 207
Mohammed Sgt Major, 78, 163
Mohammed Zubeir, 154-156
Moi President Daniel, 183
Monks E., 222, 223
Mozungullos, 89-91
Mother Teresa, 350
Mugurori, 38
Mukogodo, 54, 56
Mumuuyot, 54
Murdoch R., 392
Murphy Congressman, 256, 257
Murray M., 377
Musa L. Supt., 186, 192-195, 197
Mutha Chief Inspector, 176
Muthoni Kirima, 126, 127, 188, 226, 227
Mwami of Burundi, 47

Nairangai S., 189
Nairi J., 173
Nash J., 352
National Car Parks, 347
Nderi I., 186, 193, 197
Nderitu D., 227
Nderobo Loltianya, 173
Ndibo Molu, 405
Necho Pharoah, 86
Neumann A.H., 84
New C., 84
Ng D., 318
Nganga J.N., 177
Niyokindi A., 334
Njagi D., 126, 226
Njonjo C., 175, 185, 186, 191, 192, 222, 240
Nkasana Lekadaa, 173
Noor Abdi Ogle, 194
Northway P.E., 12
Norton-Griffiths M., 45
Nyerere President, 233, 311, 338, 385
Nzembei Major N., 173, 174

O'Hagan D., 100
O'Rourke P.J., 153
Obote M. President, 13, 124, 219

Ogiek, 47, 54
Ogutu M., 192-195, 197, 222
Olindo P., 133, 232, 233
Oliver R., 376-379
Olympio S. President, 385
Orma, 28, 83, 90, 91, 93-95, 144, 145, 153-155, 160, 227
Oswell W.C., 377, 378
Owen Bishop, 252
Owen J., 252
Oyrer H., 263

Parfet C., 175, 183, 185, 186
Parker J., 32
Parkinson A., 254, 255
Patel H., 382, 383
Patten C., 392
Patterson J.H., 43, 44
Paulsen C., 31
Peacey R.M., 347
Percival A.B., 42-44, 46, 61
Pereira R., 200
Perossi S.M., 32
Phidias, 87
Pitman C., 366
Pokomo, 64
Poole G., 165
Poon family, 290
Porto A.F. da S., 378, 379
Potgieter de Wet, 353, 355, 356
Potgieter J., 78
Potgieter Rassie, 76, 78
Powell R., 339, 343-349, 353, 355, 356, 391
Pretorius D., 31
Prince Philip, 351-353
Pringle J., 241
Prins A.H.J., 90
Purkei Lesankurukuri, 173
Pusey, 227, 250
Puttnam Lord D., 12, 340, 341, 345, 356

Quett Mosire President, 385

R & TG Consultants, 354
Rabai, 64

Rahemtullah A., 285, 287-289, 302, 329, 333, 337, 339
Rahemtullah Z., 285, 287-289, 302, 329, 333, 337, 339
Ramsden Lord, 61
Ratebold Ltd, 347, 348
Ravenscroft D., 325, 391
Reader J., 385
Reece G., 57, 163, 370
Reeve F., 70
Reeve R., 382, 383
Renamo, 337, 339, 355
Rendille, 51, 154, 155
Retherford J.M., 70
Revill J., 325, 326, 345, 346, 356
Rezende P.B. de, 90
Rhodes C., 31
Ribe, 64
Richards M., 354, 355, 360
Riebeeck J. van, 41
Ritchie A., 49, 57, 60-66, 94, 112, 163
Rodwell E., 31
Root A., 340, 398
Ross C.J., 43
Rowbotham M., 122, 124
Rupert A., 361, 362
Ryans R.W., 70

Sadruddin Aga Khan, 329-332, 336, 351
Samburu, 51, 69-71, 154-156, 163-167
San, 259
Sand P., 318
Sandeman N., 73, 100, 101, 114
Sauvage G., 69
Saw P., 7, 109
Sawyer M.J., 232
Schelde C. van der, 31
Scott Sir P., 247
Scullard H.H., 88
Seago J., 254
Sebetwane (Sebetuane), 260
Sechele Chief, 376, 378
Senghor L. President, 385
Seth-Smith A.M.D., 103, 106, 122
Shaka Zulu, 47, 259, 260

Sheldrick A., 95
Sheldrick D.L.W., 75, 76, 78-80. 83, 91-93, 95-101, 103, 104, 129, 131-138, 140, 145-147, 149, 163, 170-172, 199, 220, 253, 267, 363
Simba Singi Nzenga, 77
Simon N., 99-101, 103
Sindiyo D., 186
Skinner P.K., 347
Smith W., 310
Smuts J.C., 362
Someren V.G.L. van, 62
Soshangane, 260
Southey, 69
Spinage C., 41
St Francis, 257
Stel S. van der, 41
Stevens R., 93
Stewart D.R.M, 368
Stewart J., 368
Steyn A., 31
Steyn N., 263, 265
Stirling D., 344-350, 355, 357, 361
Sultan of Oman, 27
Swynnerton C.F.M., 44, 366, 368
Swynnerton G., 368

Tagale, 177
Takaichi Company, 277
Tanaba, 290, 299, 303, 311, 335
Tatham G., 330,331
Teita (Taita), 83, 93, 102, 404
Telemuggeh, 155
Temple-Boreham E., 67, 70, 71
Tennant L., 125
Tewfik Ismail, 208
Tharaka, 102
Thesiger W., 60
Thoms W., 70
Thoreau, 14
Torgensen D., 241
Touche G.E., 69
Touval S., 151, 160
Train Judge R., 252, 253
Transatlantic GmbH, 190

Trench C., 157, 158, 161
Turkana, 29, 51, 52, 54, 56, 70, 153, 157, 165, 230, 370, 393, 394
Turle A., 349
Turnbull R., 153, 154, 160
Turner M., 59, 251-253
Twa, 285
Twitchin B., 37

Vasco da Gama, 22, 24, 30, 91
Velde W. van der, 258, 318

Wagalla, 65
Waliangulu, 63, 83, 90, 91, 100, 101, 114
Wambua Makula, 110, 327
Wang K.T, 290.
Wanyika, 64, 90
Wapokomo, 65
Wasanye, 63, 64, 83-85, 87, 102
Wata, 7, 12, 13, 15, 21, 54, 75, 79, 81, 83-91, 93, 95, 96, 99, 100, 103, 104, 106, 110, 111, 122, 215, 305, 340, 401-407
Watson M., 132, 202
Webley M., 115
Webster J., 101
Werner A., 84
White, 333, 336, 337, 339-341, 344, 345, 356
Wickens J., 338
Wilcox B., 33
Williams J.G., 35
Wilson (trader), 377
Wilson D., 87
Wilson F. O'B., 93
Winterburn R., 114
Wissman H. von, 42
Woodley F.W. de M., 75-80, 82, 83, 93-101, 103, 110, 111, 117, 118, 129, 135, 137-147, 165, 171, 185, 199, 220, 253, 267, 363
Wooldridge M., 339
Woosnam R.B., 44-49, 51

Yutzy, 70

Cover Design: The Digital Canvas Company
                    Forres
                    Scotland
                    bookcovers@digican.co.uk

Layout: Stephen M.L. Young
                    stephenmlyoung@aol.com

Fonts: Times New Roman (10pt)

Copies of this book can be ordered via the Internet:

                    www.librario.com

or from:

                    Librario Publishing Ltd
                    Brough House
                    Milton Brodie
                    Kinloss
                    Moray IV36 2UA
                    Scotland
                    UK
                    Tel /Fax No 00 44 (0)1343 850 617